APERIODICITY AND ORDER

Volume 1

Introduction to Quasicrystals

Contributors to This Volume

Per Bak
Leonid A. Bendersky
Ofer Biham
Alan I. Goldman
T. C. Lubensky
David Mukamel
Robert J. Schaefer
Clara B. Shoemaker
David P. Shoemaker
S. Shtrikman
Michael Widom

APERIODICITY AND ORDER

Volume 1

Introduction to Quasicrystals

Edited by

Marko V. Jarić
Center for Theoretical Physics
Texas A&M University
College Station, Texas

ACADEMIC PRESS, INC.
Harcourt Brace Jovanovich, Publishers
Boston San Diego New York
Berkeley London Sydney
Tokyo Toronto

ACADEMIC PRESS, INC.
1250 Sixth Avenue, San Diego, CA 92101

United Kingdom Edition published by
ACADEMIC PRESS INC. (LONDON) LTD.
24-28 Oval Road, London NW1 7DX

Library of Congress Cataloging-in-Publication Data

Aperiodicity and order.

Bibliography: v. 1, p.
Includes index.
Contents: v. 1. Introduction to quasicrystals.
1. Metal crystals. I. Jarić, Marko V., Date-
QD921.A67 1988 530.4′1 87-11527
ISBN 0-12-040601-2 (v. 1)

Printed in the United States of America
88 89 90 91 9 8 7 6 5 4 3 2 1

Contents

Contributors

Numbers in parentheses refer to the pages on which the authors' contributions begin.

Per Bak (143), *Department of Physics, Brookhaven National Laboratory, Upton, New York 11973*

Leonid A. Bendersky (111), *Metallurgy Division, National Bureau of Standards, Gaithersburg, Maryland 20899*

Ofer Biham (171), *Department of Physics, The Weizmann Institute of Science, Rehovot, Israel*

Alan I. Goldman (143), *Department of Physics, Brookhaven National Laboratory, Upton, New York 11973*

Marko V. Jarić (preface), *Center for Theoretical Physics, Texas A&M University, College Station, Texas 77843-4242*

T. C. Lubensky (199), *Department of Physics, School of Arts and Sciences, University of Pennsylvania, Philadelphia, Pennsylvania 19104-6396*

David Mukamel (171), *IBM T. J. Watson Research Center, Yorktown Heights, New York 10598 and Department of Physics, The Weizmann Institute of Science, Rehovot, Israel*

Robert J. Schaefer (111), *Metallurgy Division, National Bureau of Standards, Gaithersburg, Maryland 20899*

Clara B. Shoemaker (1), *Department of Chemistry, Oregon State University, Corvallis, Oregon 97331*
David P. Shoemaker (1), *Department of Chemistry, Oregon State University, Corvallis, Oregon 97331*
S. Shtrikman (171), *Department of Electronics, The Weizmann Institute of Science, Rehovot, Israel*
Michael Widom (59), *Department of Physics, Carnegie-Mellon University, Pittsburgh, Pennsylvania 15213*

Preface

In many areas of science, and in particular in physics, the intuitive notion of spatial or temporal order is intimately connected with the notion of periodicity. Indeed, in many textbooks one finds that periodic crystal structures are described not only as the ultimate examples of perfect positional order but also as synonymous with such order. For example, it is commonly assumed that aperiodic spatial structures, such as structures with noncrystallographic symmetry, can exhibit order only over a limited scale. The extent of the spatial ordering, its coherence, measured in scattering experiments by the sharpness of the diffraction maxima seemed, until very recently, to agree with this assumption.

Over the last several decades we have witnessed, however, a growing awareness about the possibilities and importance of various aperiodic types of order, including quasiperiodic order, which is characterized by perfect coherence just as the periodic order is. This development was given an important impetus, on one hand, by the increasing interest in dynamical systems and deterministic chaos and, on the other hand, by the parallel expansion of our knowledge about incommensurate spatial structures. Owing to the often equivalent mathematical framework, a fruitful cross-fertilization spurred maturing of these two research areas into a broader conceptually united field. For instance, by exploiting a rigorous relationship between dynamical systems and ground states of simple interacting systems, David Ruelle conjectured the existence of certain "chaotic crystals." Although his conjecture still remains to be born out by experiments, similar analogies have already produced a number of important results pertinent to crystals with incommensurate ground states.

Ever since Roger Penrose discovered an aperiodic tiling of the plane based on a hierarchical packing of pentagons, crystallographer Alan Mackay was fascinated by the possibility that similar, especially icosahedral, structures might occur in nature but might go unrecognized if unexpected. His daring conjecture was dramatized by the recent experimental discovery of icosahedral quasicrystals by Shechtman and his collaborators. This discovery gave a driving force not only to the study of incommensurate crystals with noncrystallographic symmetries, but it stimulated an intellectual atmosphere appropriate for asking more provocative and deeper questions about the nature of order and the order of nature.

This book series will provide a forum for addressing such questions. Since the current understanding of the periodic order has its roots almost as far back as the beginnings of man's utilitarian and artistic relationship with the world, that is, the beginnings of civilization, the forthcoming volumes will focus on the more exotic and far less understood types of order.

The first volumes will be devoted to the subject of quasicrystals. They are meant to serve as an introduction for new students in the field and also as a reference for active researchers. Volume 1 is intended to give an introduction to the basic physics of quasicrystalline order and materials. An attempt has been made to present only the definitive results that are independent of any unconfirmed interpretations of experiments. Volume 1 will be shortly followed by a companion volume that will offer an introduction and a reference to the mathematical machinery necessary in studies of quasicrystals.

Chapter 1

Icosahedral Coordination in Metallic Crystals

DAVID P. SHOEMAKER AND
CLARA B. SHOEMAKER

Department of Chemistry
Oregon State University
Corvallis, OR 97331

Contents

APERIODICITY AND ORDER
Introduction to
Quasicrystals

ISBN 0-12-040601-2

1. Introduction

For many decades it has been well known to crystallographers that fivefold axes of symmetry cannot be present in crystals (i.e., materials possessing a three-dimensional translation group). Therefore, it has been an article of faith that the appearance of fivefold symmetry in any properties of ostensibly crystalline materials, including diffraction patterns, should never be expected, unless due to twinning. Twinning resulting in the appearance of five- or tenfold diffraction symmetry, or especially icosahedral diffraction symmetry which includes fivefold symmetry, would have been considered improbable by most crystallographers, although icosahedral twinning in small particles of gold and silver has been reported by Smith and Marks (1981) on the basis of lattice imaging by high-resolution electron microscopy.

Therefore, the report of Shechtman *et al.* (1984) that specimens of a rapidly quenched alloy with approximate composition $MnAl_6$ were found to exhibit icosahedral diffraction symmetry was greeted with great astonishment by crystallographers and many other scientists. The astonishment was accompanied by considerable skepticism regarding the proffered explanation that this rapidly quenched material is not crystalline in the conventional sense, but may instead be "quasicrystalline," representing a three-dimensional analog of Penrose tiling (Mackay, 1982), as described by Levine and Steinhardt (1984, 1986), Socolar and Steinhardt (1986), and others. This proposed state of matter differs from the crystalline state in possessing no lattice periodicity (i.e., no translation group), and is therefore characterized as "aperiodic." The crystallographic community preferred at the time to accept the alternative hypothesis put forward, on chemical structure considerations, by the celebrated chemist Linus Pauling (1985), namely that the material in question is actually an "icosatwin," i.e., a composite of twenty identical crystalline individuals twinned together with icosahedral symmetry (like the components of the gold and silver particles mentioned above). The idea of an icosahedral multitwin had earlier been proposed by Field and Fraser (1985), who showed that such a multitwin could produce a diffraction pattern resembling in some respects the observed one; a multitwin structural model differing from that of Pauling has been proposed by Carr (1985). Nevertheless, the condensed-matter physics community generally has preferred the quasicrystal concept, or the subsequently proposed concept of an "icosahedral glass" (Stephens and Goldman, 1986), or some compromise between the two.

We will not discuss in this chapter the experimental evidence for or against these conflicting models since this will be discussed in later chapters. In the foregoing discussion, we have sought to emphasize that the new aperiodic concepts of structure as applied to materials such as $MnAl_6$ represent a truly immense break with traditional material structure concepts. Even if crystallographers and chemists may eventually be forced to accept the quasicrystal concept, they will not be satisfied until detailed atomic structure models—with real atoms, credibly coordinated, at credible interatomic distances—have been solidly established from observed diffraction intensities. It will be among the objectives of the present chapter to illustrate structural principles of metallic materials that are crystalline (as very nearly every inorganic solid is), which may be expected to carry over, at least in considerable degree, to aperiodic materials.

Chemists, metallurgists, and crystallographers have known for many years that while fivefold and icosahedral symmetry are forbidden in the presence of two- or three-dimensional translation groups (represented respectively by two- and three-dimensional Bravais lattices), local icosahedral arrangements of atoms in crystals are possible and, in fact, for complex metallic materials fairly common. (Indeed, the crystal structure envisioned by Pauling is based on icosahedral clusters of atoms.) There is no contradiction here; the icosahedral arrangements need not have, and indeed do not have, ideal icosahedral symmetry. Each such icosahedral grouping has *local* and *approximate* icosahedral symmetry, the ideal icosahedral symmetry being broken by the requirements of the crystal lattice and the associated physical forces that must distort the icosahedral shape at least slightly. In many cases it is broken more coarsely by differences in chemical identity among atoms of the group.

Since with the quasicrystal model the detailed atomic arrangements in the so-called icosahedral phases typified by rapidly quenched i-$MnAl_x$ (where x is now believed to be somewhat less than 6) are not yet known, their relationships with icosahedral groupings in known metal alloy structures are still not known with certainty. For reasons to be discussed later, it appears likely that icosahedral groupings and linkages thereof will be important features of the structures of these materials. The fact that some icosahedral phase compositions are approximately the same as the compositions of crystalline alloys that are known to contain icosahedral groupings, as well as the fact that there have been found some correspondences between the two in observed diffraction intensities, lends some support to this view.

Accordingly, in this chapter, we will review the crystal structures of a number of alloys that exhibit icosahedral groupings, including some that we have determined over the last thirty-five years at the Massachusetts Institute of Technology and at Oregon State University. The total number of such structures is very large, and we will have to make a selection, emphasizing those that illustrate modes of linking icosahedra and propagating the icosahedral orientations throughout the crystal space.

Icosahedral groupings result from chemical forces associated with atomic *coordination* (Pauling, 1960). Icosahedral coordination (i.e., the surrounding of a central atom by twelve chemically interacting atoms in an arrangement approximating an icosahedron) is commonly found in alloys of transition metals. These possess not only *s* and *p* orbitals but also *d* orbitals and therefore form stronger bonds than the simple metals of groups I to III, which typically crystallize in the simple and well-known modes of cubic and hexagonal close packing, and body-centered cubic packing. This is shown, for example, by the higher heats of formation of the transition metal alloys. Stronger bonding favors greater efficiency in the filling of space. The cubic and hexagonal close packings are as efficient as is possible for isotropic atoms of uniform size, but these packings possess not only tetrahedral interstices but also octahedral interstices which are more wasteful of space. More efficient packing is possible, with partial or total elimination of octahedral interstices, when atoms *of different sizes* are present, as in alloys (or occasionally in elementary forms of certain metals, e.g., α-Mn (Bradley and Thewlis, 1927) and β-U (Donohue and Einspahr, 1971), where atoms of the same element are found to be present in different sizes). Since the interstices created by icosahedrally coordinating a central atom with twelve neighbors are all tetrahedral (albeit somewhat distorted), it should not be surprising that the approximately regular icosahedron is a coordination polyhedron frequently encountered in alloys containing transition metals (sometimes together with nontransition metals such as silicon and aluminum).

1.1 The Regular Icosahedron and Related Polyhedra

The *regular icosahedron,* one of the five Platonic regular solids in Euclidean three-dimensional space, is shown in Fig. 1a. It has 20 triangular faces, 30 edges, and 12 vertices. When an atom occupies the center of the icosahedron and neighboring atoms occupy all of the vertices, then the central atom has *coordination 12* (CN12). The 20 interstitial voids

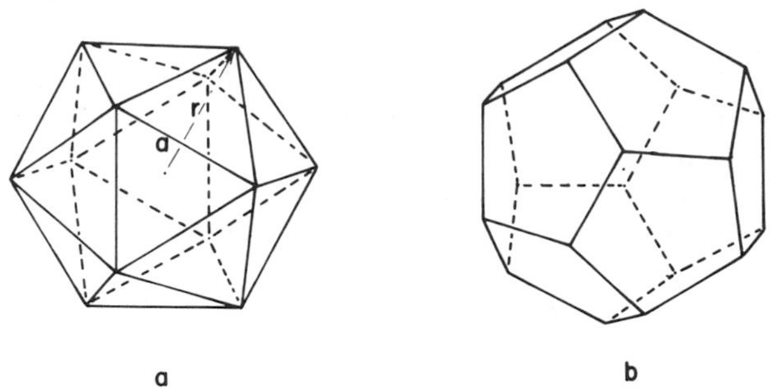

Figure 1. (a) Regular icosahedron. (b) Regular pentagonal dodecahedron.

defined by the center and the 12 vertices are all tetrahedra, slightly deformed from regularity.

The point symmetry group of the icosahedron is $m\bar{3}5$ (I_h), which includes fivefold, threefold, and twofold rotation axes and mirror planes. A very important subgroup is $m\bar{3}$ (T_h), a crystallographic point group in the cubic system, which lacks the fivefold rotation axes. When the icosahedron is found in cubic crystals, it frequently possesses this subgroup and is oriented with three of its twofold rotation axes parallel to the cube axes. The edges then divide into two groups generally of somewhat different length: six parallel to the cube axes and twenty-four inclined to these axes.

As shown by Samson (1968), the regular icosahedron can be evolved from the CN12 coordination polyhedron found in cubic closest packing, a truncated cube having point symmetry $m\bar{3}m$ (O_h). As illustrated in Fig. 2, the three equatorial squares may be distorted to rectangles having an axial ratio of τ to 1. The vertices of the three rectangles are now the vertices

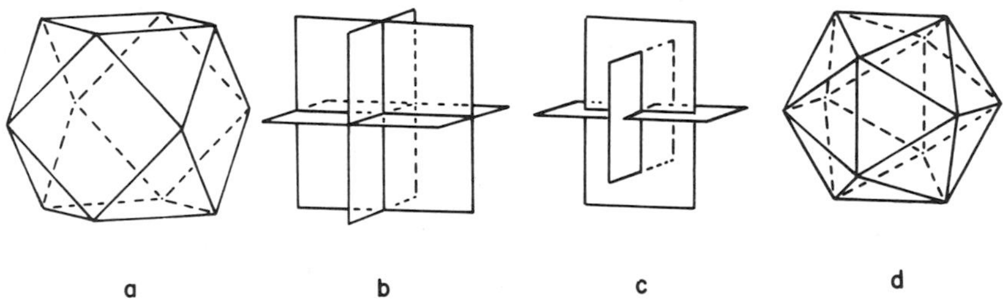

Figure 2. Evolution of regular icosahedron (d) from coordination polyhedron for cubic closest packing (a).

of a regular icosahedron. The number τ is the "golden ratio," $(\sqrt{5} + 1)/2 = 1.618033989\ldots$, a number which arises time after time in the geometry of figures with fivefold symmetry in two or more dimensions, and in the theory of quasicrystals. The distance r from the center to any of the vertices is about five percent smaller than the distance a between two neighboring vertices:

$$\frac{r}{a} = 2^{-1}5^{1/4}\tau^{1/2} = 0.951056517\ldots$$

This is one of the reasons why it is impossible to construct a structural arrangement in Euclidean space in which all atoms have icosahedral coordination. It is a key reason why icosahedral coordination is common in alloys containing atoms of different sizes, optimally where the central atom radius is about ten percent smaller than the nearest neighbor atom radius.

The "dual" of the regular icosahedron is the *regular pentagonal dodecahedron,* shown in Fig. 1b. It may be constructed by taking as vertices the centers of the faces of the original figure, and then rescaling the resulting figure until the edges intersect with those of the original figure. (In the particular case of regular polyhedra, the dual is the same as the "reciprocal" (Coxeter, 1973). The dual is related to the Voronoï or Wigner–Seitz cell, which is the figure enclosed by a set of planes that are perpendicular bisectors of the lines joining the central point to the surrounding vertices of the original figure. In the present case, the dual or reciprocal is the same as the Voronoï cell except for scale.)

The regular pentagonal dodecahedron has the same point symmetry group as the regular icosahedron, and has 12 pentagonal faces, 30 edges, and 20 vertices. Its dual is again a regular icosahedron. The pentagonal dodecahedron is found in some alloys, again somewhat distorted from regularity, as a "second coordination shell" of an atom for which the first coordination shell is an icosahedron.

An important nonregular polyhedron with icosahedral symmetry that is important in the discussion of some alloy structure types as a more distant coordination shell, and also in the discussion of quasicrystals, is a zonohedron, the *rhombic triacontahedron,* shown in Fig. 3. It has 30 rhombic faces, 32 vertices, and 60 edges. The diagonals of the rhombic faces are in the length ratio τ:1; thus the acute angle of each rhombus is $2\ \mathrm{artan}^{-1}(1/\tau) = 63.435°$, the same angle that is subtended by an edge of the regular icosahedron at the center of the icosahedron. It can be constructed as a superposition of the sets of vertices of two dual regular polyhedra, the icosahedron and the pentagonal dodecahedron.

More complex nonregular polyhedra of icosahedral symmetry exist that

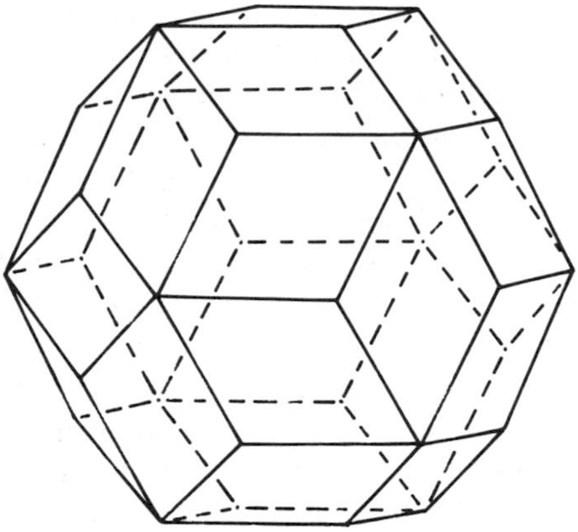

Figure 3. Rhombic triacontahedron.

will be found in the discussions of this chapter. One is a quasi-regular polyhedron, the *icosidodecahedron,* a figure with 32 faces, of which 20 are equilateral triangles and 12 are regular pentagons; 30 vertices, and 60 edges. It is the dual of the rhombic triacontahedron. A more important polyhedron will simply be referred to as a "soccer ball," as its vertices have the same relation to one another as those in that popular object, although its faces are of course plane rather than spherical. It is a truncated icosahedron. Of its 32 faces 20 are regular hexagons and 12 are regular pentagons; it has 60 vertices and 90 edges. This polyhedron is of interest to us because it exists as a fourth coordination shell in at least one alloy (namely, $Mg_{32}(Zn,Al)_{49}$; Bergman *et al.*, 1957; Section 4.5). It is the bounding convex polyhedron of the atomic cluster shown in Fig. 26c and discussed in Section 4.5.3.

1.2 Occurrence of Icosahedra and Related Polyhedra in Nonmetallic Materials

Solid-state chemists are most often aware of the icosahedron because of its occurrence in elementary boron and in several boron compounds. Boron is generally considered to behave as a nonmetal, but both in its elementary forms and in many of its compounds, it shares with metals a considerable delocalization of the valence electrons, which for boron and its compounds is well understood on the basis of multicenter bonds and/or molecular orbitals (Lipscomb, 1963). About thirty allotropic forms of elementary boron have been discovered (Donohue, 1974), most of them of unknown crystal structure. Three of them, for which detailed structures have been reported, contain B_{12} icosahedra linked together, mostly by bonds between vertices of adjacent icosahedra, but in the case of so-called β-rhombohedral-105 boron some linkages are by sharing of triangular faces (Hoard *et al.*, 1970). *The icosahedra are empty;* there are no atoms occupying the centers. In β-rhombohedral-105 boron there is an interesting 84-atom unit. It contains at its center an empty boron icosahedron to which are bonded 12 additional atoms at the vertices of a larger circumscribed icosahedron; each of these atoms forms five bonds to a circumscribed 60-atom "soccer-ball" polyhedron, described in the previous section. The combination of the soccer-ball polyhedron with the aforementioned larger icosahedron, the vertices of which are at the bottoms of the concave pentagonal dimples in an otherwise convex figure, constitutes a shell essentially identical to the outer shell of the metal cluster shown in Fig. 26c and discussed in Section 4.5.3.

The boron atoms in decaborane, $B_{10}H_{14}$, occupy ten of the twelve vertex positions of the icosahedron, the two omitted vertices being adjacent (Kasper *et al.*, 1950). All twelve of the icosahedron vertices are occupied in the dodecaborohydride anion $B_{12}H_{12}^{2-}$, as found, for example, in its crystalline dipotassium salt (Wunderlich and Lipscomb, 1960). If, in the latter, two BH^- units are replaced by CH units, an isoelectronic carborane, $C_2B_{10}H_{12}$, of approximate icosahedral shape is obtained. (Three different carboranes are possible, depending on the relative positions of the two CH units.) Presumably the dodecaborohydride ion as an isolated ion in the gas phase possesses rigorous icosahedral symmetry. However, as far as we are aware, no form of elementary boron or any of its un-ionized compounds exhibits *rigorous* icosahedral symmetry; in elementary boron and in salts containing the dodecaborohydride ion, the symmetry is broken by the crystal lattice, and in decaborane and the carboranes, it is broken through incompletion of the icosahedral shell or through substitution of carbon for some of the boron.

The pentagonal dodecahedron is also found in nonmetallic systems, for example (with slight distortions) as the figure defined by water oxygen positions in crystalline clathrate hydrates of small molecules (including noble gases). The two most important of these, from our point of view, are $Cl_2(H_2O)_8$, with a cubic unit cell of about 12 Å (Claussen, 1951b; Pauling and Marsh, 1952), and $C_3H_8(H_2O)_{17}$, with a unit cell of about 17 Å (Claussen, 1951a). In both structures, the water molecules link together with hydrogen bonds so that the water oxygen atoms form approximately regular pentagonal dodecahedra. They also form tetrakaidecahedra in the case of the hydrate with the smaller unit cell, and hexakaidecahedra in the case of the hydrate with the larger one. In the examples given, the guest molecules occupy the larger polyhedra, with the pentagonal dodecahedra remaining empty; however, in other instances, small guest molecules occupy also the pentagonal dodecahedra. Both of these arrangements are important to us here because their oxygen positional networks are the duals of two important icosahedra-containing alloy structures which will be discussed in Section 4.3, namely A15 (β-W, or Cr_3Si) and C15 ($MgCu_2$), respectively. (In each case, the dual is generated by taking as vertices of the new structure the centers of tetrahedral interstices of the original one.) In addition, the larger hydrate structure is the basis of Pauling's (1985) "icosatwin" model for rapidly cooled $MnAl_6$. He placed an icosahedral $MnAl_{12}$ group at each water oxygen position and linked them by sharing faces, generating a structure with a predicted cubic unit cell edge of 26.7 Å. As Pauling has acknowledged, the bonding of an icosahedron by faces to four tetrahedrally disposed other icosahedra requires that the arrangement around the central icosahedron be chiral. The alternation of chirality around an odd-numbered ring introduces certain problems including some strain, which Pauling considered acceptable, although similar configurations of icosahedra have not been seen previously in alloy structures. The model has been criticized on other grounds (Cahn *et al.*, 1985). Recently, Pauling (1987) has proposed a different model for the icosatwin structure of this phase.

Perhaps the only well-authenticated neutral molecule which can confidently be presumed to have rigorous icosahedral symmetry in the gas phase is pentagonal dodecahedrane, $C_{20}H_{20}$ (Paquette *et al.*, 1983). The carbon atoms occupy the vertices of a regular pentagonal dodecahedron, the bond angles of 108° approximating closely the tetrahedral angle of 109° 28′ favored for sp^3 hybridization. (The hydrogen atoms, bonded to the carbons, form a similar circumscribing pentagonal dodecahedron.) In the crystal structure (Gallucci *et al.*, 1986), the deviations from strict icosahedral symmetry forced by the lattice are very small.

Smalley and coworkers (Kroto *et al.*, 1985) have found, in the mass spectrum of the gaseous product produced when graphite is vaporized by a laser beam, a strongly predominating peak at 60 ^{12}C masses. They interpreted this as being due to a spherical carbon polymer, C_{60}, having carbon atoms at the vertices of a "soccer-ball" polyhedron, described in the previous section. They have named it "buckminsterfullerene" (it is also known as "soccerballene"). This is visualized as a sort of spherical analog of a graphite layer, with each carbon atom forming two single bonds and one double bond, thereby saturating the valence of carbon and yielding a resonating system. This interpretation is highly plausible but somewhat controversial; if such a molecule exists with this structure, it would presumably have rigorous icosahedral symmetry in the gas phase.

Any discussion of icosahedra in nonmetallic systems is necessarily incomplete but cannot omit the *icosahedral viruses,* of which poliovirus is an important example. The complete atomic structure of the protein capsid of this virus particle, which has a diameter of about 310 Å, has been determined by an X-ray diffraction study of the crystallized virus (Hogle *et al.*, 1985). The ideal local symmetry of the protein capsid of the particle (ignoring the RNA contained therein) is less than full icosahedral symmetry, owing to the chirality of the proteins of which it is constructed, and it is at best point group 235 *(I),* which lacks an inversion center.

1.3 Modes of Linking Icosahedra in Metallic Phases

A major emphasis of this chapter is the variety of ways that icosahedra can link together to propagate the icosahedral orientations through space. Here the orientations are considered to be defined by the directions of the (local and usually approximate) symmetry directions in space. These will be explored in this section. In the accompanying figures the letter *A* will be assigned to a particular orientation of an icosahedron in space; the letter *B* to an orientation derived from *A* by the operation of a mirror plane.

In Fig. 4a, we see two icosahedra linked together, along a fivefold rotation axis common to both, by sharing a vertex. There exists more than one way to orient the second one relative to the first; here we assume that the shared vertex is at an inversion center. Since the isolated icosahedron itself has an inversion center, this has two results. One is that the second icosahedron has the same orientation *(A)* as the first; the other is that another icosahedron, with orientation *(B)*, is generated with its center at the shared vertex. Note that if the two linked *A* icosahedra are perfectly regular, the intervening and interpenetrating one, *B*, cannot be regular since some of its intervertex vectors are radial vectors of the other two icosahedra. Orientations *A* and *B* are related by the mirror plane of symmetry that contains the pentagon shared by the two interpenetrating icosahedra.

As we shall see later, an icosahedron *A* may be interpenetrated simultaneously by icosahedra centered at several of its vertices, or even at all of its vertices. Since the orientations of mirror reflections depend upon the orientations of the mirror, this may generate up to six different orientations for the interpenetrating icosahedra; these we may designate with additional letters: *B, C, D, E, F, G*. The orientations of two icosahedra penetrating from diametrically opposite directions are the same.

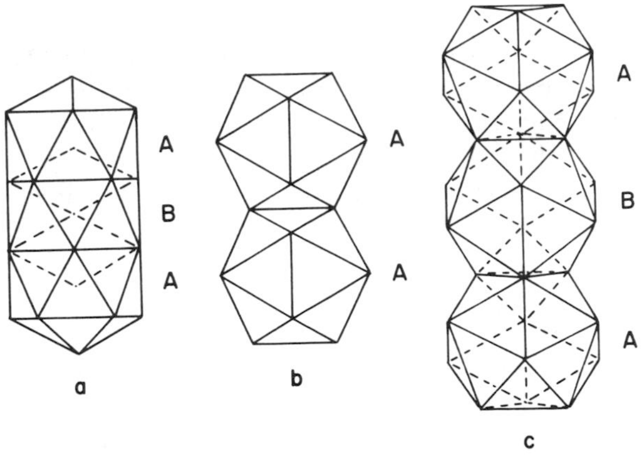

Figure 4. Modes of linking joined icosahedra. (a) Two icosahedra (both labelled *A*) share a vertex in such a way as to generate a third icosahedron *(B)* of orientation differing from the other two. (b) Two icosahedra share an edge. (c) Icosahedra join, sharing a face; the orientations alternate.

Two icosahedra joining at a vertex may also be related by a mirror plane of symmetry passing through the shared vertex, so that the adjacent vertices form a *pentagonal prism*. Since this mirror plane is not parallel to any mirror plane of either icosahedron, the two icosahedral orientations are different *(A* and *B)*. This mode, not shown here, is illustrated in the structures of some important complex alloys, such as $Sc_{57}Rh_{13}$ (Cenzual *et al.*, 1985). Other relative orientations of the two vertex-joined icosahedra are also possible through rotation around their common axis.

In Fig. 4b are shown two icosahedra joined by sharing an edge; they are related by a mirror plane of symmetry containing the shared edge. Since that mirror plane is parallel to mirror planes of the isolated icosahedra, the two icosahedra have the same orientation; if one is *A*, the other is also *A*. If both icosahedra are regular, rather short distances are generated between vertices of opposite icosahedra. In alloys containing icosahedra linked in this manner, the icosahedra must be somewhat distorted to lengthen these distances; this required distortion is accommodated by the presence of atoms of different sizes.

In Fig. 4c, icosahedra are linked by sharing triangular faces; within each pair, they are related by a mirror plane of symmetry, which is not parallel to any mirror plane of either icosahedron. Thus the orientations of the icosahedra are different. If, as shown, three icosahedra are linked together in this way along a common threefold rotation axis, the relative orientations are *A, B, A*. If several icosahedra join to an icosahedron of orientation *A* by sharing triangular faces, additional letters may be required to designate their orientations, as in the case of interpenetration.

In contrast to Fig. 4, we now turn to modes of joining *separated* icosahedra, i.e., icosahedra that do not share vertices. In Fig. 5a, icosahedra are joined through intervening tetrahedra, each of which shares opposite edges with two icosahedra. Since the tetrahedron lacks a center of symmetry but does possess a fourfold rotoinversion axis whose square is a twofold rotation axis, the sequence of relative orientations is *A, B, A*, . . . , along that axis. In Fig. 5b, two icosahedra are joined through an intervening octahedron, which shares opposite faces with two icosahedra. Since the octahedron possesses an inversion center, the two icosahedra have the same orientation *(A, A)*. The same result obtains in Fig. 5c, where the two icosahedra are joined to opposite edges of an octahedron or of a square.

In summary, icosahedral orientations along a line do not alternate if the linkages are at shared vertices (at inversion centers) or at shared edges, or (in the case of separated icosahedra) are linked through octahedra or

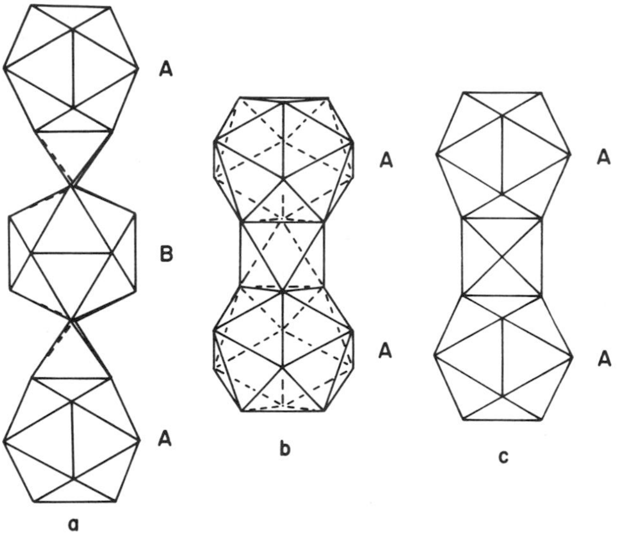

Figure 5. Modes of linking separated icosahedra. (a) Icosahedra join, sharing opposite edges with tetrahedra that separate them; the orientations of icosahedra alternate. (b) Two icosahedra join, sharing opposite faces with octahedra that separate them. (c) Two icosahedra join, sharing opposite edges with octahedra that separate them.

squares. Thus the sequence of orientations in these cases is *A*, *A*, *A*, ... On the other hand, the sequence alternates (*A*, *B*, *A*, . . .) if the icosahedra interpenetrate, are linked at shared faces, or are linked through tetrahedra. In the latter case, if the line of icosahedra is not straight, orientations other than *A* and *B* may be generated. If all of the bends in a string of linked icosahedra are at the centers of *A* icosahedra, then all *A* orientations are preserved, but the *B* orientations are variable.

As we discuss alloy structures containing icosahedra, we will point out these various modes of linkage and show in most cases how strings of linked icosahedra propagate orientations through lattice repeats, as they must. We may suspect that the sizes of unit cells are determined in many cases at least in part by requirements imposed by the linking of icosahedra.

2. Some Alloy Structures Containing Separated Icosahedra

Here we describe a few alloy structures in which the icosahedra are not linked by sharing any vertices, edges, or faces. These will be discussed with particular reference to the types of linkage and the propagation of icosahedral orientation.

2.1 $NaZn_{13}$

In Fig. 6 is shown part of the structure of this compound, determined originally by Ketelaar (1937) and Zintl and Haucke (1938) and refined by Shoemaker *et al.* (1952). This structure is found also in some other systems. It is face-centered cubic, with a_0 = 12.284 Å. The space group is $Fm\bar{3}c$. The Zn atoms are present in separated Zn_{13} clusters, each of which is an icosahedral arrangement of Zn atoms with another Zn atom at its center. This is one of the rare cases where an atom is coordinated icosahedrally by atoms of the same kind. The site symmetry at the center of each Zn_{13} cluster is $m\bar{3}$, the highest symmetry that an icosahedral cluster can possess in a crystal structure; the same site symmetry will be found in the clusters of the three alloys described in Sections 2.2–2.4. The Zn_{13} icosahedral cluster shown, which is located at the origin of the cubic unit cell, is linked to others located at the centers of the unit cell edges *through tetrahedra* in the manner described earlier, so that the orientations alternate *A*, *B*, *A*, . . . along each fourfold rotoinversion axis ($\bar{4}$), which is also a fourfold screw axis (4_2). There are eight such clusters per unit cell, four with one orientation and four with another obtained by rotating through 90° around one of the cube axes. Each of the eight Na atoms (at ¼, ¼, ¼, etc.) is surrounded by a "snub cube" of Zn atoms. The arrangement of Zn_{13} clus-

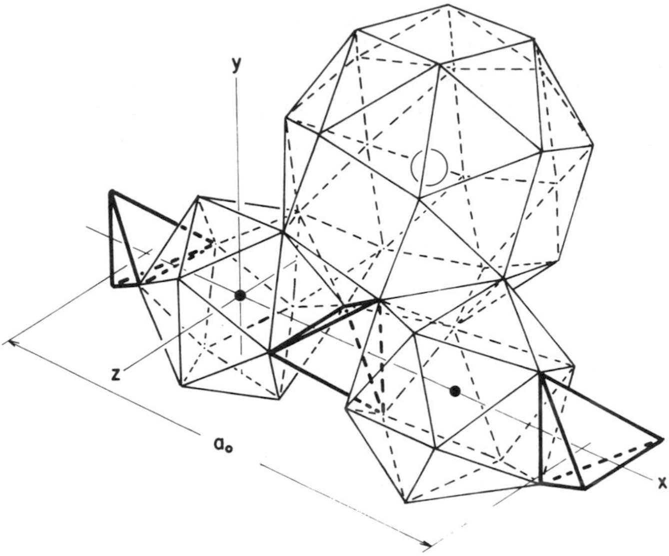

Figure 6. Part of the crystal structure of $NaZn_{13}$. Large open circle is Na at ¼, ¼, ¼, small filled circles are Zn I at 0,0,0, and 0,0,½; vertices of polyhedra are Zn II. Adapted from Shoemaker *et al.* (1952).

ters and Na atoms grossly (i.e., neglecting the different orientations of the clusters) resembles the B2 or CsCl arrangement, with a unit cell edge which is half the actual one.

2.2 Mg_2Zn_{11}

Mg_2Zn_{11} is cubic with $a_0 = 8.535$ Å, space group $Pm\overline{3}$, three formula units per unit cell (Samson, 1949b). A centered icosahedral Zn_{13} group at the unit cell center is separated from each adjacent one along a cube direction by sharing edges with a *pair of tetrahedra* having a common edge (a pair of Mg atoms). Thus, the Zn_{13} groups in adjacent unit cells have the same icosahedral orientation, as required by the translation group; the unit cell is not doubled in each direction as it is in $NaZn_{13}$, where the Zn_{13} groups are separated by single tetrahedra. The interstice at 0,0,0 is filled with a Zn_6 octahedron surrounded by a cube of Zn atoms. Additional Zn atoms (six per unit cell) lying on faces of the unit cell form a large distorted icosahedron sharing edges with similar ones in adjoining cells; it has the same orientation as the smaller one at the unit cell center. The same structure exists for $Mg_2Al_5Cu_6$ (Samson, 1949a).

This structure can be described alternatively in terms of a centered Zn_{13} icosahedral unit surrounded by a *rhombic triacontahedron,* described in Section 1.1 and shown in Fig. 3, yielding a unit of 45 atoms. (Some of the rhombic faces are bent inward at the short diagonal about 16° from planarity.) This unit is packed with other ones in a primitive cubic array, with sharing of rhombic faces at the unit cell faces. The sharing reduces the number of atoms per unit cell to 33. Each interstice at a unit cell corner is filled with six atoms forming an octahedron, so the total number of atoms per unit cell is 39, three formula units. The 45-atom unit is structurally identical to a part of the T-phase [$Mg_{32}(Zn,Al)_{49}$] structure described in Section 4.5.

2.3 $MoAl_{12}$

The crystal structure of $MoAl_{12}$, determined by Adam and Rich (1954), is shown in Fig. 7. This structure is also possessed by MAl_{12}, where M = W, Re, and Tc. It is possessed also by $MnAl_{12}$, but this phase is only metastable. The unit cell is body-centered cubic, with a_0 = 7.573 Å. The space group is $Im\overline{3}$. The $MoAl_{12}$ cluster centered at the origin is linked to the cluster at the body center by an octahedron, sharing opposite faces with both. Accordingly, the sequence along the body diagonal is *A, A,*

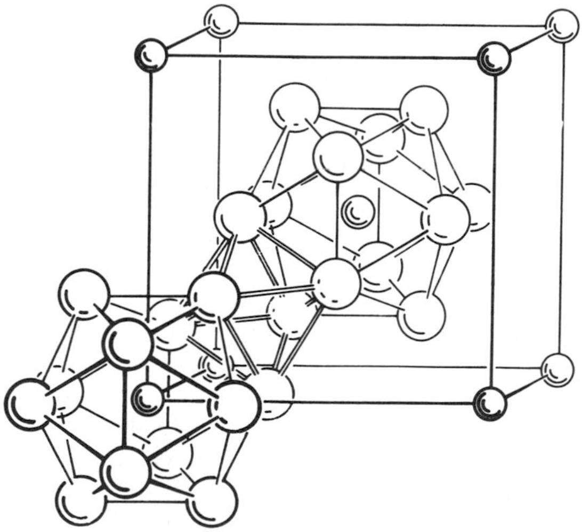

Figure 7. The crystal structure of $MoAl_{12}$. [Adapted from Linus Pauling: *The Nature of the Chemical Bond. Third Edition.* p. 425. Copyright © 1960 by Cornell University. Used by permission of the publisher, Cornell Univesity Press.]

A, ... The same sequence is found along each cubic axis, since the clusters are linked by somewhat distorted squares sharing opposite edges.

2.4 α-$Mn_{12}(Al,Si)_{57}$

The crystal structure of this phase (commonly abbreviated as α-Al-Mn-Si) is shown in Fig. 8, as redrawn from Guyot and Audier (1985); it was determined by Cooper and Robinson (1966). The (Al,Si) atoms are mostly Al. The structure is cubic, with a_o = 12.625 Å. The space group appears to be $Pm\bar{3}$, but it is very close to being the space group $Im\bar{3}$ of the nearly

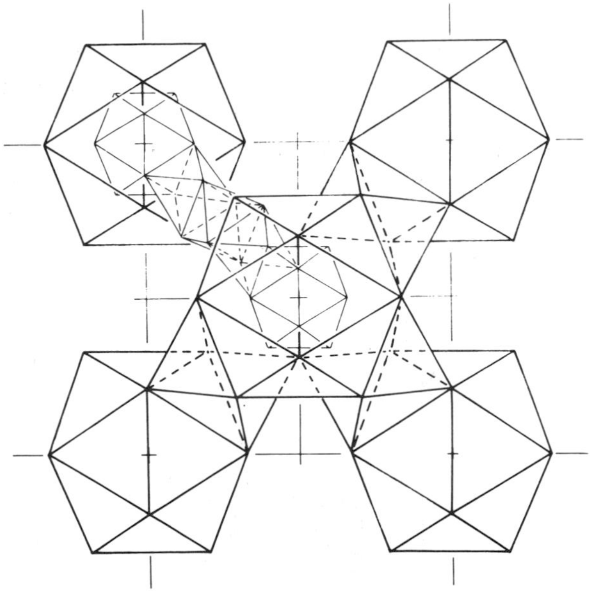

Figure 8. The crystal structure of α-$Mn_{12}(Al,Si)_{57}$. At each of the cubic unit cell corners and at the center of the unit cell is a Mackay icosahedron; this consists of an *empty* inner icosahedron of (Al,Si) atoms (light lines) surrounded by an outer icosahedron of Mn atoms (heavy lines), with additional (Al,Si) atoms near the centers of the edges of that icosahedron. The smaller icosahedra within three of the larger ones are not shown. Also not shown are some additional (Al,Si) atoms between the Mn icosahedra, as well as the (Al,Si) atoms near the centers of the edges of the Mn_{12} icosahedra (except where these are vertices of the octahedra joining the smaller $(Si,Al)_{12}$ icosahedra). Redrawn from Guyot and Audier (1985).

isomorphous α-$Fe_{12}Al_{50}Si_7$ structure (Cooper, 1967). The principal feature of the structure is the presence of a cluster of atoms known as the Mackay icosahedron, so named because it consists of the first two shells of the icosahedral shell packing described by Mackay (1962). This consists of a central atom surrounded by 12 atoms at the vertices of an icosahedron, which in turn is surrounded by 12 more atoms at the vertices of another icosahedron with 30 atoms at or near the centers of its edges. In its application to the present structure, the central atom is missing; the inner icosahedron is composed of (Al,Si) atoms. The empty (Al,Si) icosahedron recalls the empty B_{12} icosahedra which are a striking feature of elementary boron structures (Section 1.2). The outer icosahedron is composed of Mn atoms, and the atoms near the centers of the edges are (Al,Si) atoms. The composition of the cluster is $Mn_{12}(Al,Si)_{42}$. There are additional (Al,Si) atoms between the clusters which will not be discussed here. The Mn_{12} icosahedron of one cluster is linked to the Mn_{12} icosahedron of each of the eight surrounding clusters by sharing faces with a Mn_6 octahedron, while the $(Al,Si)_{12}$ icosahedron inside the cluster is linked to $(Al,Si)_{12}$ icosahedra in each of the eight surrounding clusters with a linear string of three $(Al,Si)_6$ octahedra, also with face sharing. These octahedra contain 24 (Al,Si) atoms per cluster, all of which are among the 30 (Al,Si) atoms that are near the centers of the edges of the Mn_{12} icosahedron. The linking by single octahedra or by strings of three octahedra is consistent with the fact that all icosahedra have the same orientation. (Linking by strings of even numbers of octahedra with face sharing would lead to alternation of icosahedral orientation.)

The icosahedral shell packing described by Mackay is the basis of Carr's (1985) model of a multitwin having icosahedral symmetry. The outer layer of atoms in one face of the cluster described above closely approximates close packing of spheres in a hexagonal layer. Addition of atoms to form more hexagonal layers results in a somewhat distorted cubic closest packing growing outward from that face. The resulting large cluster can be described as an icosatwin of a rhombohedral crystal structure roughly approximating face-centered cubic.

The structure is of current interest to researchers in the field of quasicrystals because the $Mn_{12}(Al,Si)_{42}$ cluster, from which the above-described crystal structure is obtained by repetition at the points of a body-centered cubic lattice (followed by addition of more Al,Si atoms), has been proposed (Elser and Henley, 1985; Guyot and Audier, 1985; Audier and Guyot, 1986) as a structural unit in a model for *i*-Mn-Al-Si, the icosahedral phase of approximately the same composition as the crystalline alpha phase.

3. Some Other Structures Containing Aluminum

3.1 Highly Distorted Icosahedra: The Stable Phase of $MnAl_6$

As already indicated, the composition $MnAl_x$ (where x was originally thought to be 6, but may be less) has been the focus of a great deal of attention. Since icosahedra are found in some alloys containing aluminum and manganese, it would be highly satisfying if such icosahedra were found in the stable crystalline phase of $MnAl_6$. Alas, the structure (as determined by Nicol, 1953, and verified and refined by Kontio and Coppens, 1981) contains no easily identifiable icosahedra, unless extremely distorted icosahedra are permitted. The structure is shown in Fig. 9. Neither the manganese atoms, shown by the smaller circles, nor the aluminum atoms, shown by the larger ones, can be said to be icosahedrally coordinated in the usual context of that term; the respective coordination numbers are ten and eleven, which may result from the removal of two or one atoms (respectively) from a roughly icosahedral coordination shell and allowing

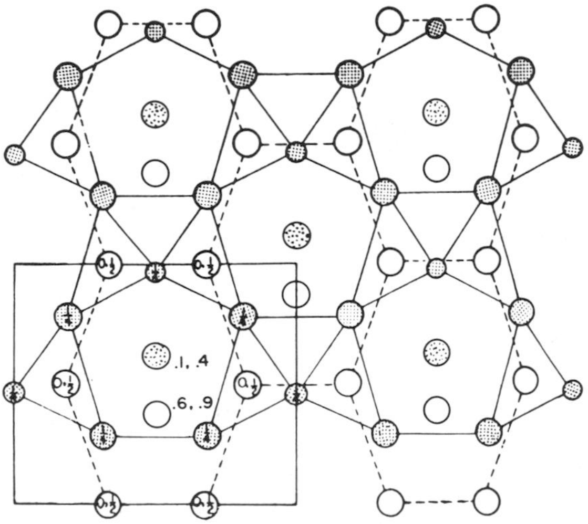

Figure 9. The crystal structure of stable $MnAl_6$. Smaller circles are Mn; larger ones are Al. Elevations of the atoms in the direction normal to the plane of the paper are indicated as fractions of the unit-cell repeat in that direction. [Adapted from Pearson (1972), p. 715. Copyright © 1972 by John Wiley and Sons, Inc.]

the structure to relax. The manganese atoms by themselves form distorted cubes. Meaningful relationships between this structure and others already mentioned are not clearly evident.

3.2 Other Aluminum-Manganese Structures

Structures related to the ones already mentioned include β-Mn_3Al_{10} (Taylor, 1959; this is essentially isostructural with β-Mn_3Al_9Si) and Mn_4Al_{11} (Kontio *et al.*, 1980). The first of these, like the alpha phase, features empty aluminum icosahedra, and the second, like $MnAl_6$, exhibits irregular coordinations (coordination of Mn by ten Al atoms). All show some Mn-Al and Al-Al distances that are verified on refinement to be abnormally short (e.g., Mn-Al = 2.40 Å in Mn_4Al_{11}, expected CN12 Mn-Al distance = 2.7 Å). These and other related structures, found in alloys of other elements, are discussed by Pearson (1972).

4. Tetrahedrally Close-Packed (t.c.p.) Structure Types

It has already been mentioned that the conventional close-packed structures for spherical atoms of equal size possess not only tetrahedral interstices but also octahedral ones, and that the presence of icosahedrally coordinated atoms favors tetrahedral interstices (the tetrahedra being slightly distorted). The alloy structures discussed up to this point do not, however, have tetrahedral interstices exclusively. We shall now discuss an important family of structure types in which all interstices are tetrahedral; we characterize these as being "tetrahedrally close-packed" (t.c.p.) (Shoemaker *et al.*, 1957; Shoemaker and Shoemaker, 1964, 1968, 1969, 1971a, 1972). Nearly all of the representatives of structure types in this family are alloys of transition metals.

An important subfamily of the t.c.p. structure types is that of the Kasper, or Frank–Kasper, phases, as described by Kasper (1956) and Frank and Kasper (1958, 1959). These may be described as layer structures. Typically, in one lattice direction (generally taken as the *c*-axis direction), there are four layers per lattice repeat: two equally spaced "main layers" (generally coinciding with mirror planes of the space group) and two "subsidiary layers" of lower atomic density, each situated approximately halfway between two main layers. The main layers consist of pentagons and/or hexagons and triangles, the atoms being indicated on many of the figures by open circles; the subsidiary layers consist of tesselations of squares or rectangles and triangles, the atoms being indicated by filled circles.

4.1 Ideal t.c.p. in Curved Space: The {3,3,5} and {5,3,3} Regular Polytopes

It is well known that a structure cannot exist in Euclidean space in which all atoms are icosahedrally coordinated. However, a structural arrangement of points, all of which are icosahedrally coordinated, may exist if the Euclidean restriction is removed, as in a Riemannian 3-space embedded in a four-dimensional manifold (Coxeter, 1973).

This particular arrangement is the {3,3,5} regular polytope, the vertices of which may be described on a 3-hypersphere. It may be regarded as the four-dimensional analog of the regular icosahedron in ordinary three-dimensional space, the vertices of which may be described on an ordinary sphere. The {3,3,5} regular polytope has 120 vertices (all coordinated by *regular* icosahedra), 720 edges, 1200 faces (equilateral triangles), and 600 "cells." (The cell is the smallest polyhedron that can be defined by a set of neighboring polytope vertices and which can pack with similar cells to fill all of the space of the structure; in this case, it is a regular tetrahedron). The indices within the curly brackets are: 3 for the number of sides of the (triangular) faces, 3 for the number of faces that come together at each vertex of the (tetrahedral) cell, and 5 for the number of cells that come together at each edge. In ordinary space, the dihedral angle between faces of a regular tetrahedron, 70.547°, differs slightly from 72°, which is exactly one fifth of 360°, so that the tetrahedra have to be slightly distorted so

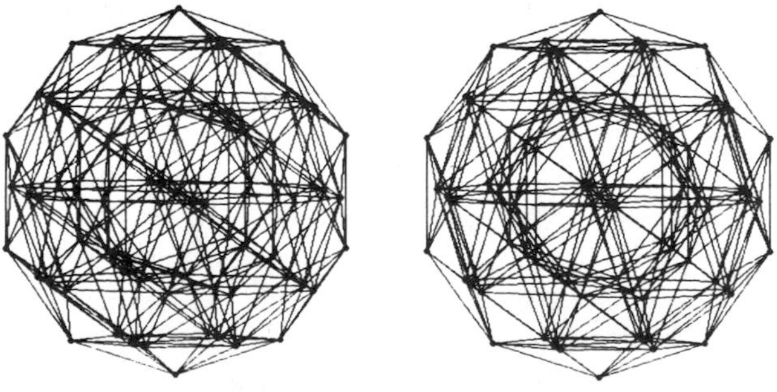

Figure 10. Stereoscopic representation of a projection of the {3,3,5} regular polytope in 4-space onto a 3-hyperplane (Euclidean space). From Shoemaker and Shoemaker (1986).

that five of them fit together around a line; in the curved space, the five regular tetrahedra fit together exactly around a line without distortion.

It is, of course, impossible to portray a four-dimensional figure directly on paper, but a projection of it onto a three-dimensional manifold can be portrayed in a stereofigure. Figure 10 is a stereoscopic representation of a particular projection of the {3,3,5} polytope. This projection is along a normal to a mirror plane of the polytope; accordingly, the 120 vertices project onto only 75 vertices in 3-space.

The {3,3,5} regular polytope may be regarded as the archtype t.c.p. structure model. To relate it to actual t.c.p. structures in Euclidean space, the curved 3-space to which the original 120 vertices are confined must be "flattened," somewhat as an orange is peeled and the peel laid out flat (see, for example, Kléman and Sadoc, 1979; Sadoc and Mosseri, 1982). This flattening must be accompanied by the insertion of additional vertices to fill the tears in the orange peel; when the flattening is complete, the number of vertices will have increased to infinity. The flattening is accomplished by the insertion of "disclination lines" which, when they pass through icosahedra, introduce additional vertices into them, increasing the coordination numbers; the icosahedra that remain become somewhat distorted. Also, the tetrahedra become distorted, and the number that may fit together around an edge increases in some cases from five to six (or more), or even occasionally decreases to four with very considerable distortion of the tetrahedra. Interstitial voids other than tetrahedra may be created. Thus may be generated, in principle, all t.c.p. structure types, plus structure types that are not t.c.p., as well as amorphous, quasiperiodic, and other aperiodic structures (Nelson, 1983). A more detailed review of this approach is given in the next chapter written by M. Widom.

As the icosahedron has its dual, so does the {3,3,5} regular polytope. Its dual is the {5,3,3} polytope which has 600 vertices, 1200 edges, 720 faces, and 120 cells. The faces are regular pentagons, and the cells are pentagonal dodecahedra. The dual of the {5,3,3} regular polytope is again the {3,3,5} regular polytope. The {5,3,3} regular polytope may be regarded as the archtype model for a froth and has been used in theoretical studies of the structures of froths (Coxeter, 1958).

4.2 *Characteristics of t.c.p. Structure Types*

The principal properties characterizing the structure types in the t.c.p. family are as follows (Shoemaker and Shoemaker, 1986): (1) all of the interstices are tetrahedral, the tetrahedra with atomic centers as vertices being only moderately distorted, with ratios of longest to shortest edges

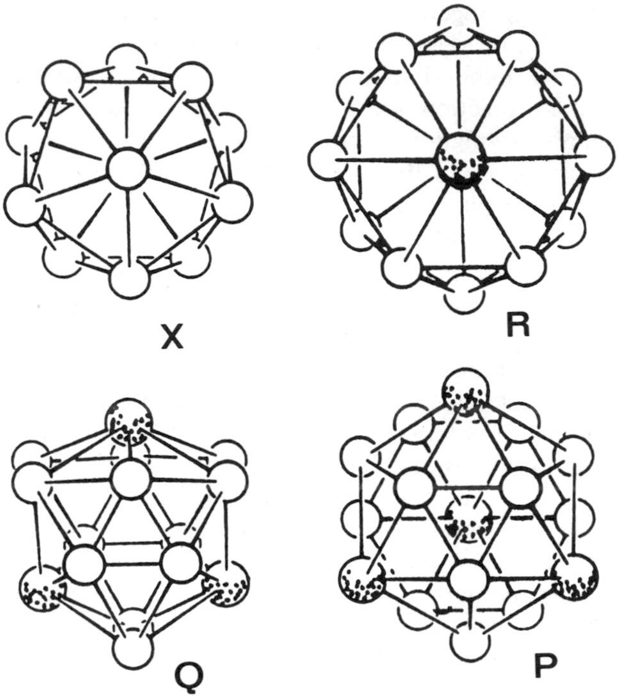

Figure 11. Coordination polyhedra found in t.c.p. structures: *X* (CN12); *R* (CN14); *Q* (CN15); *P* (CN16). [Adapted from Pearson (1972), p. 34. Copyright © 1972 by John Wiley and Sons, Inc.]

not exceeding about 4:3; and (2) the coordination types are limited to a particular set of four, which we here call *X*, *R*, *Q*, and *P*, with triangulated coordination polyhedra and with coordination numbers (CN) of 12, 14, 15, and 16, respectively. The *X*, or CN12, polyhedron is an approximately regular icosahedron. These polyhedra are shown in Fig. 11, and their characteristics are presented in Table 1. We will also use the letters *X*, *R*, *Q*, and *P* to designate the correspondingly coordinated atoms in a t.c.p. structure.

As the icosahedron, or *X* coordination polyhedron, has a dual, which is a regular pentagonal dodecahedron, the polyhedra *P*, *Q*, and *R* also have their duals, which are respectively a hexakaidecahedron, a pentakaidecahedron, and a tetrakaidecahedron having both pentagonal and hexagonal faces and having respectively 28, 26, and 24 threefold vertices compared with 20 for the pentagonal dodecahedron. These four duals are

Table 1. T.c.p. coordination polyhedra[a]

Type	Ideal point symmetry	No. of vertices 5-fold	6-fold	No. of faces[d]
P CN16	T_d-$\bar{4}3m$	12	4[c]	28
Q CN15	D_{3h}-$\bar{6}m2$	12	3[c]	26
R CN14	D_{6d}-$\overline{12}.2.m$	12	2[c]	24
X CN12[b]	I_h-$m\bar{3}\bar{5}$	12	0	20

[a]From Shoemaker and Shoemaker (1986).

[b]Regular, or approximately regular, icosahedron.

[c]Respectively disposed tetrahedrally, trigonally in plane, and digonally on axis, with respect to the center of the polyhedron.

[d]Also equal to the number of interstitial tetrahedra defined by the vertices of the faces and the center of the polyhedron.

(apart from scale) the Voronoï or Wigner–Seitz cells for the correspondingly coordinated atoms in the t.c.p. structure. Within any of these cells all points are closer to the central atom than to any other atoms in the structure.

These four coordination types are present in various combinations in the different t.c.p. structure types. The "empirical coordination formula" of a t.c.p. structure may be represented as $P_pQ_qR_rX_x$, where p, q, r, and x are integers (usually taken relatively prime) or rational fractions. The value of x may never be zero, and in Euclidean space, at least one of the other three variables must be greater than zero. The values of p, q, r, and x are not all arbitrary or independent; Yarmolyuk and Kripyakevich (1974) found empirically a restriction on these variables. Shoemaker and Shoemaker (1986) expressed this restriction algebraically in the form

$$x = 2p + 7q/6 + r/3, \qquad q \leq r, \tag{4.1a,b}$$

and established the generality of Eq. (4.1a) by arguments based on dihedral angles in the tetrahedra. In its first approximation, this treatment was based on the requirement that the relative values of p, q, r, and x be such as to cause the *average* of all tetrahedron dihedral angles in the structure to be equal to the dihedral angle in a regular tetrahedron, $\cos^{-1}(1/3) = 70.529°$. Closer approximations took account of second- and higher-order variations of the average dihedral angle in a tetrahedron caused by distortions of the tetrahedron from regularity (the first order variations being zero).

The line from an atom to a neighbor on its coordination shell may be called a "bond." If the neighbor occupies a fivefold vertex, the bond is called a "minor bond"; if it occupies a sixfold vertex, it is called a "major bond" (formerly "minor ligand" and "major ligand"; Frank and Kasper, 1959). A minor bond passes through a ring of five atoms, while a major bond passes through a ring of six. A major bond is shorter by approximately 0.4 Å than a minor bond between the same two kinds of atoms.

The aggregate of major bonds in a structure form a network which has been given the name "major network" (Frank and Kasper, 1959). The major network may be fully connected through three-dimensional space, or it may consist of a finite or infinite number of mutually noninterconnected networks which may interpenetrate or interlock. Presumably the major network corresponds to the aggregate of "disclination lines" that are required to flatten the {3,3,5} regular polytope to the t.c.p. structure in question. As we proceed in our description of typical t.c.p. structures, we will in some cases describe the major networks of these structures.

The number of known t.c.p. structure types is, at this writing, at least 23. Essential data for these 23 structures are presented in Table 2. Generally, since the space groups and lattice constants of the t.c.p. structures are given in Table 2, they will not be given in the text.

4.3 *The "Base" t.c.p. Structure Types: A15, Zr_4Al_3, C15*

Three of the most simple t.c.p. structures, layer structures with coordination formulae R_3X, $Q_2R_2X_3$, and PX_2, are of special importance because nearly all layer-type t.c.p. structures may be constructed from fragments of these three, as shown, for example, by Yarmolyuk and Kripyakevich (1974). (The reader may verify this for the layer-type structures in some of the figures that follow.) Indeed, these researchers showed that the coordination formulae of all t.c.p. structures known at that time can be recast as follows:

$$P_pQ_qR_rX_x \rightarrow (PX_2)_i(Q_2R_2X_3)_j(R_3X)_k \qquad (4.2)$$

where i, j, and k are integers, whether the structures are layer structures or not; this has also been found to be true for all t.c.p. structures discovered since that time. It was from this formulation that Eq. (4.1a) was derived by Shoemaker and Shoemaker (1986).

The three base structures are as follows:

4.3.1 ***R_3X:*** **A15 "β-Tungsten Structure," or Cr_3Si.** This cubic structure type (Borén, 1933), is illustrated in Fig. 12a,b. This structure type has the

Table 2. Summary of t.c.p. structure types[a]

Type	Example	Space group	Lattice constants				N	Eq. (4-2)			Eq. (4-1)				$CN_{av.}$	Quasicrystal claimed?[b]	Section in text
			a(Å)	*b*(Å)	*c*(Å)	β(°)		i	j	k	p	q	r	x			
A15	β-W, Cr_3Si	$Pm\bar{3}n$	4.564				8	0	0	1	0	0	3	1	13.500		4.3.1
σ	$Cr_{46}Fe_{54}$, β-U	$P4_2/mnm$	8.800		4.544		30	0	1	2	0	2	8	5	13.467		4.4.1
H	Complex	*Cmmm*	4.5	17.5	4.5		30	0	1	2	0	2	8	5	13.467		4.4.1
K′	Complex	*Pmmm*	12.5	17.1	4.5		82	0	3	5	0	6	21	14	13.463		4.4.1
F	Complex	*P6/mmm*	12.5		4.5		52	0	2	3	0	4	13	9	13.462		4.4.1
J	Complex	*Pmmm*	4.5	12.5	4.5		22	0	1	1	0	2	5	4	13.455		4.4.1
ν	$Mn_{81.5}Si_{18.5}$	*Immm*	16.992	28.634	4.656		186	6	5	10	6	10	40	37	13.441		4.4.1
Z	Zr_4Al_3	*P6/mmm*	5.433		5.390		7	0	1	0	0	2	2	3	13.428		4.3.2
P	$Mo_{42}Cr_{18}Ni_{40}$	*Pbnm*	9.070	16.983	4.752		56	1	1	1	1	2	5	6	13.428		4.4.1
δ	MoNi	$P2_12_12_1$	9.108	9.108	8.852		56	1	1	1	1	2	5	6	13.428		4.4.2
K	$Mn_{77}Fe_4Si_{19}$	*C2*	13.362	11.645	8.734	90.5	220	7	2	5	7	4	19	25	13.418		4.4.2
R	$Mo_{31}Co_{18}Cr_{51}$	$R\bar{3}$	10.903		19.342		159	8	3	2	8	6	12	27	13.396		4.4.2
μ	Mo_6Co_7	$R\bar{3}m$	4.762		25.615		39	2	1	0	2	2	2	7	13.385		4.4.1
-	K_7Cs_6	$P6_3/mmc$	9.078		32.950		26	2	1	0	2	2	2	7	13.385		4.4.1
*p*σ	$W_6(Fe,Si)_7$	*Pbam*	9.283	7.817	4.755		26	2	1	0	2	2	2	7	13.385		4.4.1
M	$Nb_{48}Ni_{39}Al_{13}$	*Pnam*	9.303	16.266	4.933		52	2	1	0	2	2	2	7	13.385		4.4.1
I	$V_{41}Ni_{36}Si_{23}$	*Cc*	13.462	23.381	8.940	100.3	228	4	1	0	4	2	2	11	13.369	Yes?	4.4.2
C	$V_2(Co,Si)_3$	*C2/m*	17.17	4.66	7.55	99.2	50	6	1	0	6	2	2	15	13.360		4.4.1
T	$Mg_{32}(Zn,Al)_{49}$	$Im\bar{3}$	14.16				162	20	3	0	20	6	6	49	13.358	Yes	4.5
X	$Mn_{45}Co_{40}Si_{15}$	*Pnnm*	12.47	15.50	4.76		74	10	1	0	10	2	2	23	13.351		4.4.1
-	Mg_4Zn_7	*C2/m*	25.96	5.24	14.28	102.5	110	16	1	0	16	2	2	35	13.345		4.4.1
C14	$MgZn_2$	$P6_3/mmc$	5.16		8.50		12	1	0	0	1	0	0	2	13.333	Yes	4.3.3
C15	$MgCu_2$	$Fd\bar{3}m$	7.080				24	1	0	0	1	0	0	2	13.333		4.3.3

[a]Adapted from Shoemaker and Shoemaker (1986). ''Almost'' t.c.p. structure types, containing some nontetrahedral interstices and/or some abnormal coordination types, are not included in this table. For references to individual structures, see the discussions of them in the text section indicated.
[b]From Kuo *et al.* (1987).

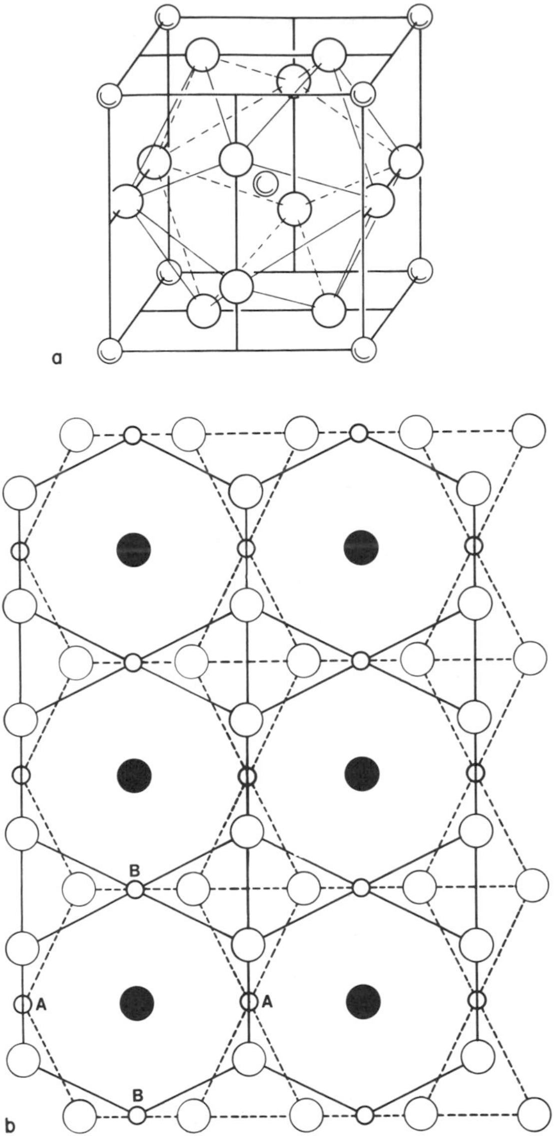

Figure 12. The A15 crystal structure (β-W; Cr_3Si). (a) Perspective drawing. Smaller circles are *X* (*e.g.*, Si); larger ones are *R* (*e.g.*, Cr). Major network is shown by lines drawn through *R* atoms; light lines define approximately regular icosahedron around the central *X* atom. (b) Projection on basal plane. Part (b) is from Shoemaker and Shoemaker (1969).

Note: In this and subsequent similar figures main layers are shown by open circles connected by solid or dashed lines, and subsidiary layers which are intercalated between the main layers are shown by filled circles; letters indicate the different orientations of the icosahedra centered by *X* atoms.

smallest fraction of X atoms of any t.c.p. structure known; in the figure, these atoms are at the unit cell corners and the body center. The icosahedra, centered by the atoms at a cell corner and at a body center, share a triangular face, resulting in an alternation of icosahedral orientation $A, B, A, \ldots$ in the body-diagonal direction. Along the cube edge, the icosahedra share an edge, resulting in the sequence $A, A, A. \ldots$ The orientations are shown by the lettering in Fig. (12b). The major network, on which all R atoms lie, consists of three nonintersecting nets of parallel lines. These are considered to play some role in the superconductivity of some alloys having this structure, such as Nb_3Sn. The structure is layered in all three directions; X and R atoms are on the main layers, and R atoms are on the subsidiary ones.

4.3.2 $Q_2R_2X_3$: Zr_4Al_3 or Z Phase. The hexagonal structure type represented by Zr_4Al_3 is illustrated in Fig. 13a,b; Hf_4Al_3 also has this structure. The main layers (open circles in Fig. 13a) are composed alternately of Q atoms (large circles) and X atoms (small circles); the subsidiary layers (filled circles) are composed of R atoms. The major network consists of an infinite stack of chicken-wire nets on the Q main layers, plus an infinite array of strings of R atoms running through the holes in the nets. There are *three* icosahedral orientations present (A, B, C), resulting from interpenetration of the icosahedra centered on neighboring atoms in the largely triangulated X main layer, as shown in Fig. 13a. Along a straight row of X atoms in the X main layer, the orientations alternate $A, B, A, B, \ldots$, or $B, C, B, C, \ldots$, or $C, A, C, A, \ldots$, in the three different directions. In the direction perpendicular to the nets, the icosahedra share edges; thus the three distinct orientations are each propagated without alternation in that direction.

The structure is also layered in projection on the $(11\bar{2}0)$ plane, as shown in Fig. 13b, but the main and subsidiary layers are differently defined than in Fig. 13a. Here the main layers consist of pentagons and triangles and contain all three kinds of atoms, while the subsidiary layers contain only X atoms. In the direction perpendicular to the paper, the icosahedra centered by X atoms (filled circles) interpenetrate and therefore alternate in orientation. The orientations of the icosahedra centered by X atoms within the subsidiary layers alternate between A and B along the heavy dash-dot line, but remain constant at A *or* B (depending on the level) along the heavy dashed line; the third orientation, C, applies to the X atoms on the main layers.

4.3.3 PX_2: C15 or $MgCu_2$ (also C14, $MgZn_2$; C36, $MgNi_2$). The structure types C15, C14, and C36 are the three so-called *Laves phases* (Laves and Witte, 1935); since the first two were actually discovered earlier by Friauf

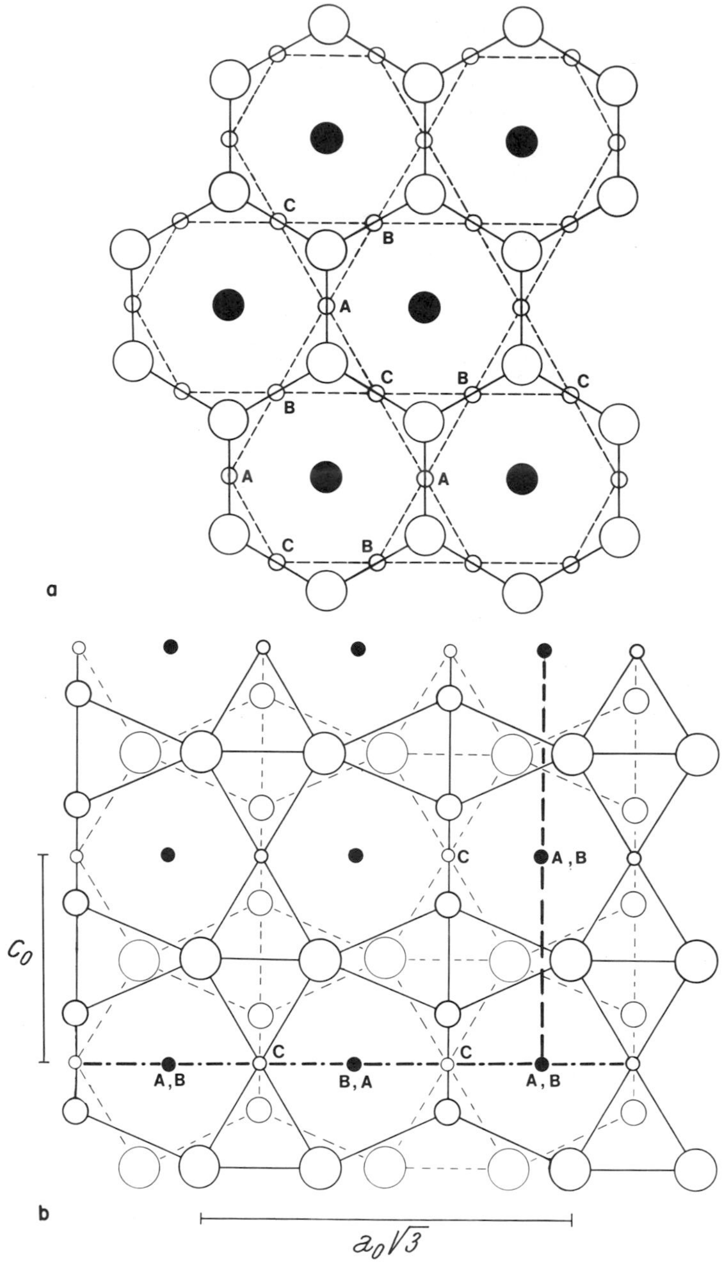

Figure 13. The Zr_4A1_3 crystal structure. (a) Projection on basal plane. Large circles are *Q;* medium circles are *R* (both Zr); small circles are *X* (A1). (b) Projection on $(11\bar{2}0)$. Atomic identification as in (a). Note the different constitutions of main and subsidiary layers in the two representations of this structure. Parts (a) and (b) are from Shoemaker and Shoemaker, (1969) and (1967), respectively.

Note: In this and simliar figures that follow, icosahedra centered by filled circles share faces (e.g.) along the heavy dash-dot line; they share edges along the heavy dashed line.

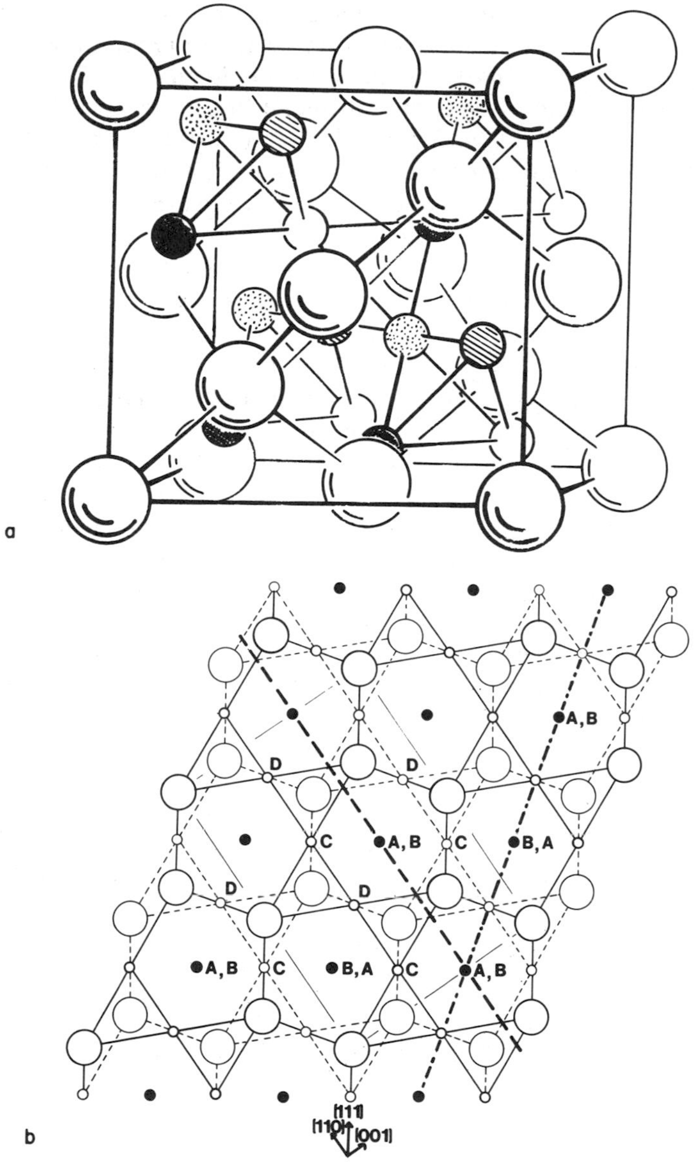

Figure 14. The C15 crystal structure (e g., $MgCu_2$). Large open circles are P (Mg); all small circles are X (Cu). (a) Perspective drawing. The four different shadings of the small circles correspond to four different orientations of the icosahedral coordination polyhedron. (b) Projection on $(1\bar{1}0)$. Part (a) is adapted from Pauling (1960) p. 426 [copyright notice in legend in Fig. 6 applies also to this figure]; part (b) is from Shoemaker and Shoemaker (1967).

(1927a, b), we prefer to call them *Friauf–Laves* phases. We turn first, and with primary emphasis, to the cubic C15 structure type, shown in Fig. 14a,b. Figure 14a shows the contents of the cubic unit cell. The *P* atoms (large circles) by themselves form a diamond-like arrangement, which is the major network of this structure; the centers of the tetrahedra of *X* atoms form another diamond-like arrangement that interpenetrates the first. The *X* icosahedra interpenetrate, and there are four orientations present; these are indicated in the figure by different shadings, to which the letters *A*, *B*, *C*, and *D* may be assigned in any order. Along the straight rows of *X* atoms, the orientations alternate between *A* and *B*, *A* and *C*, *A* and *D*, *B* and *C*, *B* and *D*, or *C* and *D* in the six different directions. Figure 14b shows a projection of the structure on a diagonal plane, $(1\bar{1}0)$. This is indeed a layer structure; the main layers (open circles) contain *P* atoms (large circles) and *X* atoms (small circles), while the subsidiary layers (filled small circles) contain *X* atoms. In this and later figures of this kind, the orientation of the icosahedra and their alternation or nonalternation in certain directions are indicated by letters and heavy dash-dot or dashed lines, as in previous figures, and will not be described in much detail.

In Fig. 15a is shown a rhombohedral unit cell decorated with *P* atoms (large dotted circles) and *X* atoms (small open circles); if the apical angle were exactly 60°, this would be the primitive unit cell of the C15 structure. Units such as this one, stacked together by repetition with the face-centered cubic translation group, would generate completely the structure type we have been discussing. However, the figure was drawn to represent an object with apical angle 63.435° (as in the rhombic faces of the rhombic triacontahedron, Section 1.1 and Fig. 3), and the figure was drawn to represent one of the two fundamental rhombohedral building blocks postulated for quasicrystals, decorated with aluminum, zinc, and magnesium,

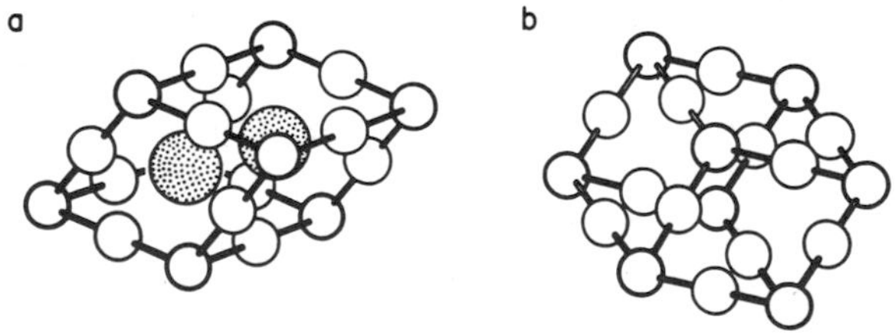

Figure 15. (a) Acute rhombohedral building unit, and (b) obtuse rhombohedral building unit for quasicrystal model of rapidly quenched Mg_{32} $(ZnAl)_{49}$ proposed by Henley and Elser (1986), with decoration proposed by these authors. Acute rhombohedral unit resembles primitive unit cell for the crystalline phase of C15 ($MgCu_2$). From Henley and Elser (1986).

as proposed by Henley and Elser (1986) for the icosahedral phase of the alloy $Mg_{32}(Zn,Al)_{49}$ (the structure of the crystalline phase of which will be discussed in Section 4.5). (In Fig. 15b is shown the complimentary quasicrystal building block, an obtuse rhombohedron, with apical angle 116.564°, decorated as proposed by Elser and Henley.) Thereby is demonstrated a connection between a known crystal structure and a postulated quasicrystal structure.

This structure type is exemplified by a large number of binary compounds of transition metals, as is also the hexagonal structure type C14, with space group $P6_3/mmc$. This structure is shown in Fig. 16. While the C15 structure type is cubic, C14 is hexagonal; the figure shows the structure projected onto the $(11\bar{2}0)$ plane. In this structure there are five icosahedral orientations, indicated by the letters *A–E*. The third Friauf–Laves structure type, C36, is a mixture of C15 and C14 slabs alternating along

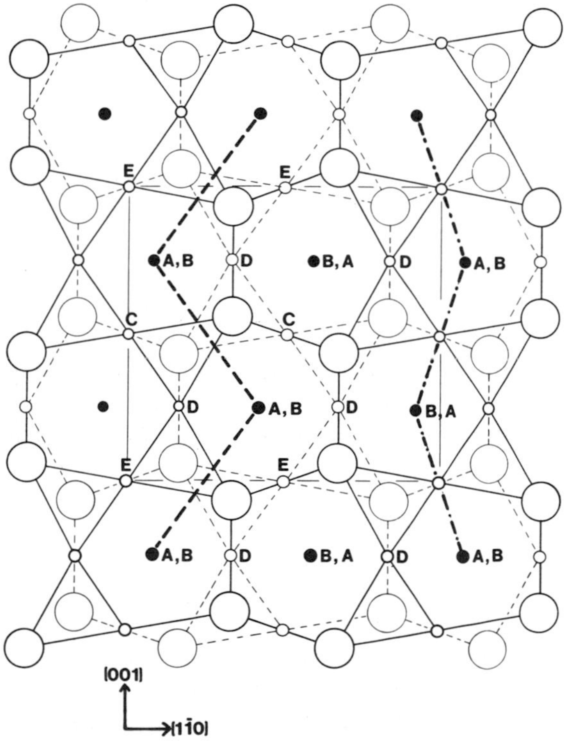

Figure 16. The C14 crystal structure (e.g., $MgZn_2$), projected on $(11\bar{2}0)$. Large circles are *P* (Mg), small ones are *X* (Zn). From Shoemaker and Shoemaker (1967).

a hexagonal axis, and hence may be regarded as a polytype; many more complex polytypes have been discovered by Komura's group (Komura and Kitano, 1977).

Incidentally, in the Mn_3Ni_2Si system an icosahedral phase has been claimed by Kuo *et al.* (1986a) which is structurally related to the Friauf–Laves phases, particularly C14 ($MgZn_2$).

The structure types C14, C36, and the Komura polytypes are not here considered base structures additional to C15, which itself suffices to represent the PX_2 formulation.

The structures of the Friauf–Laves phases, and a number of other phases, have been discussed very effectively by Samson (1968) in terms of what he calls the *Friauf polyhedron*, a truncated tetrahedron that will be described later (Section 4.5.3, Fig. 26a). It is worth noting that the primitive rhombohedral unit cell of the face-centered cubic C15 structure (resembling Fig. 15a, but with an apical angle of 60°) can be constructed of two Friauf polyhedra joined with sharing of hexagonal faces at an inversion center, plus tetrahedra at the two ends sharing triangular faces with the Friauf polyhedra.

4.4 More Complex t.c.p. Structure Types

4.4.1 Layer Structures. The sigma phase, represented by the approximate formula $Cr_{46}Fe_{54}$, occurs as a nuisance precipitate in stainless steels. This phase occurs also in many other transition metal systems. A substantially similar structure exists for the beta phase of uranium metal (Donohue and Einspahr, 1971). The structure was elucidated independently in the early 1950s by several groups, including Bergman and Shoemaker (1954). As shown in Fig. 17a, it is a tetragonal layer structure containing alternating approximately hexagonal main layers related by a 4_2 screw axis. Some of the X atoms are lettered A, B, and C along the bottom of the figure to show the four icosahedral orientations possible in that main layer. The atoms in the triangle at the lower left center icosahedra which interpenetrate; the two B icosahedra are linked by sharing an edge. It may be seen that the orientations are in accord with the lattice translation. There are similar relations among the X-atom orientations on the other main layer, but the actual orientations differ, being related to the ones previously mentioned by rotation of 90° around the c axis (perpendicular to the plane of the paper). The interconnections of icosahedral orientations between the two main layers are through B and D, and through C and E, through face sharing. Normal to the plane of the figure the icosahedra share edges, and accordingly, their orientations do not alternate in that direction.

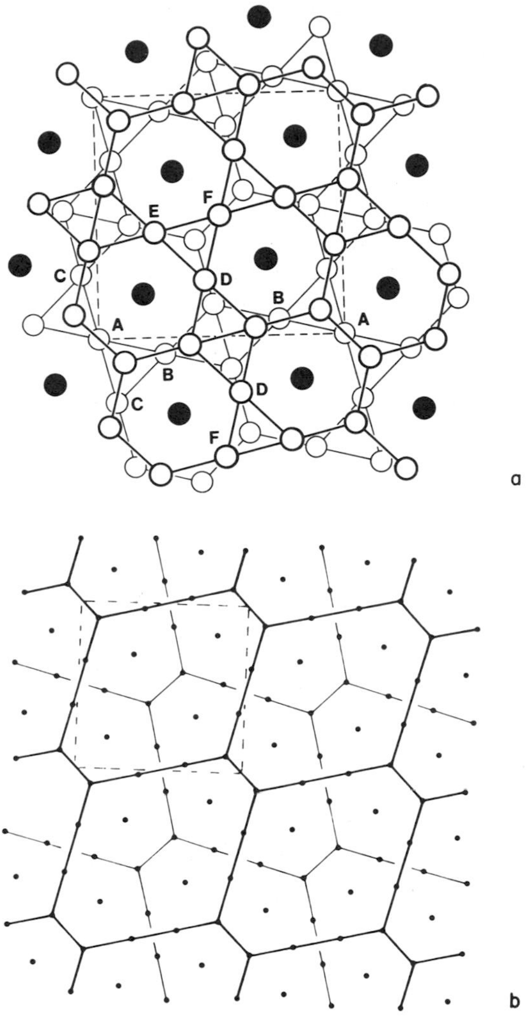

Figure 17. The σ-phase crystal structure (e.g., $Cr_{46}Fe_{54}$). (a) Projection on basal plane. (b) Major network. Part (a) is from Bergman and Shoemaker (1954). Note that the letters *A*, *B*, *C*, refer to icosahedral orientation, not to atom type as in the paper referenced.

While we have given no attention, so far, to the orientation of higher-coordinated polyhedra, we might mention that the orientations of the coordination polyhedra for the *R* atoms in the subsidiary layers alternate between two orientations in the *c*-axis direction; i.e., the orientations of the six-membered rings alternate. This is of interest in connection with a quasicrystalline phase with 12-fold rotation symmetry claimed by Ishimasa *et al.* (1985).

Figure 17b shows the major network of the sigma phase which consists of layers of Kagomé tiles alternating in orientation along the 4_2 screw axis interpenetrated by parallel rows of *R* atoms perpendicular to the layers.

An important non-t.c.p. crystal structure which is closely related to the sigma phase structure is that of alpha manganese (Bradley and Thewlis, 1927) and the isostructural chi phases. The structure is cubic, and in any of the three cubic directions, one can see warped sigma-phase-type nets. Such nets can also be seen in the D phase of Mn_5Si_2 (Shoemaker and Shoemaker, 1976).

A rather similar structure type, which we have taken the liberty of naming "pentagon sigma" ($p\sigma$), is shown in Fig. 18. This is represented

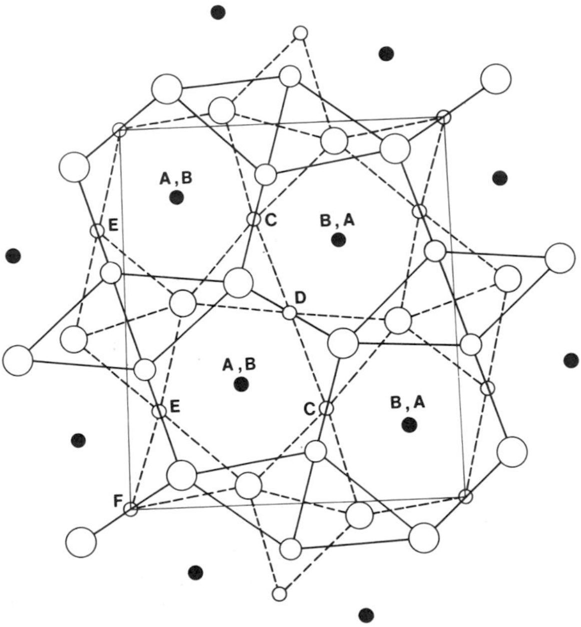

Figure 18. The $p\sigma$ crystal structure (e.g. $W_6(Fe,Si)_7$). From Shoemaker and Shoemaker (1969).

by $W_6(Fe,Si)_7$ (Kripyakevich and Yarmolyuk, 1974) and Th_6Cd_7 (Fornasini *et al.*, 1984). It is an orthorhombic structure. The six-membered rings of the sigma phase are replaced here by five-membered rings, and the atoms in rows passing through them perpendicular to the main layers (filled circles) are now *X* atoms, the icosahedral coordination polyhedra of which interpenetrate with their neighbors in the rows and therefore alternate along the rows, as in several structures previously discussed. The propagation of orientation parallel to the layers is shown in the figure. The structure can be formed from fragments of the Zr_4Al_3 and C15 structures (Figs. 13b and 14b, respectively).

The P-phase structure, represented by the composition $Mo_{42}Cr_{18}Ni_{40}$, is shown in Fig. 19. This structure, determined by Shoemaker *et al.* (1957), is orthorhombic with a unit cell nearly twice as long in one dimension as the sigma phase, resembling two sigma phase cells joined together, with half of the hexagons changed to pentagons, and therefore half of the *R* atoms in the subsidiary layers changed to *X* atoms. Icosahedral orientations and their relationships are similar to those in the sigma phase and other phases previously discussed. The *R*-atom rows (filled circles) constitute part of the major network. The remainder of the major network consists of two interlocking, mutually noninterconnected infinite three-dimensional networks.

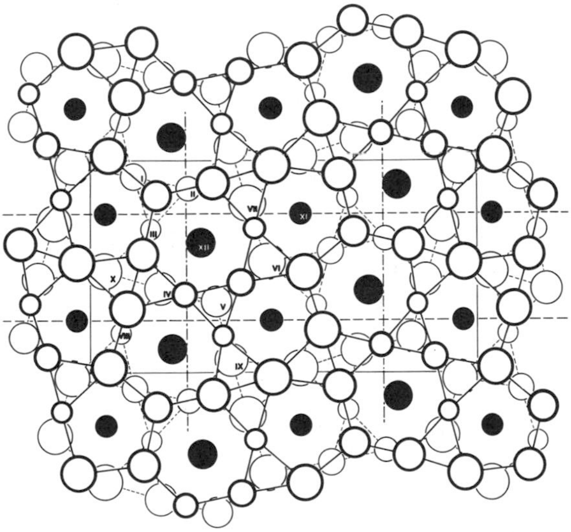

Figure 19. The P-phase crystal structure ($Mo_{42}Cr_{18}Ni_{40}$). From Shoemaker *et al.* (1957).

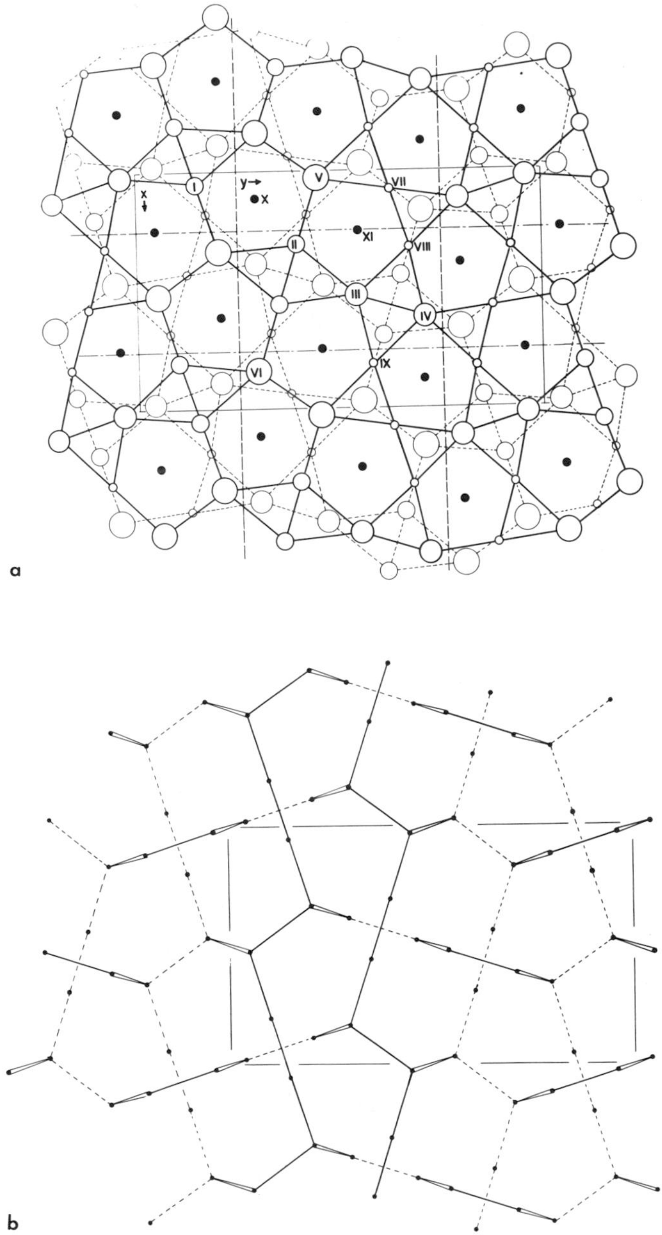

Figure 20. The M-phase crystal structure ($Nb_{48}Ni_{39}Al_{13}$). (a) The structure projected on the basal plane. (b) The major network. From Shoemaker and Shoemaker (1967).

Bearing much the same relationship to the P phase as the pentagon sigma phase does to the sigma phase is the orthorhombic M phase, represented by the composition $Nb_{48}Ni_{39}Al_{13}$ (Shoemaker and Shoemaker, 1967). As shown in Fig. 20a, the hexagons in the plane layers in the P phase are replaced by pentagons in the M phase, so that all of the atoms in subsidiary layers (filled circles) are *X* atoms. The major network is shown in Fig. 20b. It consists of two interlocking, noninterconnected three-dimensional networks somewhat similar to those in the P phase and contains no parallel rows of *R* atoms perpendicular to the layers as found in the P phase and many other layer structures.

The mu-phase structure type, represented by Mo_6Co_7 (Arnfelt and Westgren, 1935), is shown in Fig. 21, in projection on the (110) plane.

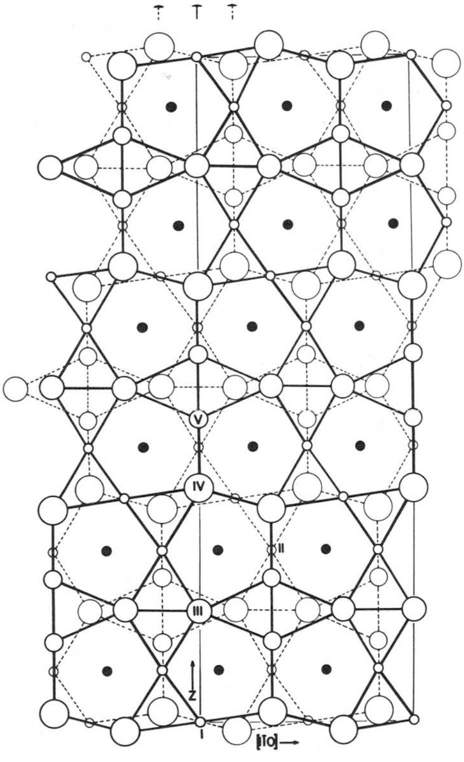

Figure 21. The μ-phase crystal structure (e.g., Mo_6Co_7), projected on the $(11\bar{2}0)$ plane. Large circles are *P*, *Q*, and *R* (Mo); small ones are *X* (Co). From Shoemaker and Shoemaker (1967).

This structure is rhombohedral. It provides another simple example of the building up of layer t.c.p. structures from fragments of base structures. The structure alternates slabs of the Zr_4Al_3 structure (Fig. 13b) and the C15 structure (Fig. 14b). Once the slabs have been identified on the figures, the orientational relationships of the icosahedra can be identified by reference to the two component structures. It is seen that there are strings of icosahedra that alternate in sharing edges and sharing faces. Note that all atoms on subsidiary layers have only two icosahedral orientations.

There exists a number of other layer t.c.p. structure types that have been determined accurately by X-ray diffraction, namely the phases named ν (Shoemaker and Shoemaker, 1971b), C (Kripyakevich and Yarmolyuk, 1970), and X (Yarmolyuk *et al.*, 1970; Manor *et al.*, 1972), and the alloys K_7Cs_6 (Simon *et al.*, 1976) and Mg_4Zn_7 (Yarmolyuk *et al.*, 1975). Recently, additional layer t.c.p. structures have been found by high-resolution electron microscopy, in the phases named H (Ye *et al.*, 1984) and F, K′ (the prime is to differentiate this phase from the phase already named K in Table 2, the structure of which was determined earlier by X-ray diffraction), and J (Li and Kuo, 1986). These phases are of complex composition; all are Ni-based (except H, which is Fe-based) and contain also Cr and several other transition metals. The structures contain main layers of hexagons and triangles and are closely related to the sigma phase. To the best of our knowledge, they have not been verified and accurately refined by quantitative diffraction methods. All of the above structure types are listed in Table 2.

4.4.2 Nonlayer Structures. These are the phases labelled δ (Shoemaker and Shoemaker, 1963), K (Shoemaker and Shoemaker, 1977), R (Komura *et al.*, 1960; Shoemaker and Shoemaker, 1978), I (Shoemaker and Shoemaker, 1981), and T (Bergman *et al.*, 1957) in Table 2. The cubic T phase, for special reasons, is treated separately in Section 4.5. The other four structure types are of low symmetry (orthorhombic, monoclinic, rhombohedral, and monoclinic, respectively) and are of considerable complexity. The monoclinic cell of the I phase in the system $V_{41}Ni_{36}Si_{23}$ contains 228 atoms of 57 crystallographically different kinds! (Incidentally, Kuo *et al.* (1987) have claimed the existence of an icosahedral phase in an alloy of the same composition as this I phase; this is now in doubt.)

The various kinds of orientational relationships among the icosahedra that are exemplified by the layer structures described are also present in the nonlayer ones, but the difficult task of describing them will not be attempted here. We shall content ourselves by presenting stereofigures of the major networks for two of them: the R phase (space group $R\bar{3}$; exemplified by $Mo_{31}Cr_{18}Co_{51}$) in Fig. 22, and the K phase (space group

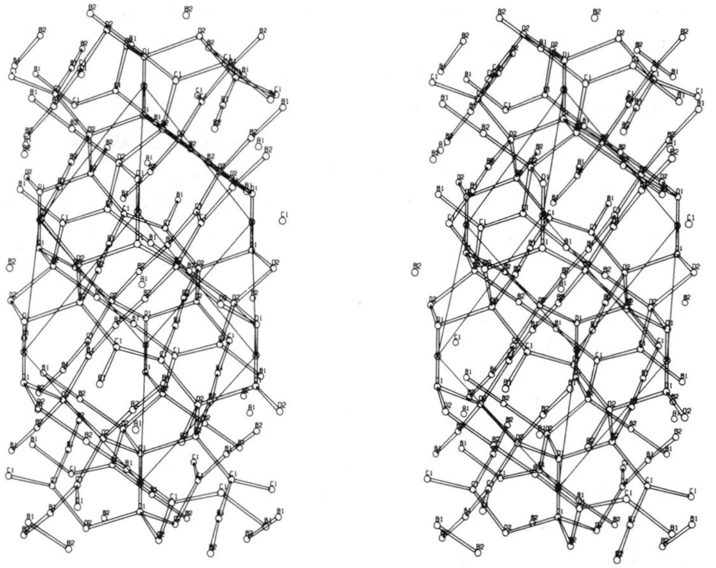

Figure 22. Stereoscopic representation of the major network of the R phase ($Mo_{31}Co_{18}Cr_{51}$ or $Mn_{85.5}Si_{14.5}$). From Shoemaker and Shoemaker (1978).

*C*2; $Mn_{77}Fe_4Si_{19}$) in Fig. 23. In both figures, the major bonds are indicated by double lines. In Fig. 22, the rhombohedral unit cell of the R-phase structure is outlined with single lines. The R-phase major network is a single infinite three-dimensional network, while the K-phase major network consists of *three* interlocking mutually noninterconnecting infinite three-

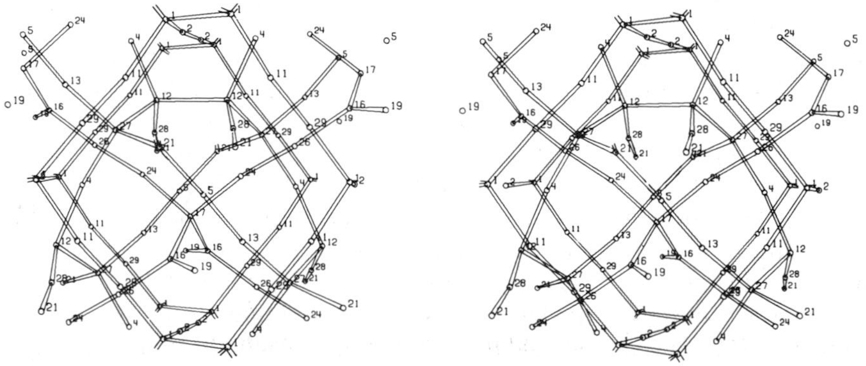

Figure 23. Stereoscopic representation of the major network of the K phase ($Mn_{77}Fe_4Si_{19}$). From Shoemaker and Shoemaker (1977).

dimensional networks. In the R phase, Friauf polyhedra (see Fig. 26a), coordinating *P* atoms (tetrahedral nodes in the major network), stack in zigzags along the edges of the rhombohedral cell. There are local regions where the major bonds form almost planar sheets; these sheets contain rings containing two parallel *P-R-R-Q* rods connected at the ends by *Q* atoms. However, the R phase cannot be considered as a layer structure except in local regions. (A distorted form of the R-phase structure, not entirely t.c.p., exists for the phase ε-$Mg_{23}Al_{30}$; Samson, 1968.)

4.5 The Structure of the T Phase, $Mg_{32}(Zn,Al)_{49}$

The cubic T-phase structure (Bergman *et al.*, 1957), with space group $Im\bar{3}$ and unit cell edge 14.16 Å, is of particular interest in the context of quasicrystals. An icosahedral phase of approximately this composition has been claimed to exist, and a quasicrystal structure based on the decorated rhombohedral building units shown in Fig. 15 has been postulated for it (Henley and Elser, 1986). Moreover, an isomorph of this phase exists as the compound $Al_5Cu\,Li_3$ (Cherkashin *et al.*, 1964); an alloy of similar composition has been found as an icosahedral phase (Dubost *et al.*, 1986) which, interestingly enough, occurs in the form of small rhombic triacontahedra! Features of the T-phase crystal structure have played a role in formulating a quasicrystal model for this phase (Audier *et al.*, 1986). This crystal structure is also of interest because it can be interpreted in several different illuminating ways, as follows.

4.5.1 The Stochastic Approach. The structure of the T phase was first proposed by Pauling (1955), using the stochastic method, which in his hands has been outstandingly successful in the formulation of structure models for many other substances. The stochastic method consists essentially of trial-and-error with educated guesses. A cluster, to be placed at the unit cell corners and body center, was formulated by Pauling as shown in Fig. 24, based on placement of atoms in successive shells around a central atom with high space-filling efficiency. The central atom is surrounded first by 12 atoms in an icosahedral shell. Then 20 atoms are placed in the dimples in that shell, forming a pentagonal dodecahedron, and 12 more atoms are placed in the resulting dimples forming a larger icosahedron. Those (20 + 12) atoms form approximately a rhombic triacontahedron, which is described in Section 1.1. (The rhombic triacontahedral shell, with the central atom and its surrounding icosahedron inside, is the same 45-atom arrangement that has been described as a feature of the

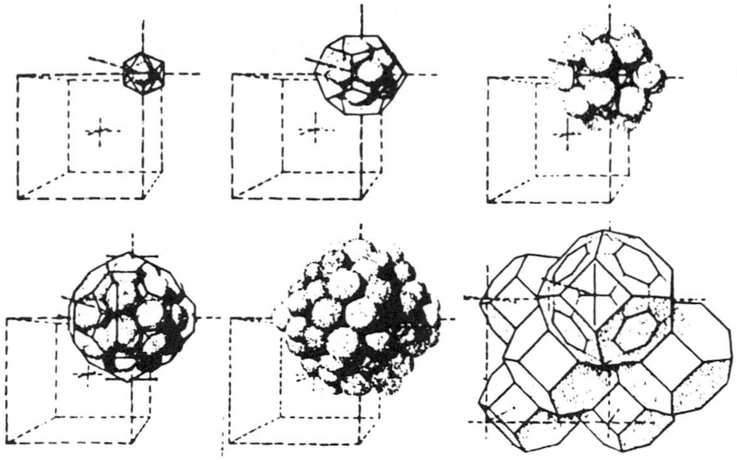

Figure 24. The T-phase crystal structure ($Mg_{32}(Zn,Al)_{49}$), illustrating the stochastic approach of Pauling (1955). From Bergman *et al.* (1957).

Mg_2Zn_{11} structure in Section 2.2, but in the T phase the rhombic faces are much more closely planar.)

These 45 atoms are followed by 60 atoms in a truncated icosahedron, similar to the "soccer ball" described in Section 1.1 and to the outer shell of the β-rhombohedral-boron structural unit described in Section 1.2. Indeed, that 84-atom unit differs from the 105-atom complex, so far developed here, in that the central icosahedron in the boron structure is empty and the pentagonal dodecahedron is missing. (The 105-atom unit, less the central atom which is assumed to be missing, figures in the icosatwin and decatwin models proposed by Pauling (1987) for the Mn-Al and Cu-Li-Al materials widely considered to be quasicrystalline. His assumed crystal structure is obtained by placing a 104-atom cluster at each of the atomic positions of the β-W structure (A15; space group $Pm\bar{3}n$), with sharing of a few peripheral atoms. This proposal has the serious difficulty that the crystallographic subgroup point symmetry $m\bar{3}$ of the cluster is incompatible with the $\bar{4}3m$ site symmetry of the positions 6(*c*) at which three quarters of the clusters are assumed to be centered.)

The 105-atom complex developed up to this point has icosahedral symmetry, which is destroyed when 12 more atoms are added to hexagonal dimples in accordance with $m\bar{3}$ symmetry. After 12 more atoms are added to the pentagonal dimples to form a large icosahedron, a 129-atom complex is obtained which is shown in the last part of the figure in the shape of a cubo-octahedron, which is the Voronoï or Wigner–Seitz cell for body-centered cubic packing, a figure which, when repeated at the positions of a body-centered cubic lattice, will fill all space. (Actually, the square faces shown are misleading; they are actually rhombs with a ratio of diagonals about 5:3. However, this will not affect the discussion that follows.) For this packing to take place, faces and corners must be shared with adjacent cubo-octahedra. The cubo-octahedron has 24 corners, each of which is shared with three other cubo-octahedra, thus three quarters of 24, or 18 atoms, must be subtracted. In addition to the corner atoms, each of the eight hexagonal faces contains a ring of six atoms, and each of the six square (rhombic) faces contains two atoms. These are shared with adjacent cubo-octahedra, so half of them, or 30, must be subtracted. The resulting number of atoms is 81, in agreement with the formula of the compound (32 + 49).

The structure turns out to be t.c.p. The Zn,Al atoms are placed in the X sites, and the larger Mg atoms are placed in the R, Q, and P sites. The central atom and the 12 surrounding it are X atoms (mostly Al), the next 20 are P atoms (Mg), and the next 12 are again X atoms (Zn,Al). The shell of 60 atoms consists of 48 X and 12 R atoms (Zn,Al, and Mg, respectively), and the final (12 + 12) atoms are Q atoms (Mg). The atoms that must be subtracted for sharing are half of the 48 X and 12 R atoms, and three quarters of the final (12 + 12) Q; this leads to the chemical formula $Mg_{32}(Zn,Al)_{49}$ and the coordination formula $P_{20}Q_6R_6X_{49}$. On applying Eq. (4.1), we see that $x = 2 \times 20 + 7 \times 6/6 + 6/3 = 49$.

The structure was refined by Bergman *et al.* (1957). They noted that the postulated central X atom of the complex may be partially absent; Samson, later (private communication), has found no evidence of its presence. This would place the alloy in the class of aluminum-containing alloys with empty icosahedra, mentioned in Section 3.2. However, in our further discussion of this structure, we will assume that this atom is present.

4.5.2 Linking of Icosahedra. The cluster described above can be described also in terms of linking or packing of icosahedra. We create a shell of twelve centered icosahedra sharing edges, as shown in Fig. 25. The twelve inner vertices of these icosahedra form a central icosahedron, to which we add (or perhaps do not add) a central atom. The 13 icosahedra all have the same spacial orientation. The vertices shared between the central and peripheral icosahedra are themselves icosahedrally coordinated, but with different orientations—six in number. The central icosahedron must be somewhat smaller than the outer ones, just as the central atom should be smaller than the ones forming the central icosahedron. (The latter circumstance may explain the suspected absence of the central atom in the

Figure 25. The cluster of 12 icosahedra sharing edges to define a larger icosahedron in the T-phase crystal structure. This figure shows the distortions required by the close approaches of atoms in icosahedra that are joined by sharing of edges. From Cenzual *et al.* (1985).

T-phase structure.) The size incompatibility could be overcome by flattening the outer icosahedra somewhat. However, in the T-phase structure, while the outer icosahedra are indeed somewhat distorted, their greater size is achieved largely by inclusion of some larger atoms in them. These atoms are the ones at the bottom of the dimples in the figure; they are the 20 *P* atoms (Mg) forming a pentagonal dodecahedron. This complex of centered icosahedra contains 117 atoms. If we now place atoms in 12 of the 20 hexagonal dimples in accordance with $m\bar{3}$ symmetry, we obtain the 129-atom unit described in the previous section.

Notwithstanding the complexity of the crystal structure built up by linking these units, it is easy to trace chains of icosahedra of the same orientation throughout the crystal structure. If we call the orientation of the central icosahedron *A*, the orientations at the atoms in the first icosahedral shell are *B, C, D, E, F, G*. The icosahedra in the next icosahedral shell (which is part of the triacontahedron, and which are the centers of the twelve icosahedra in the cluster described in this section) are again *A*, due to successive interpenetrations in the same direction. In the surrounding soccer ball, the orientations of the 48 of the 60 atoms that are icosahedrally coordinated are *B–G*. Those around a given pentagonal face of the soccer ball have the same orientations as those in the nearest parallel 5-ring of vertices in the central icosahedron. The soccer-ball *X* atoms are shared with neighboring clusters, and control the orientations in those clusters, propagating orientation to their central atoms.

4.5.3 Linking of Friauf Polyhedra. A structural unit which Samson (1968) has shown to be useful in describing t.c.p. structures is the truncated

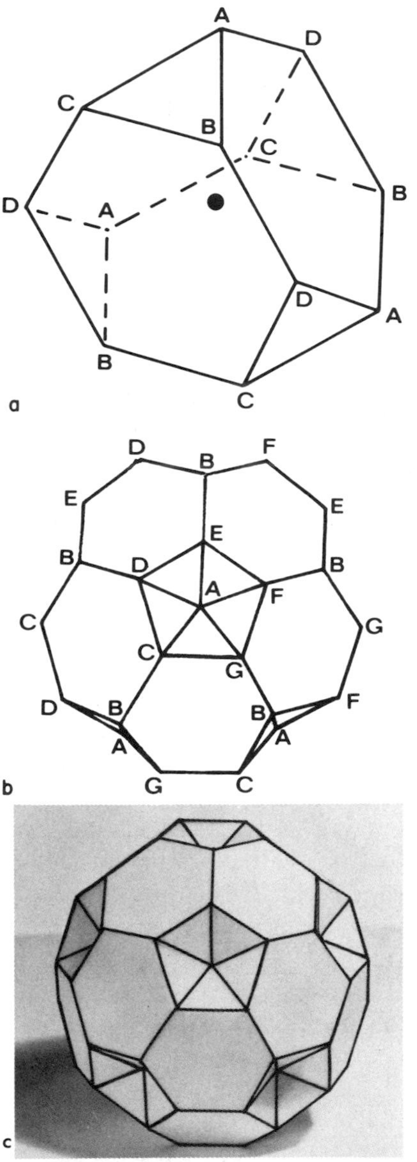

Figure 26. (a) The Friauf polyhedron (truncated tetrahedron), with a P atom at its center. Letters A–D indicate different icosahedral orientations for those vertices that are X atoms. (b) Ring of five fused Friauf polyhedra. The icosahedral orientation of the vertex in the dimple on the other side of the ring from A is B. (c) A cluster of twenty Friauf polyhedra sharing faces in the T-phase crystal structure. The vertices at the bottoms of the outer dimples all have icosahedral orientation A; the other outer vertices have orientations B–G. Part (c) is from Samson (1968).

tetrahedron, named *Friauf polyhedron* by Samson. It is defined by the 12 fivefold vertices of the *P* coordination polyhedron, and has four hexagonal faces and four triangular ones, as shown in Fig. 26a. Four different orientations *(A–D)* are possible for those vertices that are *X* atoms when their coordination shells are completed outside of the Friauf polyhedron. Since the angle included between adjacent hexagonal faces of the Friauf polyhedron is 70° 32′, which is close to one fifth of 360° or 72°, five of these Friauf polyhedra can be fused, with small distortions, through sharing of hexagonal faces, into a ring or disk, the centers of the polyhedra forming a regular pentagon, as shown in Fig. 26b. The icosahedra that are centered by vertices of this unit have all seven possible orientations *(A–G)* discussed in the previous section. The process can be continued; 20 Friauf polyhedra may be fused to form the cluster shown in Fig. 26c, in which the centers of the Friauf polyhedra (*P* atoms; Mg) form a regular pentagonal dodecahedron. The innermost triangles of the Friauf polyhedra form the central icosahedron mentioned in the descriptions of the previous sections. The 60 outermost vertices form the soccer ball previously described. This 105-atom unit is the one already described in Section 4.5.1.

4.5.4 Relation to the {5,3,3} Regular Polytope. We saw in Section 4.5.2 that when we start with an icosahedrally coordinated atom and attempt to extend that 13-atom complex by adding additional icosahedrally coordinated atoms (as if we were attempting to derive an infinite structure of exclusively icosahedrally coordinated atoms), we have some early success, but a limit is quickly reached in which both distortions and departures from icosahedral coordination are required. We have seen an essential limitation of flat (Euclidean) space. It should therefore not be surprising

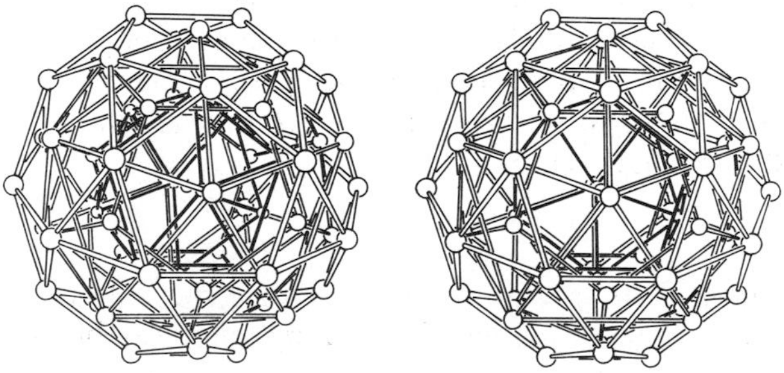

Figure 27. Stereoscopic representation of a projection of the {3,3,5} polytope onto a 3-hyperplane (Euclidean space), emphasizing concentric polyhedral shells related to those found in the T-phase crystal structure (see Fig. 23).

that the early success and subsequent failure are related to the {5,3,3} regular polytope which was discussed in Section 4.1 and shown in projection in Fig. 10. In Fig. 27, we show another stereofigure of that projection. In this figure, we show all of the projected vertices, but not all of the projected edges. In lieu of the latter, we show lines connecting neighboring vertices within successive shells of vertices: a central regular icosahedron, succeeded by a regular pentagonal dodecahedron, then a larger regular icosahedron (the latter two combining to form a *true* rhombic triacontahedron), and finally a 30-vertex *icosidodecahedron,* which is described in Section 1.1, for a total of 75 vertices in the projection. The first three shells are similar to the first three described in the previous sections. The fourth shell is different. The 60-atom shell in the T-phase structure is obtained by placing *two* atoms against each rhombic face of the triacontahedron. The 30-vertex *icosidodecahedron* in the polytope projection is obtained by placing *a single* vertex outward from the center of each (planar) rhombic face of the rhombic triacontahedron (as is understandable from the fact that the icosidodecahedron is the dual of the rhombic dodecahedron).

Close inspection of the polytope projection, in either Fig. 10 or Fig. 27, shows that, as one goes outward from the center, the icosahedra are progressively flattened. This phenomenon is entirely analogous to the compression of continental shapes in a plane projection of the earth's spherical surface. It is this flattening, already referred to in Section 4.5.2, that in the T phase must be counteracted by inclusion of larger atoms and abandonment of exclusive icosahedral coordination.

5. Structures with Giant Cubic Unit Cells

We here discuss very briefly two exceedingly complicated alloy structures determined by Sten Samson. Both are cubic; both contain over 1000 atoms per cubic unit cell. Neither is, strictly speaking, t.c.p.; however, nearly all of the interstices are tetrahedral and most of the coordination polyhedra are the ones found in t.c.p. structures. Because of the considerable complexity of the structures, we are unable here to trace the orientations of the icosahedra through the unit cells.

5.1 Cu_4Cd_3

This composition is of special interest because the existence of an icosahedral phase of approximately this composition has been claimed by Bendersky (quoted by Kuo *et al.*, 1987). The structure of the crystalline phase, as determined by Samson (1967), has space group $F\bar{4}3m$ and a unit-cell edge of 25.87 Å. There are 1124 atoms in the unit cube. Samson de-

Figure 28. Building units in the Cu_4Cd_3 crystal structure: rings of five icosahedra, and linked rings. From Samson (1968).

scribes this structure both in terms of Friauf polyhedra (centered by Cd atoms) and in terms of icosahedra (centered by Cu atoms). These descriptions are closely related, because a vertex of a Friauf polyhedron often represents the center of an icosahedron; we will be content with the icosahedral description here. The principal structural units are shown in Figs. 28 and 29. The five icosahedra in a ring (Fig. 28) are linked with shared vertices, but not in the way shown in Fig. 4a; the five-membered rings adjacent to the shared vertex of a pair of joined icosahedra form a pentagonal prism (as in $Sc_{57}Rh_{13}$; Cenzual *et al.*, 1985) rather than a pentagonal antiprism. These 5-rings of icosahedra are catenated to form larger clusters. The buildup of another type of cluster is shown in Fig. 29. A pair of interpenetrating icosahedra is shown; six such interpenetrating pairs are joined together in such a manner that, again, pentagonal prisms are

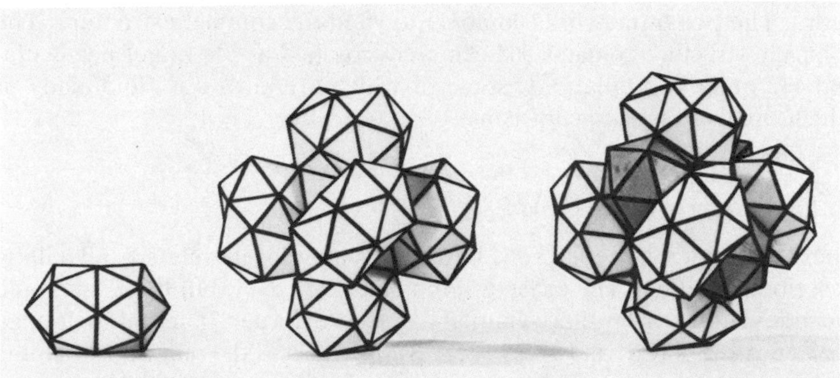

Figure 29. Building units in the Cu_4Cd_3 crystal structure: pairs of interpenetrating icosahedra, and agglomerations of such pairs. From Samson (1968).

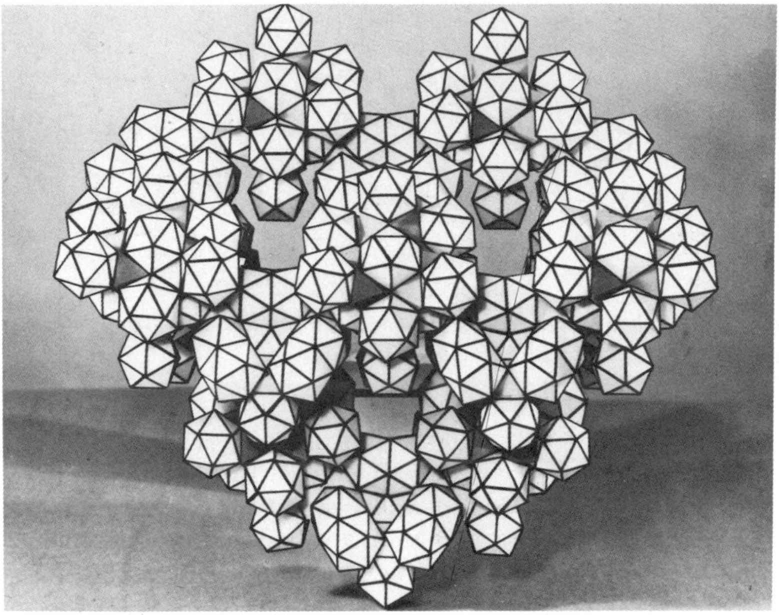

Figure 30. Part of the Cu_4Cd_3 crystal structure: the arrangement of icosahedra. From Samson (1968).

formed between adjacent icosahedra. A portion of the infinite three-dimensional framework of linked icosahedra is shown in Fig. 30. Another framework (not shown here) may be formed, based on fused Friauf polyhedra. The two frameworks combine to yield the complete structure. The complete structure contains 568 centered icosahedra, 124 Friauf polyhedra, and 432 other coordinations, some of which are not t.c.p., and many of which involve pentagonal prisms.

5.2 $NaCd_2$ and β-Mg_2Al_3

These two structures (Samson, 1962, 1965) are closely related and will be described together. The space group is ideally $Fd\bar{3}m$, but there are some local deviations from that symmetry due to disorder. The unit cell edge is 30.56 Å for $NaCd_2$ and 28.24 Å for β-Mg_2Al_3, and the numbers of atoms per unit cell are respectively approximately 1190 and 1168. The two structures largely defy any convenient description in terms of icosahedra; we will confine ourselves to a brief description in terms of Friauf polyhedra,

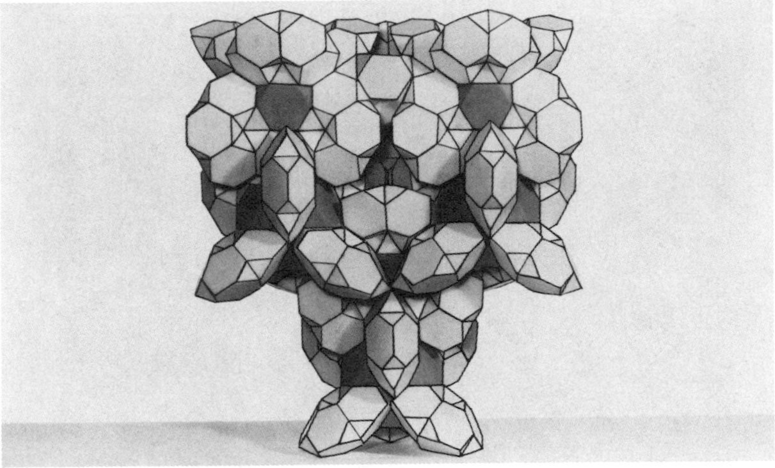

Figure 31. Part of the $NaCd_2$ or β-Mg_2Al_3 crystal structure: the arrangement of rings of five face-sharing Friauf polyhedra. From Samson (1968).

which are centered by Na in the first compound and by Mg in the second. An important structural unit consists of five Friauf polyhedra fused in a ring or disk, as described in Section 4.5.2 and shown in Fig. 26b. A portion of the structure is shown in Fig. 31, in which individual 5-rings of Friauf polyhedra may be identified, as well as some details of modes of linking together. The actual structures of these two alloys differ somewhat from this ideal model due to disorder. The unit cube contains 672 icosahedra, 252 Friauf polyhedra, and 244 other coordination polyhedra ranging from CN10 to CN16, many of which are not t.c.p.

It is a virtually hopeless task to trace the orientations of individual icosahedra through this structure. We have already seen in Section 4.5.3 that in a 5-ring of fused Friauf polyhedra there are at most seven possible orientations of icosahedra centered at vertices of that figure. Extensive linking of these 5-rings may be expected to generate additional orientations.

6. Polymicrocrystalline Materials Exhibiting Fivefold (Tenfold) Diffraction Symmetry

K. H. Kuo and collaborators in Shenyang, People's Republic of China, have investigated many transition metal alloys and have already been cited in this chapter for discovering new t.c.p. structure types by high-resolution electron microscopy, and some new quasicrystal phases by electron diffraction. In addition, they have examined by high-resolution electron mi-

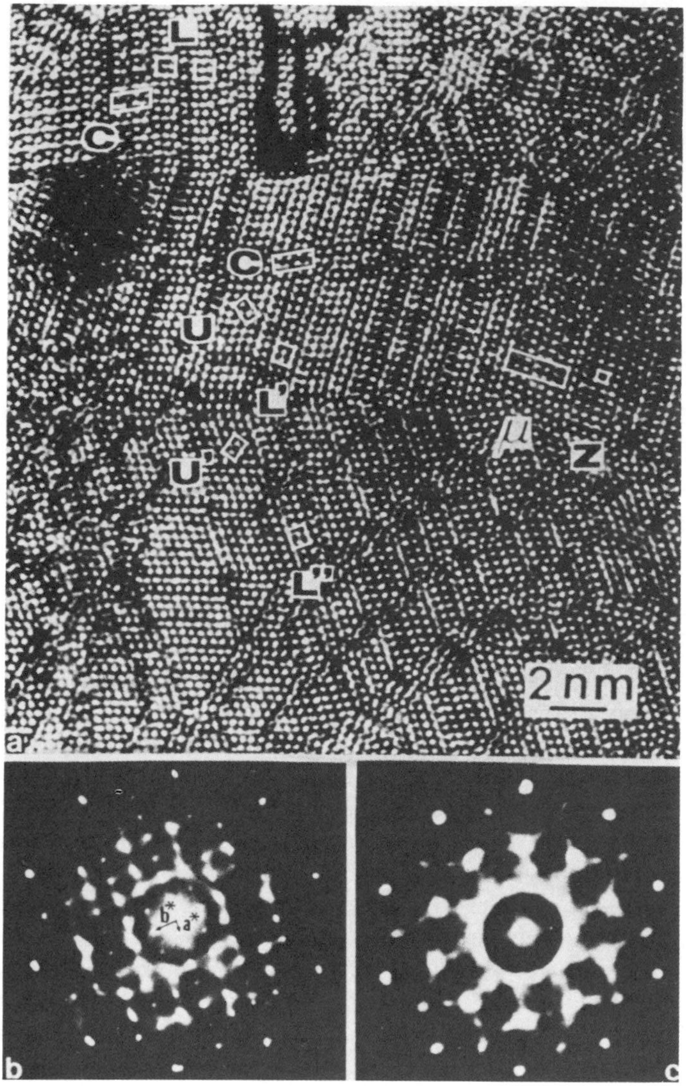

Figure 32. (a) High-resolution electron micrograph of an iron-based polymicrocrystalline alloy of complex composition, showing discrete regions of Zr_4Al_3 (Z), C14 (L), C15 (U), μ, and C, looking down approximate five fold axes of icosahedra, of which the white dots are interpreted as centers. (b) Electron diffraction pattern of a C-phase region. (c) Electron diffraction pattern of the aggregate shown in (a). (d) Calculated diffraction pattern of a single pentagonal antiprism (an icosahedron missing two opposite vertices). From Kuo *et al*. (1986b). [Copyright © (1986) by Alan R. Liss, Inc.]

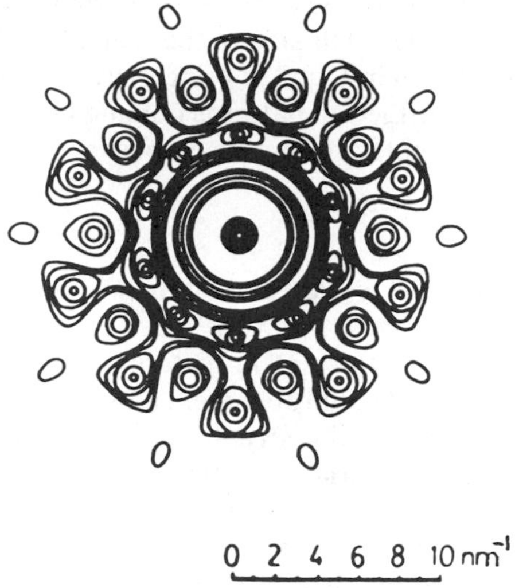

d

Figure 32. (Continued).

croscopy some alloys that show very small domains of various t.c.p. structure types existing together. An example is shown in Fig. 32a (Kuo *et al.*, 1986b) of an iron-based alloy containing substantial amounts of nickel and chromium and small amounts of other transition metals and aluminum. The authors have identified the phases present as Zr_4Al_3 (Z), C14 Friauf–Laves (L), C15 Friauf–Laves (U), μ, and C. These are all layer structures containing pentagons and triangles in their main layers, and the layers of the different structures are all parallel to the plane of the micrograph. Figure 32b shows an electron diffraction photograph of a region of C-phase structure, showing strong diffraction spots that roughly conform to decagonal symmetry, and in Fig. 32c is shown an electron diffraction pattern of the aggregate of the phases shown in Fig. 32a, also showing rough decagonal symmetry.

The bright spots in the micrograph correspond to the positions of five-membered rings forming channels through the structures in the direction perpendicular to the plane of the micrograph; these channels contain only strings of linked X atoms (in interpenetrating icosahedra) running along the centers of the channels. Figure 32d shows a calculated Fourier transform of a single pentagonal antiprism, which the diffraction pattern shown in Fig. 32c roughly resembles, particularly in the tenfold symmetry shown.

The fact that the aggregate has rough decagonal symmetry indicates that the orientations of the icosahedra forming the channels are carried across twin and domain boundaries. The diffraction peaks in Fig. 32b are rather diffuse, reflecting the small size of the individual domains, and the diffraction symmetry shown in Fig. 32c principally reflects long-range orientational order. In the diffraction patterns obtained for materials claimed to be quasicrystals, the spots are much sharper.

7. Conclusion

The discovery of fivefold and icosahedral diffraction symmetry in a number of different materials that have been (in most cases) rapidly quenched from the melt has opened a new era in materials science. The proposed quasicrystal geometries are beautiful and exciting, but chemists and crystallographers will demand that the structures be fleshed out with atomic arrangements. The source materials for constructing such atomic arrangements are the chemical and crystallographic structural data already amassed for crystalline materials of similar or related compositions. It has been a primary purpose of this chapter to provide a sampling of such data, and several researchers have already proposed quasicrystal structures with atomic arrangements that are plausible, based on some of the crystal structures described in this chapter.

However, a reliable understanding of the physical properties of these new materials will require that their detailed atomic structures be not only plausible but also well authenticated from diffraction and other data by computational procedures yet to be fully developed.

An important part of this chapter has been a consideration of the propagation of icosahedral orientations through a crystal structure. This may be also applicable in large part to quasicrystals, and perhaps also to the so-called icosahedral glasses. For example, the symmetry of an icosahedral glass may be dominated by icosahedra of one or very few orientations, linked together in a consistent manner in irregular chains or networks. Or, the units in such irregular chains or networks may be larger clusters, such as those that have been discussed in this chapter. The applicability of this consideration to the study of these new materials is not entirely clear at this time, but may become clearer as detailed atomic structures become known.

References

Adam, J., and Rich, J. B. (1954). The crystal structure of WAl_{12}, $MoAl_{12}$ and $(Mn,Cr)Al_{12}$. *Acta Cryst.* 7, 813–816.

Arnfelt, H., and Westgren, A. (1935). De intermediära fasernas kristallbyggnad och sammansättning i järn-volfram och järn-molybdenlegiringar. *Jernkontorets Ann.* 119, 185–196.

Audier, M., and Guyot, P. (1986). Al_4Mn quasicrystal atomic structure, diffraction data and Penrose tiling. *Phil. Mag. B* 53, L43–L51.

Audier, M., Sainfort, P., and Dubost, B. (1986). A simple construction of the AlCuLi quasicrystalline structure related to the $(Al,Zn)_{49}Mg_{32}$ cubic structure type. *Phil. Mag. B* 54, L105–L111.

Bergman, G., and Shoemaker, D. P. (1954). The determination of the crystal structure of the sigma phase in the Fe-Cr and the Fe-Mo systems. *Acta Cryst.* 7, 857–865.

Bergman, G., Waugh, J. L. T., and Pauling, L. (1957). The crystal structure of the metallic phase $Mg_{32}(Al,Zn)_{49}$. *Acta Cryst.* 10, 254–259.

Borén, B. (1933). Röntgenuntersuchung der Legierungen von Si mit Cr, Mn, Co und Ni. *Ark. Kemi. Min. Geol. 11A,* No. 10, 1–28.

Bradley, A. J., and Thewlis, J. (1927). Crystal structure of alpha-manganese. *Proc. Roy. Soc. A* 115, 456–471.

Cahn, J. W., Gratias, D., and Shechtman, D.; Mackay, A. L.; Bancel, P. A., Heiney, P. A., Stephens, P. W., and Goldman, A. I.; Berezin, A. A. (1985). Pauling's model not universally accepted. In Scientific Correspondence, *Nature (Lond.),* 319, 102–104.

Carr, M. J. (1985). An analysis of fivefold symmetry by microtwinning in rapidly solidified Al-Mn alloys. *J. Appl. Phys.* 59, 1063–1067.

Cenzual, K., Chabot, B., and Parthé, E. (1985). Cubic $Sc_{57}Rh_{13}$ and orthorhombic $Hf_{54}Os_{17}$, two geometrically related crystal structures with rhodium- and osmium-centered icosahedra. *Acta Cryst. C* 41, 313–319.

Cherkashin, E. E., Kripyakevich, P. I., and Oleksiv, G. I. (1964). Crystal structures of ternary compounds in the Li-Cu-Al and Li-Zn-Al systems. *Soviet Phys. Crystallogr.* 8, 681–685.

Claussen, W. F. (1951a). Suggested structures of water in inert gas hydrates. *J. Chem. Phys.* 19, 259–260, 662.

Claussen, W. F. (1951b). Second water structure for inert gas hydrates. *J. Chem. Phys.* 19, 1425–1426.

Cooper, M. (1967). The crystal structure of the ternary alloy α(AlFeSi). *Acta Cryst.* 23, 1106–1107.

Cooper, M., and Robinson, K. (1966). The crystal structure of the ternary alloy α(AlMnSi). *Acta Cryst.* 20, 614–617.

Coxeter, H. S. M. (1958). Close-packing and froth. *Illinois J. Math.* 2, 746–758.

Coxeter, H. S. M. (1973). *Regular Polytopes.* 3d ed. Dover, New York.

Donohue, J. (1974). *The Structures of the Elements.* Wiley, New York, 48–82.

Donohue, J., and Einspahr, H. (1971). The structure of β-uranium. *Acta Cryst. B* 27, 1740–1743.

Dubost, B., Lang, J. M., Tanaka, M., Sainfort, P., and Audier, M. (1986). Large AlCuLi single quasicrystals with triacontahedral solidification morphology. *Nature (Lond.)* 324, 48–50.

Elser, V., and Henley, C. L. (1985). Crystal and quasicrystal structures in Al-Mn-Si alloys. *Phys. Rev. Lett.* 55, 2883–2886.

Field, R. D., and Fraser, H. L. (1985). Precipitates possessing icosahedral symmetry in a rapidly solidified Al-Mn alloy. *Mater. Sci. and Eng.* 68, L17–L21.

Fornasini, M. L., Palenzona, A., and Manfrinetti, P. (1984). Crystal structure of the new thorium intermetallics ThIn and Th_6Cd_7. *J. Solid State Chem.* 51, 135–140.

Frank, F. C., and Kasper, J. S. (1958). Complex alloy structures regarded as sphere packings. I. Definitions and basic principles. *Acta Cryst.* 11, 184–190.

Frank, F. C., and Kasper, J. S. (1959). Complex alloy structures regarded as sphere packings. II. Analysis and classification of representative structures. *Acta Cryst.* 12, 483–499.

Friauf, J. B. (1927a). The crystal structure of magnesium dizincide. *Phys. Rev.* 29, 35–40.

Friauf, J. B. (1927b). The crystal structure of two intermetallic compounds (Cu_2Mg; $CuAl_2$). *J. Amer. Chem. Soc.* 49, 3107–3114.

Gallucci, J. C., Doecke, C. W., and Paquette, L. A. (1986). Crystal structure of the pentagonal dodecahedrane hydrocarbon, $(CH)_{20}$. Abstract PA 32, *Am. Cryst. Assn. Annual Meeting,* June 22–27, Hamilton, Ontario, Canada.

Guyot, P., and Audier, M. (1985). A quasicrystal structure model for Al-Mn. *Phil. Mag. B* 52, L15–L19.

Henley, C. L., and Elser, V. (1986). Quasicrystal structure of $(Al,Zn)_{49}Mg_{32}$. *Phil. Mag. B* 53, L59–L66.

Hiraga, K., Hirabayashi, M., Inoue, A., and Masumoto, T. (1985). Icosahedral quasicrystals of a melt-quenched Al-Mn alloy observed by high-resolution electron microscopy. *Sci. Rep. RITU A* 32, 309–314.

Hoard, J. L., Sullenberger, D. B., Kennard, C. H. L., and Hughes, R. E. (1970). The structure analysis of β-rhombohedral boron. *J. Solid State Chem.* 1, 268–277.

Hogle, J. M., Chow, M., and Filman, D. J. (1985). Three-dimensional structure of poliovirus at 2.9 Å resolution. *Science* 229, 1358–1365.

Ishimasa, T., Nissen, H. U., Fukano, Y. (1985). New ordered state between crystalline and amorphous in Ni-Cr particles. *Phys. Rev. Lett.* 55, 511–513.

Kasper, J. S. (1956). Atomic and magnetic ordering in transition metal structures. In *Theory of Alloy Phases*. American Society for Metals, Cleveland, 264–278.

Kasper, J. S., Lucht, C. M., and Harker, D. (1950). The crystal structure of decaborane, $B_{10}H_{14}$. *Acta Cryst.* 3, 436–455.

Ketelaar, J. A. A. (1937). The crystal structure of alloys of zinc with the alkali and alkaline earth metals and of cadmium with potassium. *J. Chem. Phys.* 5, p. 668.

Kléman, M., and Sadoc, J. F. (1979). A tentative description of the crystallography of amorphous solids. *J. Physique Paris* 40, L569–L574.

Komura, Y., and Kitano, Y. (1977). Long-period stacking variants and their electron-concentration dependence in the Mg-base Friauf–Laves phases. *Acta Cryst. B* 33, 2496–2501.

Komura, Y., Sly, W. G., and Shoemaker, D. P. (1960). The crystal structure of the R phase, Mo-Co-Cr. *Acta Cryst.* 13, 575–585.

Kontio, A., and Coppens, P. (1981). New study of the structure of $MnAl_6$. *Acta Cryst. B* 37, 433–435.

Kontio, A., Stevens, E. D., Coppens, P., Brown, R. D., Dwight, A. E., and

Williams, J. M. (1980). New investigation of the structure of Mn_4Al_{11}. *Acta Cryst. B* 36, 435–436.

Kripyakevich, P. I., and Yarmolyuk, Ya. P. (1970). The crystal structure of the C-phase $V_2(Co_{0.57}Si_{0.43})_3$. A new example of the structure of the homologous Zr_4Al_3-$MgZn_2$ series. *Dopov. Akad. Nauk. Ukr. R. S. R., Ser. A*, 32, 948–951.

Kripyakevich, P. I., and Yarmolyuk, Ya. P. (1974). Crystal structure of W_2FeSi = $W_6(W_{0.07}Fe_{0.465}Si_{0.465})_7$. *Dopov. Akad. Nauk. Ukr. R. S. R., Ser. A*, 36, 460–463.

Kroto, H. W., Heath, J. R., O'Brien, S. C., Curl, R. F., and Smalley, R. E. (1985). C_{60}: buckminsterfullerene. *Nature (Lond.)* 318, 162–163.

Kuo, K. H., Dong, C., Zhou, D. S., Guo, Y. X., Hei, Z. K., and Li, D. X. (1986a). A Friauf–Laves (Frank–Kasper) phase related quasicrystal in a rapidly solidified Mn_3Ni_2Si alloy. *Scripta Met.* 20, 1695–1698.

Kuo, K. H., Ye, H. Q., and Li, D. X. (1986b). Merits of high-resolution electron microscopy in the study of alloy structures. *J. Electron Microscopy Techniques* 3, 57–66.

Kuo, K. H., Zhou, D. S., and Li, D. X. (1987). Quasicrystalline and Frank–Kasper phases in a rapidly solidified $V_{41}Ni_{36}Si_{23}$ alloy. *Phil. Mag. Lett.* 55, 33–39.

Laves, F., and Witte, H. (1935). Die Kristallstruktur des $MgNi_2$ und seine Beziehungen zu den Typen des $MgCu_2$ und $MgZn_2$. *Metallwirtsch. Metallwiss. Metalltech.* 14, 645–649.

Levine, D., and Steinhardt, P. J. (1984). Quasicrystals: a new class of ordered structures. *Phys. Rev. Lett.* 53, 2477–2480.

Levine, D., and Steinhardt, P. J. (1986). Quasicrystals. I. Definitions and structure. *Phys. Rev. B* 34, 596–616.

Li, D. X., and Kuo, K. H. (1986). Some new σ-related structures determined by high-resolution electron microscopy. *Acta Cryst. B* 42, 152–159.

Lipscomb, W. N. (1963). *Boron hydrides*. W. A. Benjamin, New York, Chapters 2 and 3.

Mackay, A. L. (1962). A dense non-crystallographic packing of equal spheres. *Acta Cryst.* 15, 916–918.

Mackay, A. L. (1982). Crystallography and the Penrose pattern. *Physica (Amsterdam)* 114A, 609–613.

Manor, P. C., Shoemaker, C. B., and Shoemaker, D. P. (1972). The crystal structure of the X phase (Mn,Co,Si). *Acta Cryst. B* 28, 1211–1218.

Nelson, D. R. (1983). Order, frustration, and defects in liquids and glasses. *Phys. Rev. B* 28, 5515–5535.

Nicol, A. D. I. (1953). The structure of $MnAl_6$. *Acta Cryst.* 6, 285–292.

Paquette, L. A., Ternansky, R. J., Balogh, D. W., and Kentgen, G. (1983). Total synthesis of dodecahedrane. *J. Am. Chem. Soc.* 105, 5446–5450.

Pauling, L. (1955). The stochastic method and the structure of proteins. *Am. Sci.* 43, 285–297.

Pauling, L. (1960). *The Nature of the Chemical Bond*. 3d ed. Cornell University Press, Ithaca, N.Y.

Pauling, L. (1985). Apparent icosahedral symmetry is due to directed multiple twinning of cubic crystals. *Nature (Lond.)* 317, 512–514.

Pauling, L. (1987). So-called icosahedral and decagonal quasicrystals are twins of an 820-atom cubic crystal. *Phys. Rev. Lett.* 58, 365–368.

Pauling, L., and Marsh, R. E. (1952). The structure of chlorine hydrate. *Proc. Nat. Acad. Sci. U.S.* 36, 112–118.

Pearson, W. B. (1972). *The Crystal Chemistry and Physics of Metals and Alloys*, Wiley-Interscience, New York.

Robinson, K. (1952). The structure of β(AlMnSi)-Mn_3SiAl_9. *Acta Cryst.* 5, 397–403.

Sadoc, J. F., and Mosseri, R. (1982). Order and disorder in amorphous, tetrahedrally coordinated semiconductors. A curved space description. *Phil. Mag. B* 45, 467–483.

Samson, S. (1949a). Die Kristallstruktur von $Mg_2Cu_6Al_5$. *Acta Chem. Scand.* 3, 809–834.

Samson, S. (1949b). Die Kristallstruktur von Mg_2Zn_{11}. Isomorphie zwischen Mg_2Zn_{11} und $Mg_2Cu_6Al_5$. *Acta Chem. Scand.* 3, 835–843.

Samson, S. (1962). Crystal structure of $NaCd_2$. *Nature (Lond.)* 195, 259–262.

Samson, S. (1965). The crystal structure of the phase β-Mg_2Al_3. *Acta Cryst.* 19, 401–413.

Samson, S. (1967). The crystal structure of the intermetallic compound Cu_4Cd_3. *Acta Cryst.* 23, 586–600.

Samson, S. (1968). The structure of complex intermetallic compounds. In *Structural Chemistry and Molecular Biology* (Rich, A., and Davidson, N., eds.). Freeman, San Francisco, 687–717.

Shechtman, D., Blech, I., Gratias, D., and Cahn, J. W. (1984). Metallic phase with long-range orientational order and no translational symmetry. *Phys. Rev. Lett.* 53, 1951–1954.

Shoemaker, C. B., and Shoemaker, D. P. (1963). The crystal structure of the δ phase, Mo-Ni. *Acta Cryst.* 16, 997–1009.

Shoemaker, C. B., and Shoemaker, D. P. (1964). Interatomic distances and atomic radii in intermetallic compounds of transition elements. *Trans. A.I.M.E.* 230, 486–490.

Shoemaker, C. B., and Shoemaker, D. P. (1967). The crystal structure of the M phase, Nb-Ni-Al. *Acta Cryst.* 23, 231–238.

Shoemaker, C. B., and Shoemaker, D. P. (1969). Structural properties of some σ-phase related phases. In *Developments in the Structural Chemistry of Alloy Phases*. Giessen, B. C., ed.). Plenum Press, New York, 107–139.

Shoemaker, C. B., and Shoemaker, D. P. (1971a). Tetraedrisch dicht gepackte Strukturen von Legierungen der Übergangsmetalle. *Monatsh. Chem.* 102, 1643–1666.

Shoemaker, C. B., and Shoemaker, D. P. (1971b). The crystal structure of the ν phase, $Mn_{81.5}Si_{18.5}$. *Acta Cryst. B* 27, 227–235.

Shoemaker, C. B., and Shoemaker, D. P. (1972). Concerning systems for the generating and coding of layered, tetrahedrally close-packed structures of intermetallic compounds. *Acta Cryst. B* 28, 2957–2965.

Shoemaker, C. B., and Shoemaker, D. P. (1976). The crystal structure of Mn_5Si_2 and the D phase (V-Fe-Si). *Acta Cryst. B* 32, 2306–2313.

Shoemaker, C. B., and Shoemaker, D. P. (1977). The crystal structure and superstructure of the K phase, $Mn_{77}Fe_4Si_{19}$. *Acta Cryst. B* 33, 743–754.
Shoemaker, C. B., and Shoemaker, D. P. (1978). Refinement of an R phase, $Mn_{85.5}Si_{14.5}$. *Acta Cryst. B,* 34, 701–705.
Shoemaker, C. B., and Shoemaker, D. P. (1981). The structure of the I phase, $V_{41}Ni_{36}Si_{23}$, a pseudo superstructure. *Acta Cryst. B* 37, 1–8.
Shoemaker, D. P., and Shoemaker, C. B. (1968). Sigma-phase-related transition-metal structures with tetrahedral interstices. In *Structural Chemistry and Molecular Biology* (Rich, A. and Davidson, N., eds.) Freeman, San Francisco, 718–730.
Shoemaker, D. P., and Shoemaker, C. B. (1986). Concerning the relative numbers of atomic coordination types in tetrahedrally close-packed metal structures. *Acta Cryst. B* 42, 3–11.
Shoemaker, D. P., Marsh, R. E., Ewing, F. J., and Pauling, L. (1952). Interatomic distances and atomic valences in $NaZn_{13}$. *Acta Cryst.* 5, 637–644.
Shoemaker, D. P., Shoemaker, C. B., and Wilson, F. C. (1957). The crystal structure of the P phase, Mo-Ni-Cr. II. Refinement of parameters and discussion of atomic coordination. *Acta Cryst.* 10, 1–14.
Simon, A., Brämer, W., Hillenkötter, B., and Kullmann, H. J. (1976). Neue Verbindungen zwischen Kalium und Cäsium. *Z. anorg. allg. Chem.* 419, 253–274.
Smith, D. J., and Marks, L. D. (1981). High-resolution studies of small particles of gold and silver. *J. of Cryst. Growth* 54, 433–438.
Socolar, J. E. S., and Steinhardt, P. J. (1986). Quasicrystals. II. Unit-cell configurations. *Phys. Rev. B* 34, 617–647.
Stephens, P. W., and Goldman, A. I. (1986). Sharp diffraction maxima from an icosahedral glass. *Phys. Rev. Lett.* 56, 1168–1171.
Taylor, M. A. (1959). The crystal structure of Mn_3Al_{10}. *Acta Cryst.* 12, 393–396.
Wilson, C. G., Thomas, D. K., and Spooner, F. J. (1960). The crystal structure of Zr_4Al_3. *Acta Cryst.* 13, 56–57.
Wunderlich, J., and Lipscomb, W. N. (1960). Structure of $B_{12}H_{12}^{=}$ ion. *J. Am. Chem. Soc.* 82, 4427–4428.
Yarmolyuk, Ya. P., and Kripyakevich, P. I. (1974). Mean weighted coordination numbers and the origin of close-packed structures with atoms of unequal size but normal coordination polyhedra. *Kristallographiya* 19, 539–545 (*Sov. Phys. Crystallogr.* 19, 334–337).
Yarmolyuk, Ya. P., Kripyakevich, P. I., and Gladyshevskii, E. I. (1970). The crystal structure of the X phase in the Mn-Co-Si system. *Kristallographiya* 15, 268–274 (*Sov. Phys. Crystallogr.* 15, 226–230).
Yarmolyuk, Ya. P., Kripyakevich, P. I., and Melnik, E. V. (1975). Crystal structure of the compound Mg_4Zn_7. *Kristallographiya* 20, 538–542 (*Sov. Phys. Crystallogr.* 20, 329–331).
Ye, H. Q., Li, D. X., and Kuo, K. H. (1984). Structure of the H phase determined by high-resolution electron microscopy. *Acta Cryst. B* 40, 461–465.
Zintl, E., and Haucke, W. (1938). Metals and alloys XXV. Constitution of the intermetallic phases $NaZn_{13}$, KZn_{13}, KCd_{13}, $RbCd_{13}$ and $CsCd_{13}$. *Z. Elektrochem.* 44, 104–111.

Chapter 2

Short- and Long-Range Icosahedral Order in Crystals, Glass, and Quasicrystals

MICHAEL WIDOM

Department of Physics
Carnegie-Mellon University
Pittsburgh, PA 15213

Contents

ISBN 0-12-040601-2

1. Introduction

The discovery by Shechtman *et al.* (1984) of sharp diffraction patterns with icosahedral symmetry demands an explanation in terms of the local icosahedral order known to be present in related crystal structures. The difficulty with such an explanation is that local icosahedral order does not extend to fill flat three-dimensional space without defects. In fact, the tendency towards local icosahedral order may be an important ingredient in the formation of metallic glass. This paper discusses the theory of local icosahedral order in glass in order to clarify the relationship between the local order and long-range disorder characteristic of glass, and the long-range noncrystallographic order of icosahedral quasicrystals.

The role of icosahedral frustration in determining the structure of rapidly quenched metals is the main theme of this chapter. Icosahedral frustration results from a conflict between geometrical properties of the icosahedron and the geometrical constraints of filling flat three-dimensional space with atoms. These constraints are most important at large distances, so small clusters of atoms, either isolated in space or embedded within a bulk material, may satisfy their preference for icosahedral order. On the other hand, the formation of local icosahedral domains inhibits the formation of conventional simple crystals such as FCC, allowing liquids to cool below their freezing points without crystalizing.

Competition between the local preference for icosahedra and the global requirements of filling space results in a rich variety of stable and metastable phases in rapidly quenched metallic systems. This chapter discusses features of crystalline, quasicrystalline, and glassy order resulting from this competition. I shall focus in particular on the interplay of icosahedral order, and disclination line defects caused by icosahedral frustration. I shall discuss the origins of the preference for local icosahedral order. This preference for icosahedra has its roots in a preference for tetrahedral close packing of atoms. The icosahedron, a packing of 20 tetrahedra around a single point in space, inherits the high density, rigidity, and low energy of the tetrahedron. This packing is not perfect, however. Icosahedral frustration ultimately prevents the formation of an icosahedral crystal. This frustration results from the absence of curvature in flat three-dimensional space. One can form a perfect icosahedral crystal in the curved three-dimensional surface of S^3, a sphere in four dimensions. The frustration arises when we flatten the sphere to fill flat space. Certain defects efficiently relieve the frustration without violating the local preference for tetrahedral packing. Consideration of defect networks leads to the identification of local coordination shells known as Voronoi polyhedra.

Evidence abounds for tetrahedral and icosahedral ordering in nature. In Section 2, I shall present examples of icosahedral and tetrahedral ordering, including the sequence of magic numbers in atomic clusters, the Voronoi statistics in supercooled liquids, the dense random packing model of glass, and the Frank–Kasper phases of crystals. All these systems also betray the influence of icosahedral frustration.

Polytope {3,3,5} provides an idealized model of the icosahedral order in these systems. This polytope is a perfect icosahedral crystal in S^3, the curved surface of a sphere in four-dimensional space. Section 3 of this paper discusses the mathematical subject of regular polytopes with special emphasis on the icosahedral polytope {3,3,5}. I shall discuss the peculiar geometry of curved space, and also the symmetry groups of the polytopes. Introducing the idealized icosahedral crystal allows discussion of the physical implications of icosahedral order in glass presented in Section 4. The frustration-induced defects, for example, play an important role in the dynamics at the glass transition. The same defects broaden the peaks in the glassy-structure function. I also consider the influence of icosahedral order on the electronic and acoustic properties of glass.

Finally, we address the interesting problem of relating the local icosahedral order in glass to the local icosahedral order in crystals and the long-range icosahedral order in quasicrystals. We find that consideration of frustration induced by the lack of spatial curvature provides a unifying theme. Different schemes for flattening the polytope produce any of these three structures. Section 5 presents a possible explanation for long-range icosahedral order arising from the local preference for icosahedral structures.

1.1 Origins of Icosahedral Order

Consider placing atoms in three-dimensional space. Assume the atoms interact through a pair potential such as the Lennard–Jones potential

$$V = \varepsilon_0[(\sigma/r)^{-12} - 2(\sigma/r)^{-6}], \tag{1.1}$$

where σ is the position of the pair potential minimum, and ε_0 is the strength of each bond. Place each atom so as to minimize the total energy of the system. Thus, the second atom forms a line segment of length σ with respect to the first. The third one creates an equilateral triangle, and the fourth sits above the equilateral triangle to form a regular tetrahedron (Fig. 1a). Each atom in the tetrahedron sits exactly the distance σ from the other three.

Now the analysis becomes more difficult. We can no longer place atoms in simultaneous contact with every other atom. Two possibilities for five-atom clusters are a square pyramid and a triangular bipyramid. Inspecting Figs. 1b and 1c reveals eight nearest-neighbor bonds in the square pyramid

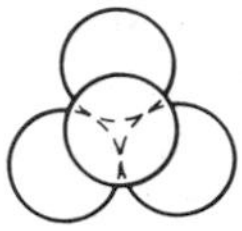

(a)

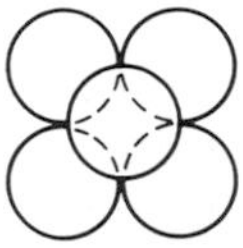

(b)

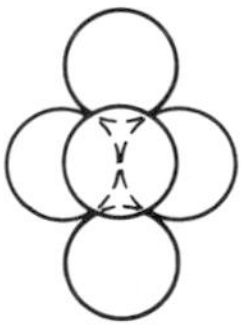

(c)

Figure 1. Small clusters of atoms. (a) Tetrahedron. (b) Square pyramid. (c) Triangular bipyramid.

and nine nearest-neighbor bonds in the triangular bipyramid. Thus, energetics favor the triangular bipyramid, which can be described as two tetrahedra joined on their face. Including further neighbor interactions does not alter this conclusion. Energetics (for interactions with a strong potential minimum such as Lennard–Jones) favors the structure with triangular faces and tetrahedral cells.

Consider the problem of packing many atoms so as to maximize the number of tetrahedral cells. Every time we place an atom above a triangular face, we create at least one additional tetrahedral cell. Noting that the dihedral angle (the angle between two faces) of a tetrahedron is $\cos^{-1}(1/3) = 70.529°$, we observe that 5 tetrahedra may share a common bond to form a pentagonal bipyramid (Fig. 2a). Two intersecting pentagonal bipyramids form an icosahedral cluster of 13 atoms containing 20 tetrahedral cells (Fig. 2b).

The fit is not perfect. There is actually room for $360°/70.529° = 5.1043$ regular tetrahedra around a bond. Since tetrahedra are indivisible units, the actual number must be an integer, most likely 5. Similarly, there is a

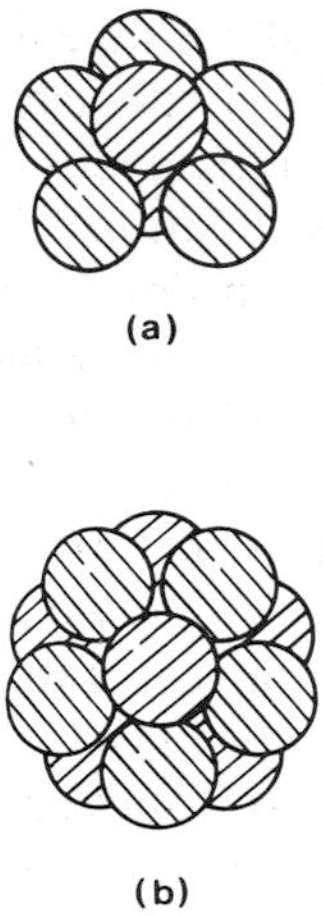

Figure 2. Efficient packings of tetrahedra. (a) Seven atom pentagonal bipyramid. (b) Icosahedron. Note the small gaps between surface atoms.

little play in the icosahedral packing of 12 atoms around a central atom. The vertex–vertex distance is $\sqrt{2-2/\sqrt{5}} = 1.05146$. One of the great outstanding problems of geometry is to prove that no more than 12 identical hard spheres may simultaneously touch a central sphere (or else find a counter example). This looseness, which is visible in Fig. 2, results from what is called icosahedral frustration and is a central theme in our discussion of icosahedral order in materials.

Let us consider the stability of the icosahedral cluster. We find 12 center–vertex bonds, and 30 vertex–vertex bonds. The total binding energy, counting only nearest-neighbor bonds, is thus $-39.972\varepsilon_0$. If we allow the structure to relax so that the vertex–vertex bonds are closer to their minima, the new center–vertex distance is 0.9697σ and the new binding energy is $-41.144\varepsilon_0$. In contrast, the FCC structure has 12 center–vertex bonds but only 24 vertex–vertex bonds for a binding energy of just $-36\varepsilon_0$. Even if we include the 12 second-neighbor bonds of length $\sqrt{2}$, the binding energy is only $-38.813\varepsilon_0$. 13-atom FCC or HCP clusters spontaneously deform into the icosahedron (Hoare, 1976). We have now understood the origin of the preference for local icosahedral order. Clearly, small clusters of atoms (at least those well described by Lennard–Jones and similar potentials) will tend to form icosahedral clusters.

Further theoretical and experimental evidence concerning icosahedral order in clusters is presented in Section 2.1. Since small numbers of atoms prefer to form icosahedral clusters, systems rapidly quenched from a liquid state may rearrange to form small icosahedral domains satisfying the local preference without regard to global energetic considerations (Frank, 1952). Section 2.2 presents evidence of local icosahedral order in numerical simulations of supercooled liquids. Section 2.3 examines models of metallic glass for evidence of icosahedral order. The most extensive experimental evidence of local icosahedral order in metallic systems is provided by the so-called Frank–Kasper phases of metal alloys. Section 2.4 reviews the structure of these materials. The article in this volume by the Shoemakers presents a more detailed analysis of Frank–Kasper phases.

Given a system of atoms, how can we determine the degree of icosahedral ordering? If the system is a real, experimentally produced structure, we do not know where each atom lies. Structures must be determined indirectly by comparing structure functions and radial distribution functions with theoretical models. Once a theoretical model is obtained, we can employ a powerful technique known as Voronoi analysis to study its local structure. The idea behind Voronoi analysis is to represent the first coordination shell of each atom by the convex polyhedron formed by the perpendicular bisecting planes of each atomic bond. The number and types of vertices and faces on the Voronoi polyhedron reflect the local environment of each atom. For instance, three faces meeting at a vertex represent a tetrahedron formed by the atom and three of its neighbors. A

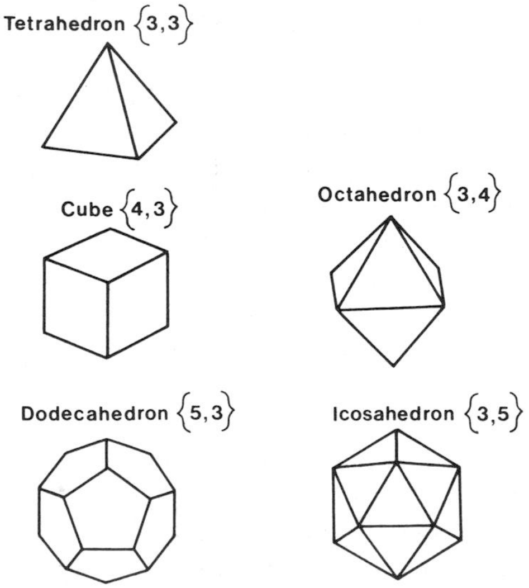

Figure 3. The five Platonic solids and their Schlafly symbols.

pentagonal face indicates five tetrahedra sharing a bond. In general, each face of the Voronoi polyhedron corresponds to an atom of the first coordination shell.

We can predict the structure of an average Voronoi polyhedron in an idealized glass containing only perfect tetrahedral cells. There exist precisely five regular, convex polyhedra (Fig. 3). These are the five platonic solids: the tetrahedron, with Schlafly symbol {3,3}; the octahedron, with Schlafly symbol {3,4}; the cube, with Schlafly symbol {4,3}; the icosahedron, with Schlafly symbol {3,5}; and the dodecahedron, with Schlafly symbol {5,3}. The symbol $\{p,q\}$ means that the solid has the polygon $\{p\}$ for its faces, and q of these meet at each vertex.

The two values p and q in a polyhedron's Schlafly symbol $\{p,q\}$ determine the numbers of vertices, edges, and faces. This relationship follows from Euler's formula

$$V - E + F = 2, \tag{1.2}$$

where V is the number of vertices, E is the number of edges, and F is the number of faces. Since the polyhedron is regular, we have the relationship

$$qV = 2E = pF. \tag{1.3}$$

Solving these Eqs., we find

$$\begin{aligned} V &= \frac{4p}{2q + 2p - qp}, \\ E &= \frac{2qp}{2q + 2p - qp}, \\ F &= \frac{4q}{2q + 2p - qp}. \end{aligned} \tag{1.4}$$

The requirement that V, E, and F be positive integers restricts the allowed set of values of p and q to those of the Platonic solids.

Now try to construct an ideal glass (Coxeter, 1969, and Nelson, 1983b). We wish to fill space with atoms so as to create only perfect tetrahedral cells. Consider the Voronoi polyhedron of any atom. All faces of this polyhedron must meet in threes, since by hypothesis the system has only tetrahedral cells. Thus, the Schlafly symbol for the Voronoi polyhedron is $\{p,3\}$, where p should be 5 if the glass were to have perfect icosahedral coordination (and thus only dodecahedral Voronoi polyhedra). But icosahedra cannot pack to fill flat space. There is actually room for 5.104 perfect tetrahedra around each bond. Thus, we search for a packing with Voronoi polyhedron {5.104,3}.

No such packing exists, but one can consider packings where each bond is surrounded by p tetrahedra, and p may be any integer. We assign an energy cost for p not close to 5.104. Thus most bonds will have $p = 5$, the next most common will be $p = 6$. In terms of coordination shells, the icosahedron {3,5} will be the most common with its 12 vertices corresponding to the 12 pentagonal faces of the dodecahedral Voronoi polyhedron {5,3}. The average number of atoms in the first shell, however, is given by Eq. (1.4) with $q = 3$,

$$F = \frac{12}{6 - p}. \tag{1.5}$$

Thus, setting $p = 5.104$, we find $F = 13.398$ atoms on average arranged around the central atom to form $V = 22.796$ tetrahedral cells. Voronoi statistics for supercooled liquids, dense random packing models of metallic glass, and Frank–Kasper crystals agree quite well with this prediction, as shown in Sections 2.2, 2.3, and 2.4, respectively.

2. Local Icosahedral Order

Many biological, chemical, and physical systems display icosahedral order. Viruses, for example, frequently possess icosahedral shells (Caspar and Klug, 1962), as do many radiolarians (Thompson, 1952). The biological preference for icosahedra arises from the need to efficiently form a strong shell around a volume in space. The complex phase diagrams of elemental boron and compounds rich in boron arise from competition between a tendency to form icosahedral B_{12} packing units and a chemical version of icosahedral frustration (Hoard and Hughes, 1967).

This section focuses on the occurrence of local icosahedral order and the influence of frustration in atomic clusters, supercooled liquids, metallic glass, and crystals. I begin in 2.1 by discussing clusters of noble gas atoms, because a wide body of experimental data and theoretical calculations clearly demonstrates the importance of icosahedra and frustration. I shall consider supercooled liquids in the next section. Frank (1952) postulated that supercooling of liquid metals occurs as a result of the nucleation of icosahedral clusters of atoms. We reexamine his theory and also mention recent computer simulations (Steinhardt *et al.* 1983).

Section 2.3 explores experiments and computer simulations which show the presence of icosahedral order in glass. I consider dense random packing and other aggregation models of glass and review the radial distribution functions and structure functions of these models. In Section 2.4, I inspect the Frank–Kasper phases of intermetallic compounds. Voronoi statistics provide convincing evidence of icosahedral order in supercooled liquids, metallic glass, and Frank–Kasper crystals.

2.1 *Atomic Clusters*

Our theory of local icosahedral order in glass is based on an assumed tendency for small numbers of atoms to form icosahedral clusters. In this section, I review experimental observations which support this assumption. The experimental data I discuss is the mass spectrum of Ar^+ clusters produced in a low-temperature free jet expansion. I choose to discuss inert gas clusters because these are likely to better approximate the true interatomic forces between metal atoms in a bulk system than an actual metallic cluster would. This is because of finite-size effects in which the conduction electrons in metallic clusters form closed shells (Knight *et al.* 1984).

Although many researchers have discussed spectra and so-called magic numbers of neutral inert gas clusters, interpretation of the results is hampered by the possibility of fragmentation during the ionization of the cluster necessary for mass spectrography. Thus, I focus on the experiment of Harris *et al.* (1984) which starts with an ionized seed. Figure 4 shows their mass spectrum. Note the many prominent peaks, which are labeled in the figure by the numbers 13, 19, 23, 26, 29, 32, 34, 43, 46, 49, 55, ... These numbers, called magic numbers, correspond to highly stable structures which can be built out of the given number of atoms. The sequence of magic numbers may be understood with the following pictures (Fig. 5).

Figure 5 should be interpreted as follows. Start with a 13-atom icosahedral cluster. Then add atoms at the positions shown in Fig. 5A. The first 5 atoms each sit above a triangular face of the icosahedron, forming 5 new tetrahedra consisting of the 3 atoms on each face of the icosahedron,

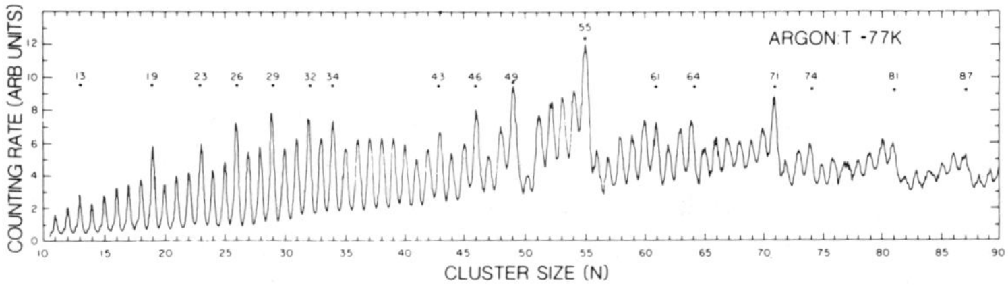

Figure 4. Experimental mass spectrum for charged argon clusters. From Harris *et al.* (1984).

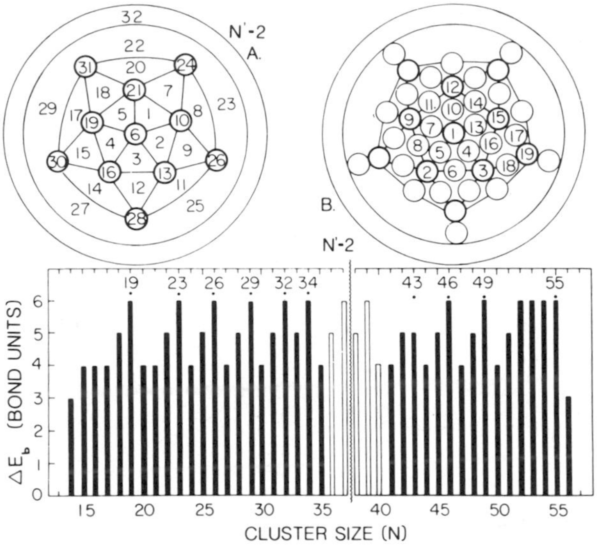

Figure 5. Model for magic numbers. (a) growth of endcaps on icosahedron. (b) deletion of atoms from 55-atom Mackay icosahedron. ΔE_b is the difference in binding energies of N atom clusters and N-1 atom clusters. From Harris, *et al.* (1984).

and the new atom. These 5 atoms together form a pentagonal ring. Place the sixth atom above the center of this ring, directly above a vertex of the icosahedron, completing a pentagonal bipyramid.

We may describe this highly stable 19-atom structure as a 13-atom icosahedral cluster with a 6-atom endcap. Alternatively, we may think of it as 2 interpenetrating icosahedra, just as the icosahedron itself may be thought of as 2 interpenetrating pentagonal bipyramids. Subsequent atoms are added so as to form additional endcaps in as efficient a manner as possible. The histogram below 5A displays the expected peak intensity. The magic-number sequence is reproduced perfectly. Computer simulations confirm these geometrical interpretations (Honeycutt and Andersen, 1987). The identical structure occurs in the metal carbonyl cluster $[Pt_{19}(CO)_{22}]4^-$ which is important in catalysis (Haggin, 1987).

The intersecting icosahedra model breaks down at about 40 atoms. Beyond this point, the Mackay icosahedra (Fig. 6) probably model the system better. In fact, Fig. 5B shows a sequence of atoms removed from a 55-atom Mackay icosahedron which seems to reproduce the magic number sequence from 43 to 55. Hoare (1976) discusses icosahedral order in atomic clusters which he calls "amorphons" due to their relationship to metallic glass. In the limit, as N goes to infinity, it is believed that an FCC crystal

Figure 6. Mackay icosahedra are multiply twinned FCC microcrystals.

should have stability greater than these icosahedral structures. The crossover point to FCC has never been determined. Figure 7 shows that icosahedral structures dominate up to $N = 2000$, the largest systems studied (Allpress and Sanders, 1970). In fact, for metallic clusters the crossover point may be around 10^5 to 10^7 (Phillips, 1986).

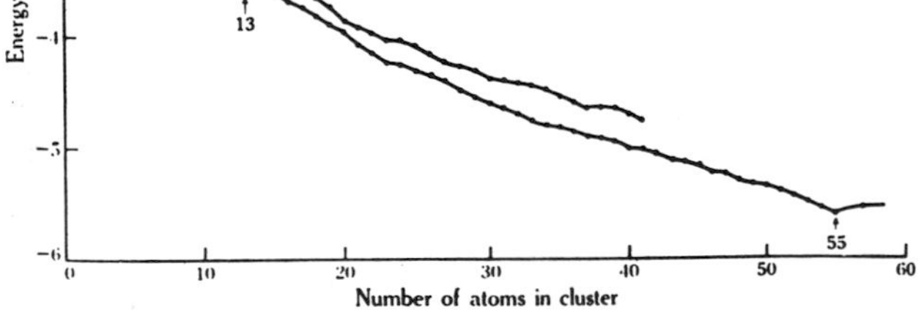

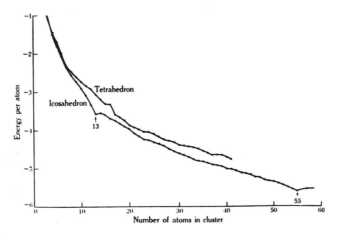

Figure 7. Binding energies of Mackay icosahedra and FCC single microcrystals. From Allpress and Sanders (1970).

2.2 Supercooled Liquids

Frank (1952) proposed that icosahedra should be prevalent in supercooled liquid metals. He argued that the extent of supercooling attainable in liquid metals (Turnbull, 1952) required that small clusters of atoms in a supercooled liquid form icosahedral structures which cannot form the seed of crystal nucleation. Only larger and less prevalent clusters take the structure of the bulk crystal phase. This notion has been tested and confirmed through molecular dynamics simulations of liquid argon (Rahman, 1966) and for a supercooled Lennard–Jones system (Steinhardt *et al.*, 1983).

The simulation by Rahman calculated Voronoi statistics in the liquid phase of argon. Even in the normal liquid, one observes the characteristic distribution of local environments dominated by five-sided faces. The anisotropy favoring six-sided faces over four-sided faces is also apparent. In the liquid phase, the Voronoi polyhedra have on average 14.45 faces with 5.169 sides per face. Recent simulations by Andersen (1987) on Lennard–Jones liquids reveal enhancement of the number of five-sided faces accompanying the transition from supercooled liquid to glass.

The simulation by Steinhardt *et al.* (1983) focusses on the possibility of long-range icosahedral bond-orientational order in supercooled liquids. The chief test for icosahedral order in this simulation was the projection of bond angles onto spherical harmonics. They define the local bond-orientational order parameter

$$Q_{lm}(\mathbf{r}) = Y_{lm}(\theta(\mathbf{r}),\phi(\mathbf{r})), \tag{2.1}$$

where the Y_{lm} are the usual spherical harmonics, and θ and ϕ denote the angles of an atomic bond at point **r**. This quantity's usefulness as a measure of icosahedral order results from the fact that icosahedral symmetry prohibits nonzero values of Q_{lm} for $l < 6$, whereas cubic and lower symmetries have nonzero projection onto $l = 4$. For a perfect icosahedron oriented along the z axis, one finds

$$Q_{60}^2 = \frac{11}{7} Q_{6\pm5}^2 . \tag{2.2}$$

Darby and Evans (1978) note a near vanishing of the Q_{lm}, except for $l = 6$, $m = 0, \pm 5$, in a model of metallic glass.

Because Steinhardt *et al.* were interested in the question of long-range bond-orientational order, they focussed on bulk quantities such as

$$Q_l = \left\{ \frac{4\pi}{2l+1} \sum_{m=-l}^{l} |< Q_{lm} >|^2 \right\}^{1/2}, \tag{2.3}$$

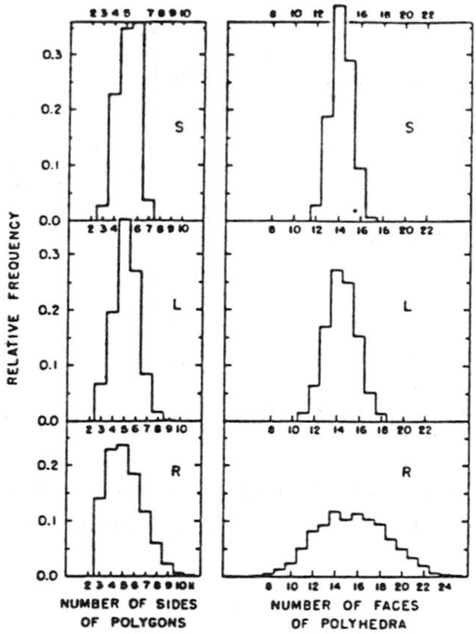

Figure 8. Voronoi statistics for solid (S) and liquid (L) argon and a random set of points (R). From Rahman (1966).

where the brackets < > denote averaging over many bonds. Figure 9 shows histograms of Q_l for the 13-atom icosahedral cluster (a), the 13-atom FCC cluster (b), and the result of a molecular dynamics simulation on an 864-atom system (c). The pronounced peak at $l = 6$ in Fig. 9c reveals a tendency towards icosahedral order. The fact that the peak survives averaging out to several atomic radii suggests that there may actually be long-range bond-orientational order, although it could be just a finite-size effect. Since even the FCC cluster has a peak at $l = 6$, a more sensitive test is needed to ensure that the observed order is actually icosahedral.

One such measure is provided by the orientational correlation function

$$G_l(\mathbf{r}) = [4\pi/(2l + 1) \sum_{m=-l}^{l} < Q_{lm}(\mathbf{r})Q_{lm}(\mathbf{0}) >]/G_0(\mathbf{r}), \qquad (2.4)$$

where

$$G_0(\mathbf{r}) = 4\pi < Q_{00}(\mathbf{r})Q_{00}(\mathbf{0}) >, \qquad (2.5)$$

and the angle brackets indicate an average over all atoms separated by **r**. Figure 10 shows the $l = 4$ and 6 correlation functions. At high tem-

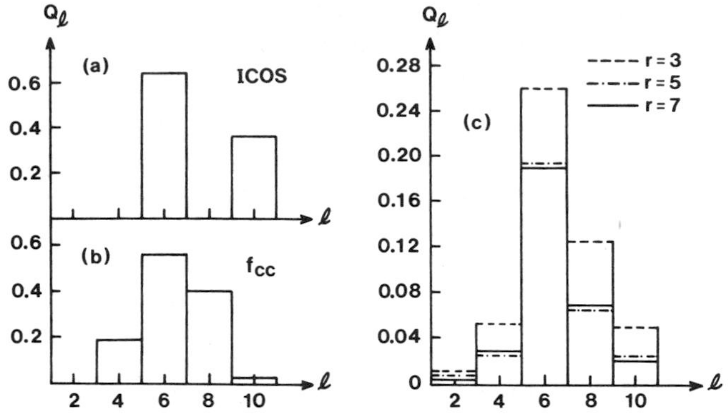

Figure 9. Quadratic invariants for (a) 13-atom icosahedral cluster, (b) 13-atom FCC cluster, and (c) computer simulation. From Steinhardt *et al.* (1983).

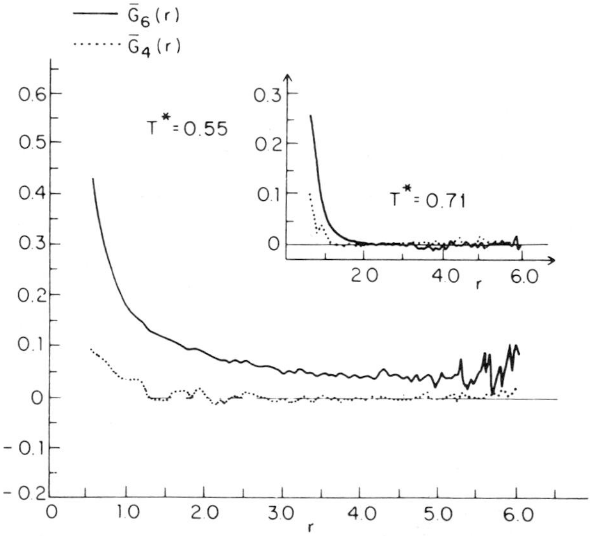

Figure 10. $l = 4$ and $l = 6$ orientational correlation functions at high temperature and low temperature. From Steinhardt *et al.* (1983).

peratures, both fall off rapidly as expected in the liquid phase. At low temperatures, however, the icosahedral ($l = 6$) bond-orientational order parameter suggests long-range order. This effect also could be simply a finite-size effect.

2.3 *Metallic Glass*

It is often suggested that the structure of metallic glass is related to that of a supercooled liquid. The idea is that as the icosahedral order parameter in the liquid grows large, the defects lines forced in by icosahedral frustration become entangled. This effect may slow down the dynamics leading to a long-lived, disordered, metastable state. To test this idea, we must know the actual locations of atoms in the glass. Unfortunately, little is known experimentally about the actual atomic positions in glass because the usual techniques of crystallography depend on the orderly repetition of a crystal unit cell. The experimentally measured structure factors and radial distribution functions do not uniquely determine the structure. However, we do gain some insight from these measurements. For instance, microcrystalline models of amorphous metals may be ruled out because of qualitative discrepancies between amorphous and crystalline diffraction patterns.

Computer simulations provide a bridge between theory and experiment. Simplified models of glass may be constructed by computer algorithms of hard sphere packing which take into account the preference for tetrahedral clustering (Finney, 1970, and Bennett, 1972). These models yield good agreement with experimentally measured diffraction patterns and radial distribution functions. Because the computer has a record of every atom's location, the actual structure is known and can bc analyzed. Early models of this sort, known as dense random packing models, were even constructed and analyzed by hand (Bernal, 1964).

Bernal's theory is actually a theory of liquid structure. He proposes that liquids may be modeled as a dense random packing of atoms in which there is no empty cell large enough to contain an extra atom. The theory of metallic glass discussed in this chapter considers the glass to be a supercooled liquid which has slow dynamics. Indeed, dense random packing models are the most successful class of models of metallic glass (Cohen and Turnbull, 1964, Cargill, 1970, Chaudhari and Turnbull, 1978, and Zallen, 1983). Figure 11 illustrates Bernal's dense random packing structure. How can we describe such a structure quantitatively? The radial distribution function, structure function, and Voronoi statistics have all been calculated for computer-generated dense random packings. Discussion of

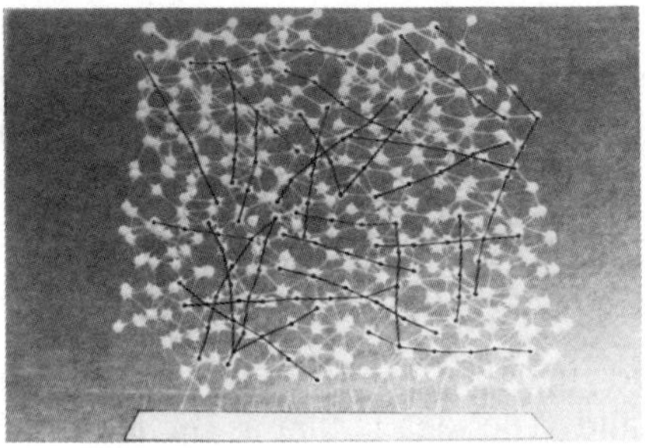

Figure 11. Dense random packing model of Bernal (1964). Black lines mark approximate colineations.

the structure function is deferred until Section 4. We begin by discussing the Voronoi statistics.

Bernal noted that the structure is dominated by tetrahedral cells, and that the Voronoi polyhedra have predominantly pentagonal faces. Figure 12 shows the Voronoi statistics for the Ichikawa–Bennett model, both before and after relaxation in a soft sphere potential. Note the peak at 5 tetrahedra per bond and also the asymmetry favoring 6 over 4. This asymmetry is a consequence of icosahedral frustration which requires slightly more than 5 tetrahedra per bond on average. Compare Fig. 12 with the data of Rahman (1966) for liquid argon (Fig. 8).

We can understand qualitative features of the radial distribution function (Fig. 13) by consideration of tetrahedral packing geometry. First of all, each atom has a number of near neighbors in its first coordination shell. For hard sphere systems, no atoms have separation less than σ, so the first coordination shell extends from σ to wherever the first minimum occurs. The first peak is therefore extremely asymmetrical, with a discontinuous rise and more gradual falloff. The second coordination shell is somewhat controversial, consisting either primarily of atoms on opposite sides of triangular biprisms (Sadoc *et al.* 1973) leading to a smooth symmetrical peak at 1.63σ, or else of pairs of coplanar equilateral triangles leading to an asymmetrical peak with discontinuous drop at 1.73σ (Bennett, 1972). The third peak arises from colineations of three atoms creating an asymmetrical peak with a discontinuous drop at 2.0σ.

The key difference between the models is that the Sadoc model explicitly constructs tetrahedra, while the Finney and Bennett models seeks

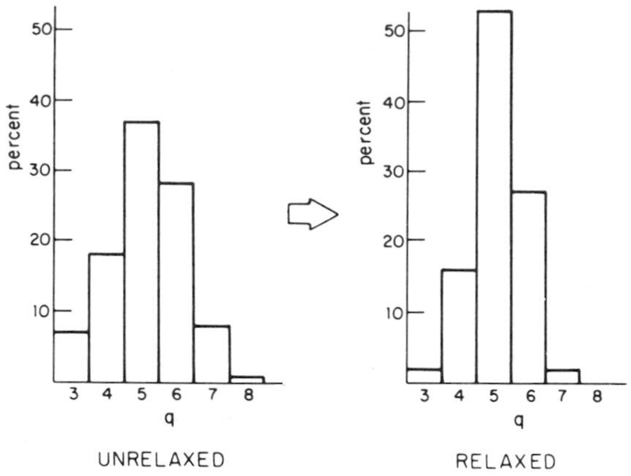

Figure 12. Voronoi statistics for unrelaxed and relaxed Ichikawa–Bennett dense random packing model (Nelson, 1983).

instead to maximize global density. Note that the Finney and Bennett models predict that the third peak should have greater weight than the second peak in apparent contradiction of the experiment (Fig. 13). The Finney and Bennett models can be improved by statically relaxing the configuration in a soft sphere potential. However, the Voronoi statistics (Fig. 12) suggest that this amounts to increasing the degree of tetrahedral ordering, which is already present in the Sadoc model.

2.4 *Crystals and Quasicrystals*

The clearest evidence of icosahedral order in metallic solids occurs in a family of intermetallic compounds whose crystal structures boast large unit cells with extensive icosahedral coordination. Frank and Kasper (1958 and 1959) classified crystal phases which contain only tetrahedral cells. Most Frank–Kasper phases arise in metallic alloys where the smaller atoms occupy the icosahedral sites and the larger atoms accommodate the icosahedral frustration by occupying sites of non-icosahedral coordination. Many authors predict (Elser and Henley, 1985, Guyot and Audier, 1985) or observe (Ma, Stern, and Gayle, 1987) similar short-range order in crystalline and quasicrystalline phases.

Because these materials are discussed in detail in this volume by Shoemaker and Shoemaker, I only present a few facts pertinent to the present

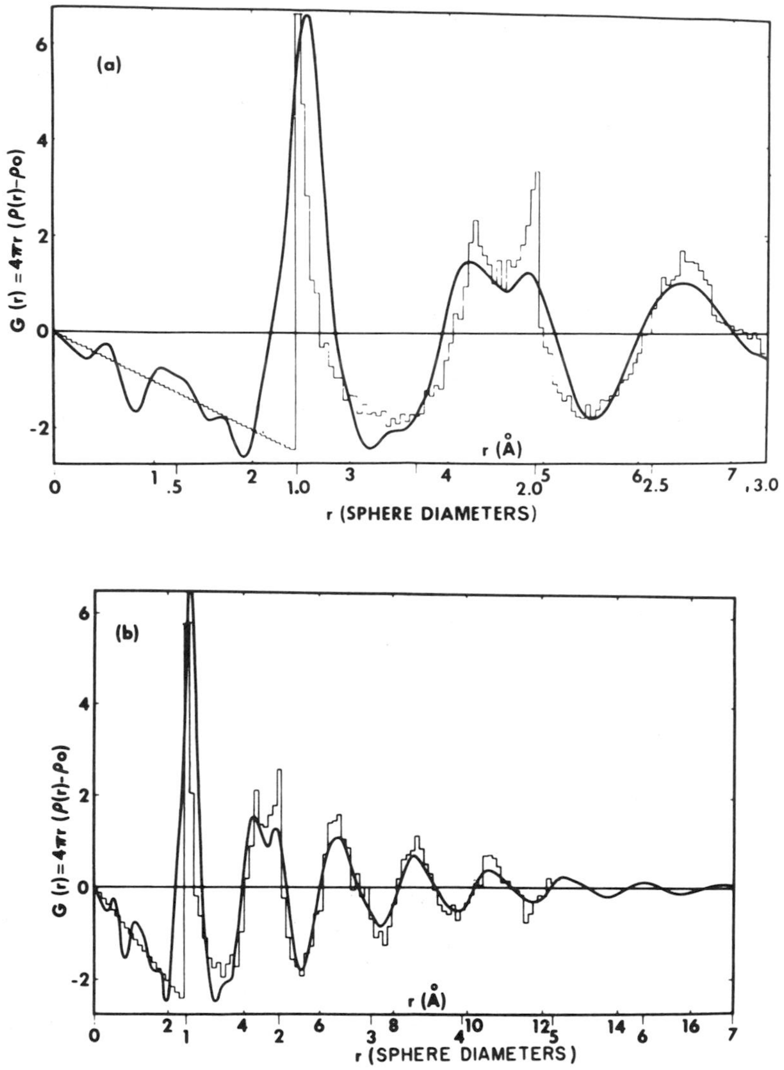

Figure 13. Radial distribution function of Finney's (1970) dense random packing model of hard spheres compared with experimental data for amorphous $Ni_{76}P_{24}$ (Cargill, 1975).

discussion. First of all, because crystals have long-range orientational order, there are long-range orientational correlations between icosahedral sites, just as in quasicrystals. Mechanisms for propagating icosahedral orientational order are discussed in the final section (Section 5) of this article. Secondly, the Voronoi statistics approach the values for the ideal glass. The larger unit-cell crystals typically come closer than the smaller unit-cell crystals. Figure 14 shows a plot of mean coordination number versus number of tetrahedra per bond, comparing liquids, glasses, and Frank–Kasper phases.

3. Mathematics of Polytopes

Icosahedra pack perfectly in S^3, the curved three-dimensional surface of a sphere in four dimensions. The resulting structure is known as polytope {3,3,5}. In general, polytopes are high-dimensional analogues of polyhedra and polygons. Just as polygons and polyhedra have varying degrees of

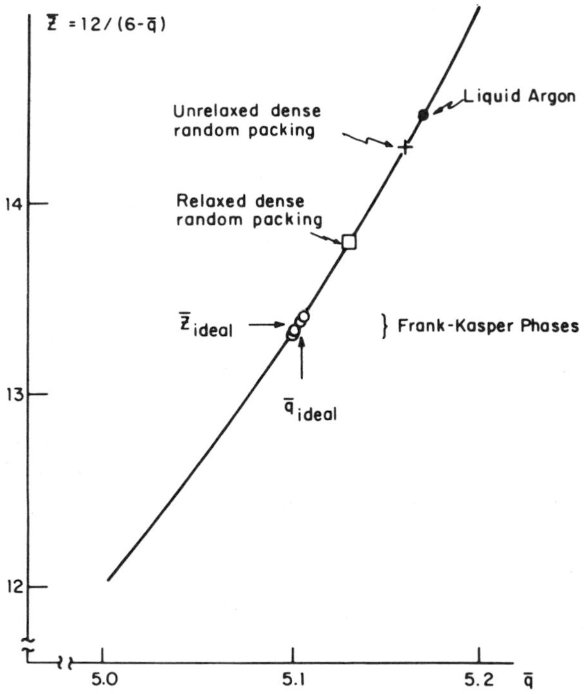

Figure 14. Coordination number z vs. q for different particle configurations (Nelson, 1983b).

complexity and symmetry, so the set of polytopes is rich and intriguing. This section presents a survey of polytope structures and their mathematical properties. Section 3.1 defines the regular polygons, polyhedra, and polytopes in two, three, and four dimensions. Section 3.2 introduces the group of unit quaternions as the natural coordinate system for studying polytopes. Section 3.3 exploits a correspondence between quaternions and three-dimensional rotations to identify the polytopes' vertices and to calculate the four-dimensional rotational symmetry group of polytope {3,3,5}.

3.1 Regular Polytopes

The essence of metallic-glass formation rests on icosahedral frustration (Kleman and Sadoc, 1979, Nelson, 1983a, Sethna, 1983). The local preference for icosahedra demonstrated in the previous section is thwarted at large-length scales. When a metal is rapidly quenched, this large-distance effect is unable to influence the local order. Thus, a metallic glass may be thought of as a highly defected icosahedral crystal. The defects are a remnant of the frustration.

In order to exploit this view of a glass as a defected icosahedral crystal, we must describe the underlying defect-free icosahedral crystal. That is, we must find a way to adjust the degree of icosahedral frustration or even remove it entirely. Remarkably, such a feat is possible. Icosahedral crystals are prohibited only in three-dimensional flat space. By curving space suitably, one can reduce or altogether eliminate icosahedral frustration. The origin of icosahedral frustration is thus the absence of curvature in physical space. This observation underlies the remainder of this article. The main idea is to remove the frustration by curving space, to understand the resulting structure (polytope {3,3,5}), then to reintroduce the frustration by flattening space back out in a controlled fashion.

Polytopes may model systems quite different from metallic glasses as well. A polytope known as polytope 240, which is related to polytope {3,3,5} in the same manner as the diamond structure is related to FCC, models tetracoordinated glasses (Kleman and Sadoc, 1979, Sadoc and Mosseri, 1982). The frustration of double-twist vector fields in $\mathbf{R}^3$ is relieved on S^3, giving rise to polytope-like models of the blue phases of cholesteric liquid crystals (Hornreich, Kugler, and Shtrikman, 1982, Sethna, 1983 and 1985, Nelson, 1983b). Twisted vector fields on S^3 also serve as a model for molten polymers (Kleman, 1985).

This section describes those structures known as polytopes. Polytopes may be thought of as crystals in a non-Euclidean three-dimensional space.

Alternatively, we may consider them embedded in a higher-dimensional flat space. I adopt this second view primarily because it is better suited to my intuition. Lower-dimensional analogues of polytopes are polyhedrons and polygons which may be regarded as, respectively, two- and one-dimensional non-Euclidean crystals embedded in three-dimensional and two-dimensional flat space.

Consider first the set of regular polygons. Each regular polygon can be identified by its Schlafly symbol $\{p\}$, which is simply the number of vertices. Of course, p also equals the number of edges. There exist infinitely many regular polygons—p may be any integer greater than or equal to 3. To view a regular polygon in a curved one-dimensional space, simply inscribe it in a circle (S^1). The vertices of the polygon form a regular lattice on the circle.

Polyhedra are somewhat more complicated. We already have seen the five Platonic solids (Fig. 3). Recall that the symbol $\{p,q\}$ means that the solid has the polygon $\{p\}$ for its faces (called, in general, "prime faces"), and q of these meet at each vertex. If we intersect the polygon with a plane so as to cut off one corner, the truncated surface is a polygon $\{q\}$ known as the "vertex figure." Two polyhedra are said to be dual to each other if their Schlafly symbols are the reverse of each other. Thus, the cube {4,3} and octahedron {3,4} are dual to each other, as are the icosahedron {3,5} and dodecahedron {5,3}. Dual polyhedra share identical symmetry groups. The tetrahedron {3,3} is self dual. Each regular polyhedron may be inscribed inside a sphere (S^2) and thus considered as a non-Euclidean two-dimensional crystal. In the limit of infinite radius of curvature, we have the Euclidean crystals known as the triangular lattice {3,6} and its dual, the hexagonal lattice {6,3}, as well as the self-dual square lattice {4,4}.

Now go up one dimension and consider polytopes (Coxeter, 1973). In general, the polytope with Schlafly symbol $\{p,q,r\}$ divides the three-dimensional surface of S^3 into prime faces which are cells bounded by polyhedra $\{p,q\}$. The vertex figure (intersection of polytope with flat three-dimensional hyperplane) forms the polyhedron $\{q,r\}$. There exist six regular, convex polytopes (Fig. 15): the self dual regular simplex ("pentatope") {3,3,3}; the hypercube ("tesseract") {4,3,3} and its dual the 16-cell (cross-polytope) {3,3,4}; the self-dual 24-cell {3,4,3}; the 120-cell {5,3,3}, and its dual the 600-cell {3,3,5}. All vertices of the regular polytopes sit on the hypersphere S^3 in four dimensions. Thus the polytopes form non-Euclidean three-dimensional crystals. Detailed description of the icosahedral polytope {3,3,5} follows after discussion of coordinate systems for S^3. In all higher dimensions, there exist just three regular polytopes: the

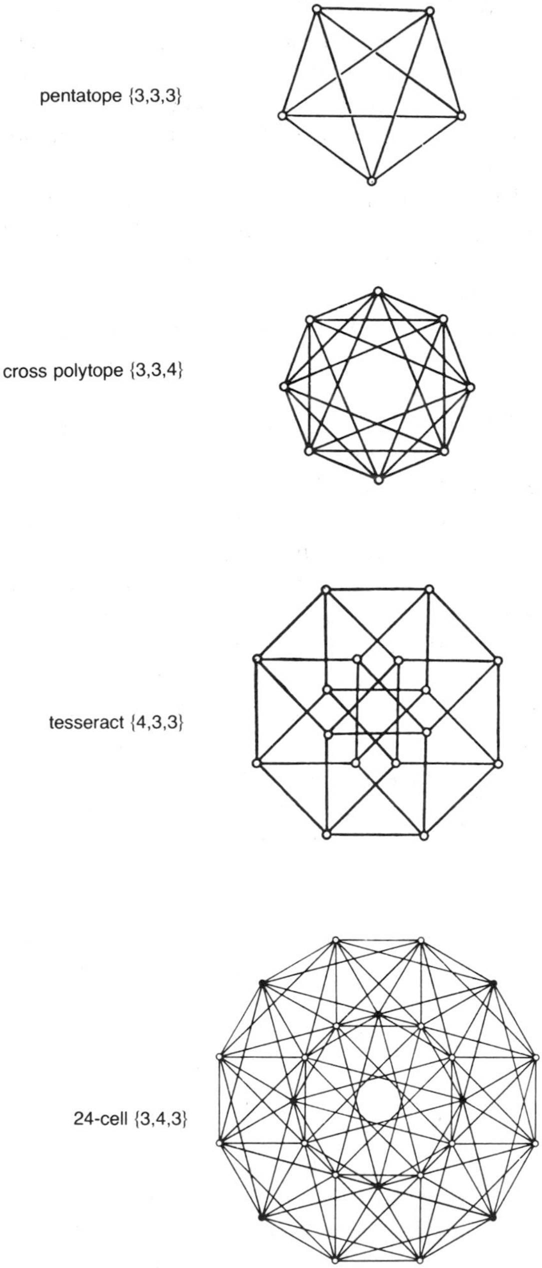

Figure 15. The six regular polytopes, projected into two dimensions. From Coxeter (1969 and 1973).

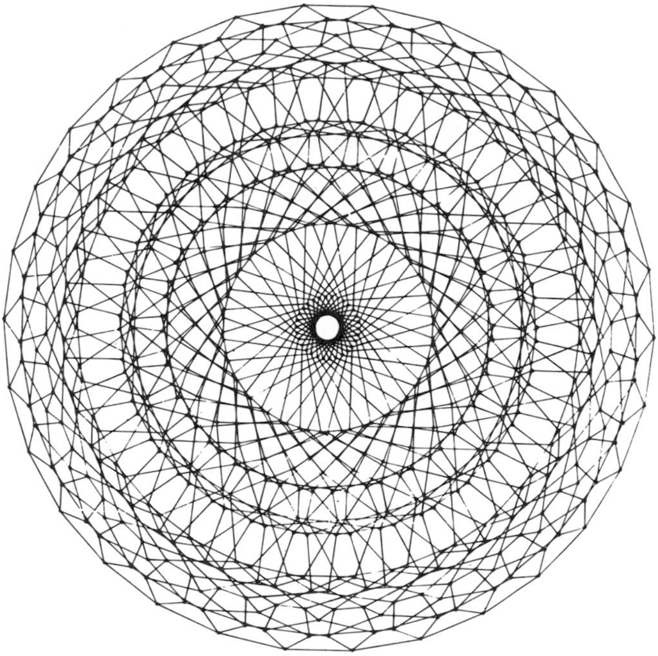
120-cell {5,3,3}

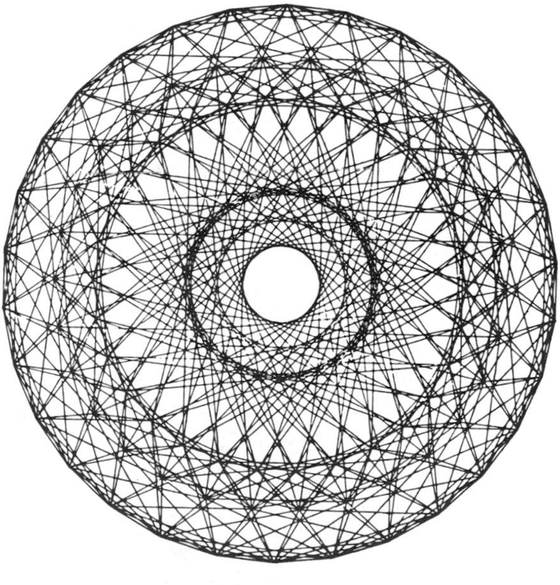
600-cell {3,3,5}

Figure 15. (Continued)

self dual simplex $\{3^d\}$; the hypercube $\{4,3^{d-1}\}$; and its dual, the cross-polytope $\{3^{d-1},4\}$.

3.2 Quaternions and Coordinate Systems

Because the non-Euclidean space S^3 represents the surface of a sphere in Euclidean four-dimensional space, we may do analytic geometry in S^3 using four-dimensional Cartesian coordinates. That is, any point in $\mathbf{R}^4$ may be expressed as

$$\mathbf{u} = u_\alpha \hat{\mathbf{e}}_\alpha \tag{3.1}$$

where $\hat{\mathbf{e}}_\alpha$ ($\alpha = 0, 1, 2, 3$) are the four basis vectors of $\mathbf{R}^4$. Any four-dimensional vector belongs to the field of real quaternions. The algebraic properties of the quaternions turn out to be of great use in studying the polytopes (DuVal, 1964).

Multiplication between two quaternions is defined by

$$\mathbf{uv} = (u_o v_o - u_j v_j)\hat{\mathbf{e}}_0 + (u_0 v_j + v_0 u_j)\hat{\mathbf{e}}_j + \varepsilon_{ijk} u_j v_k \hat{\mathbf{e}}_i, \tag{3.2}$$

where $i,j,k = \{1,2,3\}$. Note that quaternion multiplication is noncommutative. The inverse conjugate of $\mathbf{u}$ is

$$\mathbf{u}^{-1} = u_0 \hat{\mathbf{e}}_0 - u_i \hat{\mathbf{e}}_i. \tag{3.3}$$

The scalar product of a quaternion $\hat{\mathbf{u}}$ and a real number s is defined by

$$s\mathbf{u} = (s u_\alpha)\hat{\mathbf{e}}_\alpha. \tag{3.4}$$

Finally, quaternions may be added together by

$$\mathbf{u} + \mathbf{v} = (u_\alpha + v_\alpha)\hat{\mathbf{e}}_\alpha. \tag{3.5}$$

Not all points in $\mathbf{R}^4$ lie on S^3. Thus it is convenient to restrict attention to unit quaternions which satisfy

$$|\hat{\mathbf{u}}|^2 = u_\alpha u_\alpha = 1. \tag{3.6}$$

Clearly, the unit quaternions are in one-to-one correspondence with points on S^3. It turns out that unit quaternions themselves form a group, denoted $\mathbf{Q}$. Remarkably, the group $\mathbf{Q}$ is isomorphic to the familiar group $SU(2)$ through the mapping

$$\hat{\mathbf{u}} = u_\alpha \hat{\mathbf{e}}_\alpha \in \mathbf{Q} \leftrightarrow u = \begin{Bmatrix} u_0 + iu_3 & iu_1 + u_2 \\ iu_1 - u_2 & u_0 - iu_3 \end{Bmatrix} \in SU(2). \tag{3.7}$$

The relationship is perhaps clearer using a different representation of quaternions. Any unit four vector may be written in the form

$$\hat{\mathbf{u}} = (\cos \psi, \hat{\mathbf{n}} \sin \psi), \tag{3.8}$$

where ψ is the geodesic distance between $\hat{\mathbf{u}}$ and the "north pole" (1,0,0,0), and $\hat{\mathbf{n}} \in S_2$ is a unit three-vector. This quaternion $\hat{\mathbf{u}}$ is isomorphic to the $SU(2)$ matrix

$$u = e^{i\psi \hat{\mathbf{n}} \cdot \boldsymbol{\sigma}}, \tag{3.9}$$

where $\boldsymbol{\sigma}$ is the vector of Pauli matrices.

We can easily calculate the geodesic separation between any two points $\hat{\mathbf{u}}$ and $\hat{\mathbf{v}}$ on S^3 by expressing them as $SU(2)$ matrices u and v. The geodesic separation of points on a unit sphere is simply the angle between them,

$$\psi(\hat{\mathbf{u}},\hat{\mathbf{v}}) = \cos^{-1}(1/2 \; Tr \; uv^{-1}). \tag{3.10}$$

In particular, the point $\hat{\mathbf{u}}$ defined above (Eq. 3.8) lies a geodesic distance ψ from the north pole of S^3.

Finally, we observe that $SU(2)$ is the universal covering group of $SO(3)$, the group of rotations in three dimensions. That means any $SU(2)$ matrix (and therefore any unit quaternion) denotes a unique rotation in ordinary three-dimensional space. Furthermore, any three-dimensional rotation is generated by precisely two elements of the group $SU(2)$ (two unit quaternions). Formally, to generate a three-dimensional rotation using $SU(2)$ matrices, write an arbitrary three-dimensional unit vector $\hat{\mathbf{m}} \in S^2$ as an $SU(2)$ matrix

$$v = e^{i\phi \hat{\mathbf{m}} \cdot \boldsymbol{\sigma}}, \tag{3.11}$$

where ϕ is arbitrary. Then act on v with u defined in Eq. 3.9 by

$$u\colon v \rightarrow uvu^{-1} = e^{i\phi \, R_u(\hat{\mathbf{m}}) \cdot \boldsymbol{\sigma}}, \tag{3.12}$$

where $R_u(\hat{\mathbf{m}})$ denotes a rotation of $\hat{\mathbf{m}}$ by the angle 2ψ around the axis $\hat{\mathbf{n}}$. The fact that $SU(2)$ is a double covering of $SO(3)$ arises from the fact that Eq. (2.12) is quadratic in u and therefore u and $-u$ induce the same three-dimensional rotation.

We now have a coordinate system for S^3 in which any point $\hat{\mathbf{u}}$ of S^3 written in the form (3.8) corresponds to a three-dimensional rotation through Eq. (3.12). This turns out to be a useful labeling of points in S^3 for describing polytopes.

3.3 Polytope Symmetries

Describing the symmetry groups of the Platonic solids as sets of points on S^3 provides an amusing and instructive exercise. Consider, for instance, the rotational symmetry group $\mathbf{Y} \subset SO(3)$ of the icosahedron. The lift $\mathbf{Y}$

$\subset SU(2)$ of this group has 120 elements contained in 9 conjugacy classes corresponding to different types of rotational symmetries of the icosahedron (see Table 1). List the elements of **Y** in terms of increasing angle of rotation. The first class

$$\mathscr{C}_0 = \{\mathbf{u} = (1, 0, 0, 0)\} \tag{3.12a}$$

contains only the identity. The second class

$$\mathscr{V}_1 = \{\hat{\mathbf{u}} = e^{i\pi/5\,\hat{\mathbf{n}}\cdot\boldsymbol{\sigma}} : \hat{\mathbf{n}} \text{ points to vertices of an icosahedron}\} \tag{3.12b}$$

contains rotations by $2\pi/5$ around axes passing through the 12 vertices of an icosahedron. The third class

$$\mathscr{F}_2 = \{\hat{\mathbf{u}} = e^{i\pi/3\,\hat{\mathbf{n}}\cdot\boldsymbol{\sigma}} : \hat{\mathbf{n}} \text{ points to faces of an icosahedron}\} \tag{3.12c}$$

contains rotations by $2\pi/3$ around axes passing through the 20 faces of an icosahedron. The fourth class

$$\mathscr{V}_3 = \{\hat{\mathbf{u}}^2 : \hat{\mathbf{u}} \in \mathscr{V}_1\} \tag{3.12d}$$

contains rotations by $4\pi/5$ around the vertices of an icosahedron. The fifth class

$$\mathscr{E}_4 = \{\hat{\mathbf{u}} = e^{i\pi/2\,\hat{\mathbf{n}}\cdot\boldsymbol{\sigma}} : \hat{\mathbf{n}} \text{ points to edges of an icosahedron}\} \tag{3.12e}$$

contains rotations by π around axes passing through the 30 edges of an icosahedron. The remaining four classes are simply the negatives of the first four and therefore do not correspond to new rotations.

$$\mathscr{V}_5 = \{-\hat{\mathbf{u}} : \hat{\mathbf{u}} \in \mathscr{V}_3\} \tag{3.12f}$$

$$\mathscr{F}_6 = \{-\hat{\mathbf{u}} : \hat{\mathbf{u}} \in \mathscr{F}_2\} \tag{3.12g}$$

$$\mathscr{V}_7 = \{-\hat{\mathbf{u}} : \hat{\mathbf{u}} \in \mathscr{V}_1\} \tag{3.12h}$$

and

$$\mathscr{C}_8 = \{\hat{\mathbf{u}} = (-1,0,0,0)\}. \tag{3.12i}$$

What positions do these points take on the sphere S^3? The identity $\mathscr{C}_0$ is the northpole of S^3. The 12 members of the class V_1 each sit a geodesic distance $\pi/5$ from the northpole. A set of points a fixed distance from the northpole of a hypersphere S^3 is simply an ordinary sphere S^2, which surrounds the northpole in the non-Euclidean three-dimensional space just as the sphere S^2 surrounds its center. Therefore, the set $\mathscr{V}_1$ sit at the vertices of an icosahedron surrounding the identity $\mathscr{C}_0$. The geodesic distance

Table I. Character of table of the icosahedral double group Y′ C SU(2).

Y	$1\mathscr{C}_0$	$12\mathscr{V}_1$	$20\mathscr{F}_2$	$12\mathscr{V}_3$	$30\mathscr{E}_4$	$12\mathscr{V}_5$	$20\mathscr{F}_6$	$12\mathscr{V}_7$	$1\mathscr{C}_8$
A	1	1	1	1	1	1	1	1	1
E_1	2	Ω	1	Ω^{-1}	0	$-\Omega^{-1}$	-1	$-\Omega$	-2
E_2	2	$-\Omega^{-1}$	1	$-\Omega$	0	Ω	-1	Ω^{-1}	-2
F_1	3	Ω	0	$-\Omega^{-1}$	-1	$-\Omega^{-1}$	0	Ω	3
F_2	3	$-\Omega^{-1}$	0	Ω	-1	Ω	0	$-\Omega^{-1}$	3
G_1	4	1	-1	-1	0	1	1	-1	-4
G_2	4	-1	1	-1	0	-1	1	-1	4
H	5	0	-1	0	1	0	-1	0	5
I	6	-1	0	1	0	-1	0	1	-6

from $\mathscr{C}_0$ to points of $\mathscr{V}_1$ is $\pi/5$, precisely half the angle of rotation of symmetry axes passing through vertices of an icosahedron.

Figure 16a shows the conjugacy classes $\mathscr{C}_0$ and $\mathscr{V}_1$ projected into flat space $\mathbf{R}^3$ from curved space S^3. This is precisely the 13-atom icosahedral cluster observed in so many physical contexts in Section 2. The next conjugacy class $\mathscr{F}_2$ forms a new shell surrounding $\mathscr{C}_0$ and $\mathscr{V}_1$ at a distance $\pi/3$ from the northpole in S^3. Figure 16b illustrates the projection of this structure into $\mathbf{R}^3$. Because the atoms in $\mathscr{F}_2$ sit on the triangular faces of the shell $\mathscr{V}_1$, the number of tetrahedral cells is maximized. The next conjugacy class $\mathscr{V}_3$ occupies the pentagonal rings of atoms in $\mathscr{F}_2$ to form pentagonal bipyramids. Figure 16c shows the 45-atom structure consisting of conjugacy classes $\mathscr{C}_0$, $\mathscr{V}_1$, $\mathscr{F}_2$, and $\mathscr{V}_3$.

Although the growing icosahedral frustration is clearly visible in the gaps between the atoms, the structure in Fig. 16c actually occurs in nature (Bergman *et al.*, 1957) within the unit cell of the Frank–Kasper phase of $Mg_{32}(Al,Zn)_{49}$. In addition, the growth of clusters by forming an icosahedron $\mathscr{V}_1$, adding atoms above the faces in $\mathscr{F}_2$, then capping the pentagonal rings with atoms in $\mathscr{V}_3$, is precisely the explanation for the magic number sequence of Ar^+ clusters (Harris *et al.* 1984). The final structure (Fig. 16d), which includes the 30 atoms in the conjugacy class $\mathscr{E}_4$, is not known in nature.

What is the structure in S^3 obtained from the entire icosahedral group **Y**? That is the icosahedral polytope {3,3,5}! The 120 elements of the symmetry group **Y** of an icosahedron expressed as quaternions form the 120 vertices of polytope {3,3,5}. Figure 17 shows the radial distribution function of polytope {3,3,5}, which can be calculated immediately once the relation

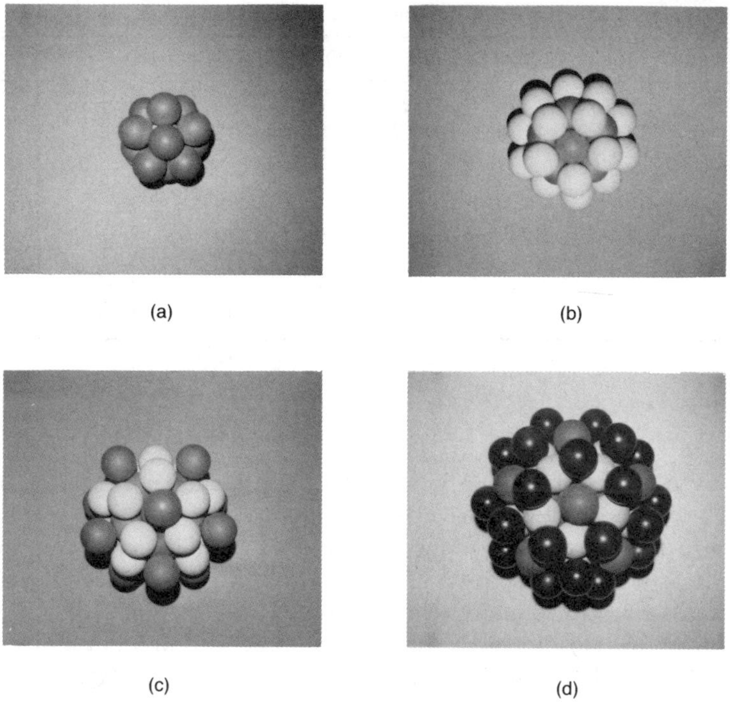

Figure 16. Fragments of polytope {3,3,5}, projected into flat space. The outer coordination shell is $\mathcal{V}_1$ in (a), $\mathcal{F}_2$ in (b), $\mathcal{V}_3$ in (c), and $\mathcal{E}_4$ in (d).

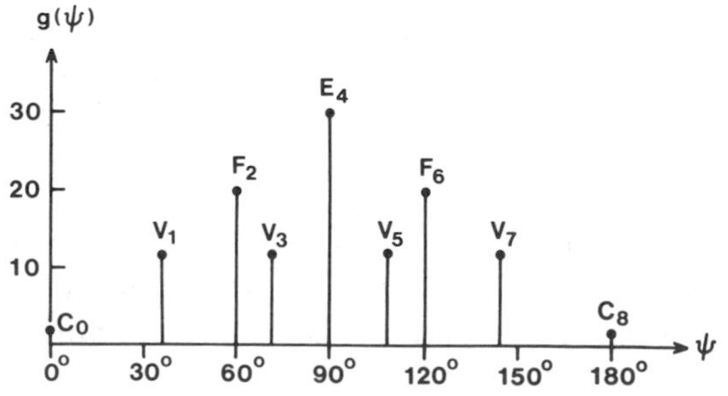

Figure 17. Radial distribution function of polytope {3,3,5}.

is known between polytope structure and symmetries of an icosahedron.

Can we generalize this statement to other symmetries? DuVal (1964) answers this question. He shows that the 24 elements of the symmetry group **T** of a tetrahedron form the 24 vertices of the polytope {3,4,3}. The 8-element dihedral group $\mathbf{D}_2$ forms the 8 vertices of the hyperoctahedron {3,3,4}. The 16 elements of the 24-element tetrahedral group **T** which are not in $\mathbf{D}_2$ form the 16 vertices of the hypercube {4,3,3}. Finally, the 600 vertices of polytope {5,3,3} arise from cosets of the icosahedral symmetry group **Y** of the form $\hat{\mathbf{p}}^j\hat{\mathbf{t}}\mathbf{Y}$, where $\hat{\mathbf{t}}$ is any element of the coset of **T** in the octahedral group **O**, $\hat{\mathbf{p}}$ is any element of order 5 in **Y**, and j runs from 0 through 4.

The relationship between geometrical properties of polytopes in S^3 and rotational symmetry groups (subgroups of $SU(2)$) of polyhedrons in S^2 leads immediately to the four-dimensional rotational symmetry groups (subgroups of $SO(4)$) of the polytopes. First consider the group $SO(4)$ itself. Because S^3, the sphere in four dimensions, is isomorphic to the group of unit quaternions, we may rotate S^3 by multiplying it on the left and right by unit quaternions,

$$(\hat{\mathbf{l}},\hat{\mathbf{r}}) : S^3 \rightarrow \hat{\mathbf{l}}\, S^3\, \hat{\mathbf{r}}^{-1}. \tag{3.13}$$

Every four-dimensional rotation may be expressed in this manner. Noting that the group of unit quaternions is simply $SU(2)$, we find

$$SO(4) = SU(2) \times SU(2) / Z_2, \tag{3.14}$$

where the mod Z_2 arises because $(\hat{\mathbf{l}},\hat{\mathbf{r}})$ and $(-\hat{\mathbf{l}},\hat{\mathbf{r}})$ denote identical rotations. Similarly, the symmetry group G of polytopes {3, 3, 5} and {5, 5, 3} is related to the direct square of Y, the symmetry group of an icosahedron,

$$G = Y \times Y / Z_2. \tag{3.15}$$

Certain four-dimensional rotations may be understood in terms of familiar three-dimensional concepts. For example, the rotation $(\hat{\mathbf{l}},\hat{\mathbf{l}})$, where $\hat{\mathbf{l}} = e^{i\psi\hat{\mathbf{n}}\cdot\vec{\sigma}}$, leaves the set of points on S^3 corresponding to $e^{i\phi\hat{\mathbf{n}}\cdot\boldsymbol{\sigma}}$ invariant. Imagine sitting at the northpole of S^3. During the rotation, the northpole remains fixed, as does a straight line (actually a geodesic) of points running out of the northpole along the axis $\hat{\mathbf{n}}$. From the point of view of the northpole, then, a rotation of the type $(\hat{\mathbf{l}},\hat{\mathbf{l}})$ appears to be a conventional three-dimensional rotation through the angle 2ψ around the axis $\hat{\mathbf{n}}$. In contrast, rotations of the type $(\hat{\mathbf{l}},\hat{\mathbf{l}}^{-1})$ appear as three-dimensional translations of the northpole along the axis $\hat{\mathbf{n}}$ a distance 2ψ, because they replace $e^{i\phi\hat{\mathbf{n}}\cdot\boldsymbol{\sigma}}$ with $e^{i(\phi+2\psi)\hat{\mathbf{n}}\cdot\boldsymbol{\sigma}}$.

4. Physical Properties of Icosahedral Materials

The importance of polytope {3,3,5} lies in its utility in calculations of physical properties of amorphous materials. True amorphous materials have no crystal symmetries. Therefore we cannot apply Bloch's theorem or other calculational tools which exploit the regularity of a crystal lattice. On the other hand, Section 2.3 demonstrates that glass may have a high degree of short-range icosahedral order. The polytope provides a natural way of applying boundary conditions to these icosahedral domains, which offers great computational tractability.

In this section, I describe some of these calculations including a Landau free energy in Section 4.1 which incorporates both the local preference for icosahedral order and geometrical frustration. This Landau theory determines the structure function of metallic glass. One finds a set of defect-broadened peaks at wavenumbers determined by polytope symmetries. These peaks fit experimental data quite well. I use the polytope's symmetries in Section 4.2 to simplify the discussion of the defects which arise when the polytope is flattened. I also extend Bloch's theorem to the sphere S^3 and use the polytope's symmetries to derive electronic (Section 4.3) and vibrational (Section 4.4) densities of states.

4.1 Landau Theory

The Ginzburg–Landau model of frustrated icosahedral order (Nelson and Widom, 1984) provides a phenomenological description of supercooled liquids, glasses, and solid phases which may be crystalline or quasicrystalline (Widom, 1987). I define the Landau free energy so as to encourage polytope-like ordering at short-length scales. Icosahedral frustration, which appears when propagating the polytope ordering through space, is included in the form of a covariant derivative (Sethna, 1983) in the gradient terms of the Landau free energy. Sachdev and Nelson (1984) calculated the structure function of supercooled liquids using this model and found good agreement with the structure function of metallic glass.

To construct a Landau free energy for a liquid about to undergo a change of phase into a solid, we must first identify an order parameter which grows large after the phase transition. For example, if the solid has a set of reciprocal lattice vectors $\{\mathbf{G}\}$, then the Fourier transform of the mass distribution at those wavenumbers provides a convenient set of order parameters. The order parameter need not be uniform in space. One defines a local coarse-grained order parameter

$$\rho_G(\mathbf{r}) = \frac{1}{V}\int d^3\mathbf{r}\; e^{i\mathbf{G}\cdot\mathbf{r}}\,\rho(\mathbf{r}) \tag{4.1}$$

where the volume V must be larger than a lattice constant to effectively coarse-grain, but is much smaller than the macroscopic system size.

To construct a translationally and rotationally invariant free energy, one must consider the transformation properties of the order parameter. Combining a spatial translation $\mathbf{r} \to \mathbf{r} + \mathbf{u}$ and a rotation $\mathbf{r} \to \mathbf{r} + \boldsymbol{\theta} \times \mathbf{r}$, the order parameter

$$\rho_{\mathbf{G}}(\mathbf{r}) \to e^{i\mathbf{G}\cdot\mathbf{u} + i\mathbf{G}\cdot\boldsymbol{\theta}\times\mathbf{r}} \rho_{\mathbf{G}}(\mathbf{r}). \tag{4.2}$$

The Landau free energy associated with ρ is (Nelson and Toner, 1981)

$$\mathcal{F} = \left(\frac{K_0}{2}\right) \sum_{\mathbf{G}} |(\boldsymbol{\nabla} - i\mathbf{G} \times \boldsymbol{\theta})\rho_{\mathbf{G}}|^2 + \frac{1}{2} r \sum_{\mathbf{G}} |\rho_{\mathbf{G}}|^2 + \mathcal{O}(\rho^3), \tag{4.3}$$

where the gradient term is defined to maintain rotational invariance.

In order to define an order parameter of this type for polytope-like ordering, we must find an analogue of reciprocal lattice vectors for polytope {3, 3, 5}. Projection onto hyperspherical harmonics (Bander and Itzykson, 1966) plays the role on S^3 of Fourier transformation in $\mathbf{R}^3$. Thus for a mass density $\rho(\hat{\mathbf{u}})$ on S^3, we define

$$Q_{n,m_1m_2} = \int d\Omega_{\hat{\mathbf{u}}}\, Y_{n,m_1m_2}(\hat{\mathbf{u}})\rho(\hat{\mathbf{u}}). \tag{4.4}$$

The index n may be identified with the wavevector $\mathbf{G}$ through the relationship

$$|\mathbf{G}|^2 = \kappa^2\, n(n + 2), \tag{4.5}$$

where κ is the curvature of S^3. Inspecting Eq. (3.12b), we find $\kappa = \pi/5d$, where d is the average interatomic distance in the icosahedral domains. The angular-averaged structure function is thus (Nelson and Widom, 1984)

$$S_n = \frac{1}{(n + 1)^2} \sum_{m_1m_2} < |Q_{n,m_1m_2}|^2 >, \tag{4.6}$$

where $< >$ denotes thermal average. Symmetries of the polytope force S_n to vanish except for a special set of values of $n \in B$, where

$$B = \{0, 12, 20, 24, 30, 32, \ldots]. \tag{4.7}$$

We identify the set B as the set of Bragg wavenumbers of the polytope.

Now that the reciprocal lattice vectors of the polytope have been identified, we define local icosahedral order parameters

$$Q_{n,m_1m_2}(\mathbf{r}) = \int d\Omega_{\hat{\mathbf{u}}}\, Y_{n,m_1m_2}(\hat{\mathbf{u}})\rho(\hat{\mathbf{u}}), \tag{4.8}$$

where the mass distribution $\rho(\hat{\mathbf{u}})$ is the projection from a small region of $\mathbf{R}^3$ onto a tangent sphere S^3. The index n denotes the magnitude of the wavevector while the indices m_1 and m_2 denote angular information. Often I denote the matrix Q_{n,m_1m_2} simply as $\mathbf{Q}_n$.

For the polytope, the analogue of spatial translations in the direction $\hat{\mathbf{n}} \in \mathbf{R}^3$ are four-dimensional rotations which leave the directions perpendicular to $\hat{\mathbf{n}}$, and the extra fourth dimension w invariant. The analogue of spatial rotations around the axis $\hat{\mathbf{n}}$ are four dimensional rotations which leave $\hat{\mathbf{n}}$ and w themselves invariant. An arbitrary combination of rotation and translation $(\hat{\mathbf{l}},\hat{\mathbf{r}}) \in SO(4)$ may be expressed as a combination of the six generators of four-dimensional rotations $\mathcal{L}_\alpha$

$$\mathbf{Q}_n \rightarrow e^{i\sum_\alpha \theta_\alpha \mathcal{L}_\alpha^{(n)}} \mathbf{Q}_n. \tag{4.9}$$

In analogy with Eq. (4.3) the rotationally and translationally invariant gradient term in the Landau free energy is

$$\mathcal{F}_{\text{grad}} = \frac{1}{2} \sum_{n \in B} \mathrm{K}_n \; |\mathbf{DQ}_n|^2, \tag{4.10}$$

where

$$(\mathbf{DQ}_n)_{m_1m_2} = \nabla Q_{n,m_1m_2} - i\kappa \sum_{j=1}^{3} \hat{e}_j \sum_{\substack{m_1' \\ m_2'}} \mathcal{L}_{0j}^{(n)})_{m_1m_2;m_1'm_2'} Q_{n,m_1'm_2'} \tag{4.11}$$

is chosen so that configurations separated by a distance $\mathbf{d}$ are related by

$$\mathbf{Q}_n(\mathbf{r} + \mathbf{d}) = e^{i\kappa \sum_\mu \mathcal{L}_{0\mu}^{(n)} d_\mu} \mathbf{Q}_n(\mathbf{r}). \tag{4.12}$$

The full free energy consists of this gradient term plus "potential" terms which begin at quadratic order in $\mathbf{Q}_n$ and include third-order invariants, so that the transition to the true ground state is likely to be first order.

$$\mathcal{F} = \frac{1}{2} \sum_n \left\{ K_n \; |\mathbf{DQ}_n|^2 + r_n \; |Q_n|^2 \right\} + \mathcal{O}(\mathbf{Q}^3). \tag{4.13}$$

Peaks in the structure function occur for wavenumbers corresponding to negative coefficients of the quadratic term. Assuming that the material is forming polytope-like ordering at the phase transition, these are the values of n in the set of Bragg wavevectors B. Thus, although the peak height and width are determined by parameters of the Landau theory, the peak positions are determined primarily by symmetries of the polytope. Indeed, the peaks in the structure functions of metallic glasses show approximately universal positions close to those predicted by the polytope. The ground state of the free energy equation (4.13) is unknown at present, but quite likely corresponds to a Frank–Kasper type crystal. Jarić (1985) has considered a general fourth-order Landau theory combining translational and orientational ($l = 6$) order parameters and shown that quasi-

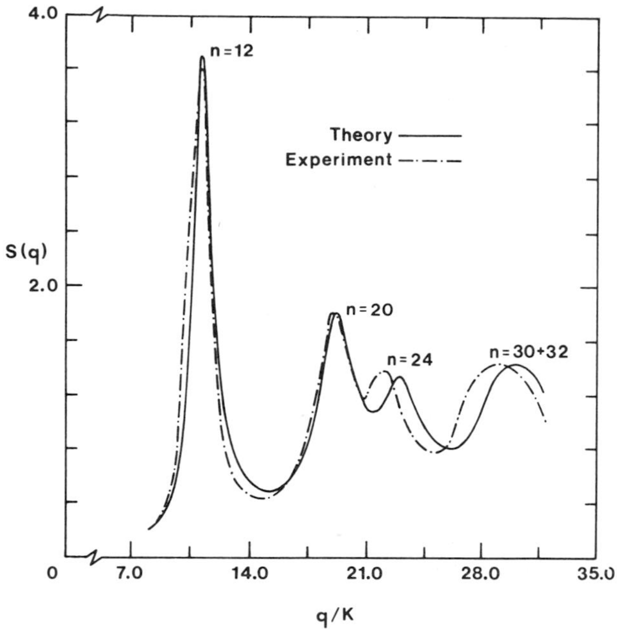

Figure 18. Structure function of metallic glass predicted by Landau theory compared with experimental data for amorphous cobalt (Sachdev and Nelson, 1984).

crystalline and icosahedral liquid-crystalline phases occupy a large portion of the phase diagram.

4.2 *Defects*

We defined the gradient term of the free energy to be minimized when the local order parameter $\mathbf{Q}_{ll}$ is obtained by rolling the polytope along a path. We have not yet considered whether this minimum can be obtained in practice. In fact, it cannot! The free energy $\mathcal{F}$ given by Eq. (4.13) is frustrated (Nelson and Widom, 1984, and Sethna, 1983). Although perfect polytope order may propagate along a line via Eq. (4.12), rolling the polytope around a closed loop reveals the inevitable presence of defects. Mathematically the frustration arises because of the noncommutativity of rotations. Physically the frustration arises because of the impossibility of filling space with icosahedra.

Consider rolling the polytope around the plaquette shown in Fig. 19. On each leg of the circuit, the order parameter transforms as in Eq. (4.12).

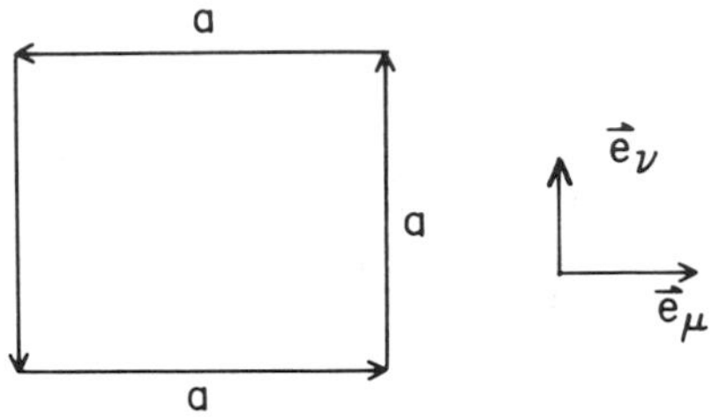

Figure 19. Plaquette (μ, υ) with area a^2.

Thus around the entire plaquette, there is a net transformation of the order parameter

$$\mathbf{Q} \rightarrow e^{i\kappa a \mathcal{L}_{0\mu}} e^{i\kappa a \mathcal{L}_{0\nu}} e^{-i\kappa a \mathcal{L}_{0\mu}} e^{-i\kappa a \mathcal{L}_{0\nu}} \mathbf{Q} = e^{i\kappa^2 a^2 \mathcal{L}_{\mu\nu}} \mathbf{Q}, \quad (4.14)$$

where we used the commutation relations of the $SO(4)$ angular momentum operators (Biedenharn, 1961)

$$[\mathcal{L}_{0\mu}, \mathcal{L}_{0\nu}] = i\mathcal{L}_{\mu\nu}. \quad (4.15)$$

Of course, $\mathcal{L}_{\mu\upsilon}$ induces a rotation of the $\mu\nu$ plane. Such a rotation can be accommodated by threading a finite density κ^2 of $-72°$ wedge disclination lines oriented normal to the plaquette.

Mermin (1979) and Michel (1980) described how to calculate the algebra of defect lines—the manner in which defects transform when lines cross one another. The first step is to find the symmetry group of the order parameter. This is simply $SO(4)$ modulo the symmetry group G of the polytope. The algebra of defect lines is given by the universal covering group

$$\pi_1(SO(4)/G) = \mathbf{G}. \quad (4.16)$$

The covering group of $SO(4)$ is $SU(2) \times SU(2)$, so

$$\mathbf{G} = \mathbf{Y} \times \mathbf{Y}. \quad (4.17)$$

Thus any defect in an icosahedral medium can be labelled by a pair $(\hat{\mathbf{l}},\hat{\mathbf{r}})$ of $SU(2)$ matrix charges. Curiously, disclinations labelled by $(\hat{\mathbf{l}},\hat{\mathbf{l}})$ and dislocations labelled by $(\hat{\mathbf{l}},\hat{\mathbf{l}}^{-1})$ belong to the same class.

The icosahedral frustration forces disclinations into an otherwise well-ordered medium. These disclination lines tend to become entangled because the defect algebra is noncommutative (Mermin, 1979). Thus, as the degree of icosahedral order in a supercooled liquid increases, the dynamics slow down. The glass transition may be a purely dynamical phenomenon in which the medium is unable to reach the true ground state of the free energy because of defect entanglement. Straley (1986) has demonstrated the phenomenon convincingly by showing that 120 atoms in S^3 readily freeze into a perfect polytope even though 108 atoms in R^3 are unable to reach a crystalline configuration in short simulations. Caflish *et al.* (1986) report Monte Carlo simulations explicitly of the free energy (Eq. 4.13) also with glassy dynamics.

It is interesting to consider explicitly what defects look like in an icosahedral medium. Because icosahedral frustration forces a minimum density of $-72°$ disclination lines, these must be the most common type of defect in well-ordered icosahedral structures. Indeed, the Frank–Kasper phases contain only this type of defect. Physically, this defect is favored because there are small gaps between the atoms in the pentagonal ring of a pentagonal bipyramid, allowing an occasional sixth atom to squeeze in, creating a six-fold ring in the first coordination shell of an atom.

One can show (Frank and Kasper, 1958) that the number of six-fold rings in an otherwise perfect icosahedral shell is never equal to 1. The same is true for four-fold rings. This means a disclination line never terminates at an atom. Rather, disclination lines thread through the medium, ending only in intersections with other disclinations. Figure 20 shows co-

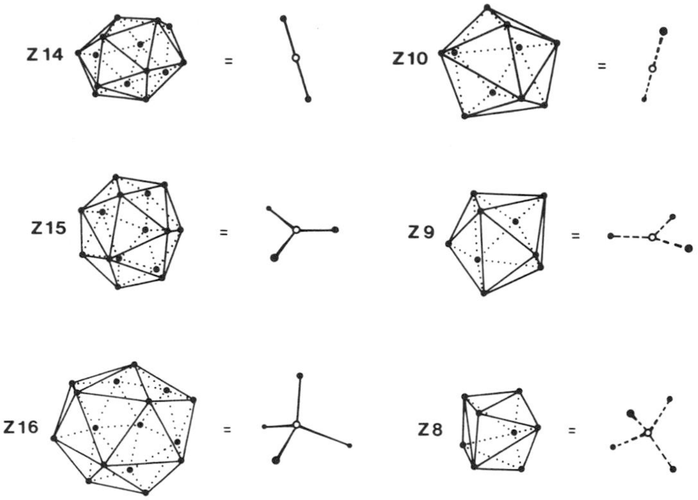

Figure 20. Coordination shells associated with $\pm 72°$ disclination lines. Dashed lines indicate four-fold coordinated bonds. Solid lines indicate six-fold coordinated bonds. The numbers $Z\ N$ indicate the total number of atoms in each shell. From Nelson (1983b).

ordination shells of atoms with two, three, or four six-fold disclinations or four-fold disclinations. Not shown are the intersection of two six-fold and one four-fold disclination on a Z13 atom and the intersection of two four-fold and one six-fold disclination on a Z11 atom. These Z13 and Z11 atoms may be described in terms of an interstitial and vacancy, respectively.

In order to study the influence of defects on physical properties of icosahedral materials, it is important to be able to place defects into the polytope in a controlled fashion. Straley (1985) and Nicois, Mosseri, and Sadoc (1986) have shown how to insert isolated disclination lines. Represent the sphere S^3 with the coordinates (θ,λ,ϕ), so that

$$\begin{aligned} w &= \cos\theta\cos\lambda \\ x &= \sin\theta\cos\lambda \\ y &= \cos\phi\sin\lambda \\ z &= \sin\phi\sin\lambda, \end{aligned} \tag{4.18}$$

where the polytope {3, 3, 5} results from the choices

$$\begin{array}{lll} \lambda = 0 & \theta = (2\mu + 1)\pi/10 & \phi = \text{arbitrary} \\ \lambda = \tan^{-1} 1/2 & \theta = 2\mu\ \pi/10 & \phi = 2\nu\ \pi/10\ (\mu + \nu \text{ even}) \\ \lambda = \tan^{-1} 2 & \theta = 2\mu\ \pi/10 & \phi = 2\nu\ \pi/10\ (\mu + \nu \text{ odd}) \\ \lambda = \pi/2 & \theta = \text{arbitrary} & \phi = (2\nu + 1)\ \pi 10, \end{array} \tag{4.19}$$

and μ and ν take on integer values between 0 and 9.

In order to insert a six-fold disclination line around the equator, allow μ to run from 0 to 11 and step θ in multiples of $\pi/12$. The result is a polytope with 142 atoms, all of which have perfect icosahedral coordination except for a string of Z14 atoms along the equator. Similar alterations of Eqs. (4.19) lead to a $+72°$ disclination line around the equator with a total of 98 atoms, + or $-72°$ disclination lines along great circles passing through the north pole, linked six-fold lines (168 atoms), or linked four- and six-fold lines (116 atoms).

Small numbers of $-72°$ disclination lines reduce the mean curvature of the polytope slightly. To approximate flat space, we need to insert large numbers of disclinations and create polytopes with arbitrarily large numbers of atoms. Mosseri and Sadoc (1984) have proposed an iterative scheme for flattening the polytope which produces a hierarchical structure of defects in a sequence of polytopes of decreasing mean curvature. The idea is to take the dual of polytope {3,3,5}, namely polytope {5,3,3}, and to create a new polytope by filling in the 120 dodecahedral cells of polytope {3,3,5} with 13-atom icosahedral clusters and placing an atom at each of the 600 vertices of polytope {5,3,3}. The new structure contains $120 \times 13 = 1560$ icosahedrally coordinated atoms and 600 Z16 atoms. The Z16 atoms are those at vertices of polytope {5,3,3}, so the $-72°$ disclination network runs along near-neighbor bonds of polytope {5,3,3} (see Fig. 21).

At each step of the flattening, near-neighbor bonds of the dual of the old polytope become a disclination network for the new polytope. Each Z12 vertex gets replaced by thirteen Z12 vertices and five Z16 vertices. Each Z14 vertex gets replaced by twelve Z12 vertices, three Z14 vertices, and six Z16 vertices. Each Z16 vertex gets replaced by twelve Z12 vertices, four Z14 vertices, and eight Z16 vertices. Each iteration can thus be described by the transfer matrix (Sadoc and Mosseri, 1985)

$$\begin{bmatrix} 13 & 12 & 12 \\ 0 & 3 & 4 \\ 5 & 6 & 8 \end{bmatrix} \tag{4.20}$$

which relates the numbers of 12, 14, and 16 coordinated vertices at each step to the number at the previous step. The largest eigenvalue of this matrix, $\lambda = 20$, gives the asymptotic rate of increase of the total number of atoms. The corresponding eigenvector yields the relative numbers of 12, 14, and 16 coordinated vertices in the flat space limit, and therefore the mean coordination number. The value of the mean coordination number $40/3 = 13.333$ for the hierarchical structure is not far from the ideal value 13.397.

4.3 Electronic Properties

The symmetry group of polytope {3,3,5} allows exact solution of many problems of physical interest. In this section, I shall discuss the generalization of Bloch's theorem to curved space in connection with solution of simple electronic tight-binding models. The simplest tight-binding model describes electrons hopping between *s* orbitals of neighboring atoms. The

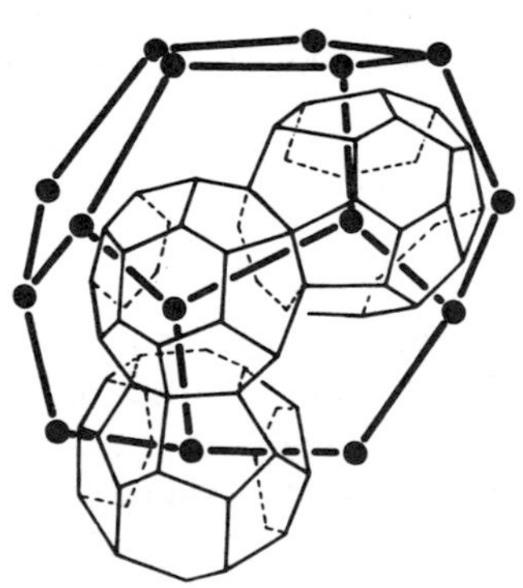

Figure 21. Fragment of disclination network. Dark lines represent edges of polytope {5,3,3}. Lighter lines illustrate near-neighbor bonds between atoms inserted into polytope. From Mosseri and Sadoc (1984).

hopping matrix element has no angular dependence for s orbitals, so the Hamiltonian takes the simple form (Nelson and Widom, 1984)

$$H = - \sum_{\hat{\mathbf{v}} \in \mathcal{V}1} V_1 \ |\hat{\mathbf{u}}><\hat{\mathbf{v}}\hat{\mathbf{u}}| \ , \tag{4.21}$$

where the second sum over $\hat{\mathbf{v}} \in \mathcal{V}_1$ sums over all near neighbors of $\hat{\mathbf{u}}$. The hopping matrix element is normalized to $V_1 = -1$.

Bloch's theorem states that eigenfunctions of H form the basis functions for irreducible representations of the symmetry group of H. But the symmetry group of H is precisely the symmetry group of polytope {3,3,5}, namely $G = \mathbf{Y} \times \mathbf{Y}/Z_2$. Because the symmetry group G is related to the direct square of the icosahedral group $\mathbf{Y}$, we can conveniently express the irreducible representations of G in terms of the irreducible representations of $\mathbf{Y}$ (Widom, 1984 and 1985). In particular, if α and β label two irreducible representations of $\mathbf{Y}$, then the ordered pair (α,β) labels an irreducible representation of G.

Not all irreducible representations of G need to be eigenfunctions of H. We are only interested in defining scalar-valued single electron wavefunctions on the vertices of polytope {3,3,5}. All such wave functions must be contained within the 120-dimensional representation $\mathcal{R}$ generated by the tight-binding basis functions $|\hat{\mathbf{u}}>$ centered at each of the 120 vertices $\hat{\mathbf{u}}$ of the polytope. It is easy to determine which irreducible representations (α,β) are contained in $\mathcal{R}$. The number of times (α,β) is contained in $\mathcal{R}$ is given by (Landau and Lifshitz, 1965)

$$A(\alpha,\beta) = \frac{1}{\mathcal{O}(G)} \sum_{(\hat{\mathbf{l}},\hat{\mathbf{r}}) \in G} \chi_{\mathcal{R}}^{G}(\hat{\mathbf{l}},\hat{\mathbf{r}})\, \chi_{\alpha\beta}^{G}{}^{*}(\hat{\mathbf{l}},\hat{\mathbf{r}}), \tag{4.22}$$

where χ^G are the characters of representations of G.

We already know the characters $\chi_{\alpha\beta}$ of irreducible representations of G because these are simply the products of characters of the icosahedral group $\mathbf{Y}$

$$\chi_{\alpha\beta}^{G}(\hat{\mathbf{l}},\hat{\mathbf{r}}) = \chi_{\alpha}^{\mathbf{Y}}(\hat{\mathbf{l}})\chi_{\beta}^{\mathbf{Y}}(\hat{\mathbf{r}}). \tag{4.23}$$

The problem then is to determine the characters of the reducible representation $\mathcal{R}$. Note that for any four-dimensional rotation which maps polytope vertices onto polytope vertices (i.e., any element of G), the representation matrix in the basis provided by $\mathcal{R}$ is a sparse matrix of zeros and ones giving the permutations of vertices. The character of a representation is the trace of the representation matrix. Therefore, the character of the rotation in the basis $\mathcal{R}$ is the number of ones which fall on the diagonal of the matrix. But a "1" on the diagonal indicates a vertex which remains unmoved under the rotation.

How many vertices are stationary under $\hat{\mathbf{u}} \rightarrow \hat{\mathbf{l}}\,\hat{\mathbf{u}}\,\hat{\mathbf{r}}^{-1}$? Note that if $\hat{\mathbf{u}}$ is stationary, then

$$\hat{\mathbf{l}} = \hat{\mathbf{u}}\,\hat{\mathbf{r}}\,\hat{\mathbf{u}}^{-1}, \tag{4.24}$$

so $\hat{\mathbf{l}}$ and $\hat{\mathbf{r}}$ must belong to the same class. The number of vertices $\hat{\mathbf{u}}$ satisfying Eq. (4.24) for a given $\hat{\mathbf{l}}$ and $\hat{\mathbf{r}}$ is

$$\chi^{G}_{\mathcal{R}}(\hat{\mathbf{l}},\hat{\mathbf{r}}) = \delta_{\mathcal{A}\mathcal{B}}\,\theta(Y)/\theta(\mathcal{A}), \tag{4.25}$$

where $\mathcal{A}$ and $\mathcal{B}$ are the conjugacy classes containing $\hat{\mathbf{l}}$ and $\hat{\mathbf{r}}$. Combining Eqs. (4.21), (4.23), and (4.24), we obtain

$$A^{G}_{\mathcal{R}}(\alpha,\beta) = (1/\mathcal{O}(\mathbf{Y})) \sum_{\hat{\mathbf{l}}\in\mathbf{Y}} \chi_{\alpha}{}^{\mathbf{Y}(\hat{\mathbf{l}})}\,\chi_{\beta}{}^{\mathbf{Y}(\hat{\mathbf{l}})}, \tag{4.26}$$

which is just the dot product of rows α and β in the character table of **Y**, and therefore

$$A^{G}_{\mathcal{R}}(\alpha,\beta) = \delta_{\alpha\beta}. \tag{4.27}$$

So we see that the eigenfunctions of H are basis functions for diagonal irreducible representations of G and can therefore be labelled with a single icosahedral symmetry group irreducible representation.

The energies can be calculated quite easily also. Because G is the direct square of the icosahedral group **Y**, irreducible representation basis functions of G transform as the product of irreducible representation basis functions of **Y**,

$$(\hat{\mathbf{l}},\hat{\mathbf{r}}):\ \psi_{\alpha\beta,m_1m_2}(\hat{\mathbf{u}}) \rightarrow D^{\alpha}_{m_1m_1'}(\hat{\mathbf{l}})\,\chi_{\alpha\beta,m_1'm_2'}(\hat{\mathbf{u}})\,D^{\beta}_{m_2'm_2'}(\hat{\mathbf{r}}^{-1}). \tag{4.28}$$

Denoting the eigenfunctions of H which transform as the (α,α) irreducible representation of G by $|\,\alpha\ m_1m_2 >$, we find

$$<\alpha\ m_1m_2\,|H|\,\alpha\ n_1n_2> = \delta_{m_2n_2} \sum_{\hat{\mathbf{v}}\in\mathcal{V}_1} D^{\alpha}{}_{n_1m_1}(\hat{\mathbf{v}})\,. \tag{4.29}$$

Note that $\mathcal{V}_1$, the set of nearest neighbors of the north pole is a conjugacy class of G. This means that the sum in Eq. (4.29) commutes with $D^{\alpha}(\hat{\mathbf{u}})$ for all $\hat{\mathbf{u}} \in \mathbf{Y}$. Applying Schur's lemma yields the resulting spectrum of H (Nelson and Widom, 1984, see also Widom, 1985, Mosseri *et al.*, 1985, and Warner, 1982),

$$E_{\alpha} = -\mathcal{O}(\mathcal{V}_1)\,\chi^{\mathbf{Y}}_{\alpha}(\mathcal{V}_1)/d_{\alpha}, \tag{4.30}$$

with degeneracy $d_{\alpha}{}^2$, where d_{α} is the dimension of the α irreducible representation of **Y**. We can generalize this result to include further neighbor

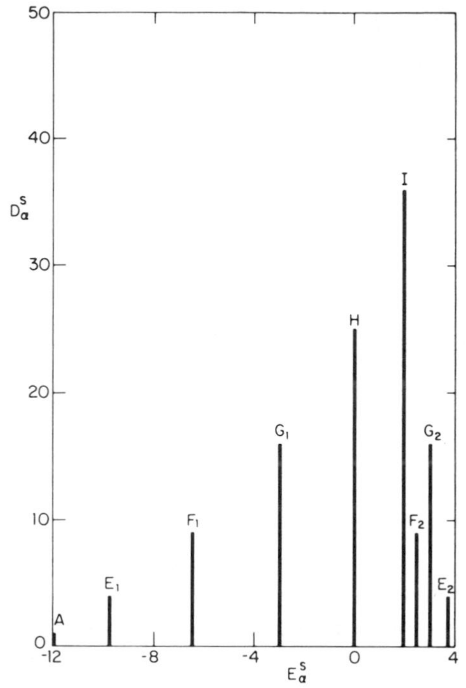

Figure 22. Density of states of the Hamiltonian (4.21). From Widom (1985).

interactions by replacing the class of nearest neighbors $\mathcal{V}_1$ with other conjugacy class of **Y**.

Figure 22 shows the resulting density of states. Because of the finite number of atoms, we obtain only a histogram instead of a continuous curve. It is an interesting exercise to associate wavenumbers with each energy level. Figure 23 plots the repeated zone spectrum obtained in this fashion (Widom, 1985). Note how the lowest-lying energy levels reach their minimum whenever the associated wavenumber belongs to the set of Bragg wavevectors B. Splittings of the dispersion relation occur when the wavenumbers are half of a Bragg wavevector. This is evidence for pseudo-Brillouin zones which have been observed in electronic and vibrational properties of dense random packing models of glass (Rehr and Alben 1977, Von Heimendahl 1979, Hafner 1983, and Grest *et al.*, 1984) and has been postulated in frustrated systems (Kleman, 1982). Usually, in realistic models, the dispersion curve fails to reach its absolute minimum value, presumably because the Bragg peaks are not perfectly sharp.

Although the energy spectrum of the polytope can be easily calculated

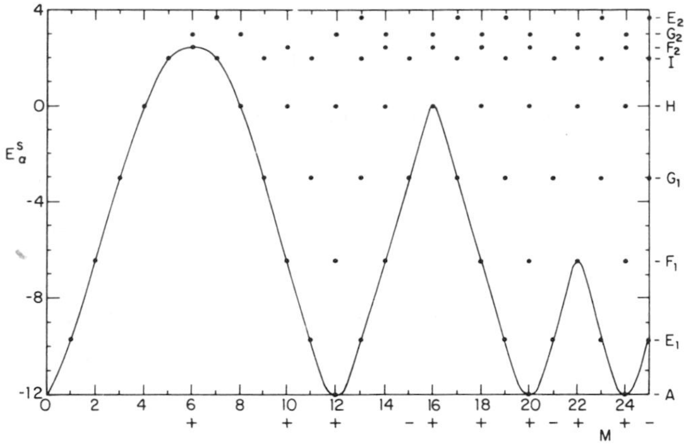

Figure 23. Repeated zone spectrum of H. Energy levels are labeled by irreducible representations of Y'. Even (odd) values of M for which $2M \in B$ are labeled with a + (−) sign.

and provides an approximation to the true energy spectrum of metallic glass, we really need to calculate the properties of a defected icosahedral crystal. Selinger and Nelson (1985) calculate the densities of states of Straley's defected polytopes and find qualitative agreement with the true dense random packing density of states which, like polytope {3,3,5}, has a peak near the upper band edge (Gaspard, 1976). In addition, realistic models of transition metal glasses must consider the electronic d band in addition to the s band. An important feature of icosahedral symmetry is that the atomic d shell is not split by the crystal field. This enhances the d band density of states near the band center (Gaspard, 1976, and Widom, 1985). Atomic f levels are split into a doublet with degeneracies 3 and 4 which may have relevance for metallic glasses or quasicrystals containing rare earth compounds. Electronic properties of polytope 240 have also been studied (Mosseri *et al.*, 1985), and optical selection rules have been determined (Brodsky and DiVincenzo, 1983) for amorphous tetracoordinated semiconductors.

One can calculate the moments of the density of states of the ideal glass, polytope {3,3,5.104}, by performing calculations for the polytope $\{3,3,q\}$ and interpolating to $q = 5.104$. The idea is that the rth moment of the density of states is simply equal to the number of closed paths of length r (Cyrot-Lackmann, 1967), because

$$\mu_r = \int E^r n(E)\, dE = Tr\{H^r\}. \tag{4.31}$$

One finds

$$\mu_2 = F, \tag{4.32}$$

where $F = 12/(6-q)$ is the coordination number of every atom,

$$\mu_3 = Fq, \tag{4.33}$$

and

$$\mu_4 = F^2 + Fq^2 + 3Fq + F. \tag{4.34}$$

Thus, setting $q = 5.104$, we obtain $\mu_2 = 13.397$, $\mu_3 = 68.384$, and $\mu_4 = 747.090$.

4.4 Melting

The techniques developed in Section 4.3 to analyze the electronic properties of polytope {3,3,5} can be adapted to study the vibrational properties of the polytope (Widom, 1986). Because the vibrational wavefunctions are vector fields instead of scalar fields, certain off-diagonal as well as diagonal irreducible representations must be considered. Just as the high polytope symmetry prevented splitting of the *d* bands at zero wavevector, so the polytope is acoustically isotropic in continuum elastic theory. As a result of this isotropy, the polytope is more stable against melting than an FCC crystal with the same interatomic potential.

Assuming a simple repulsive inverse twelfth potential between atoms, one can calculate the mean square atomic displacement in the harmonic approximation. One finds (Deng and Widom, 1987)

$$< u^2 > = 0.0027 k_B T \tag{4.35}$$

for an FCC crystal, and

$$< u^2 > = 0.0023 k_B T \tag{4.36}$$

on the polytope, where the potential is $\phi(r) = 4/r^{12}$ and the equilibrium atomic spacing is taken to be 1, so that the curvature $\kappa = \Omega^{-1}$. Thus, at a given temperature, the atoms in an FCC crystal oscillate more violently than the atoms in polytope {3,3,5}. The discrepancy arises because FCC crystals have soft transverse modes in the [110] direction, whereas the polytope is elastically isotropic. It is interesting to note that the harmonic approximation is in excellent agreement with data from Monte Carlo simulations by Straley (1986) all the way up to the melting temperature.

Using the mean square atomic displacement at melting obtained from Eq. (4.36), combined with the melting temperature $T = 16$ obtained by Straley (1986), we can test the Lindeman theory of melting. The ratio of RMS atomic displacement in the polytope before melting to the lattice constant is 0.19, as compared to 0.15 in an FCC crystal. Thus, not only does the icosahedral structure oscillate less at a given temperature than an FCC crystal, but also the polytope can tolerate more violent oscillations without melting than an FCC crystal (Deng and Widom, 1987). As a result, the melting temperature of the polytope is double the melting temperature of an FCC crystal. Stability of the polytope is discussed also in the density functional theory of Singh, Stoessel, and Wolynes (1985).

5. Long Range Icosahedral Order

The requirements of filling flat space frustrate a rapidly quenched metal's preference for icosahedral order. Frustration produces defects, the defects tend to entangle, and glass formation results. In this final section, I shall discuss the structures which result when the defect lines do not entangle and the system reaches, or at least approaches, equilibrium.

The stable Frank–Kasper phases exemplify the structures which emerge. In these structures, the defect lines arrange themselves in a regular lattice. The atoms form unit cells of rather high complexity, and the unit cells pack in a crystalline fashion. The Frank–Kasper phases display long-range icosahedral order in the sense that each unit cell contains icosahedral structures, and these unit cells themselves pack with long range translational and orientational order. Crystallography does not forbid this type of icosahedral ordering, since the crystal lattice itself lacks icosahedral symmetry.

If we describe icosahedral quasicrystals as crystals with infinitely large unit cells, then the key question is how icosahedral order persists across a single unit cell in the Frank–Kasper crystals. I propose that minimizing the free energy equation (4.13) leads to structures with long-range orientational order, including Frank–Kasper phases and quasicrystals.

Rolling the polytope along paths through three-dimensional flat space maps the local order of the polytope from S^3 to $\mathbf{R}^3$. Everywhere an atom of the polytope touches three-dimensional space, place an ordinary three-dimensional atom. I shall show that long-range orientational and translational order result from rolling the polytope along a straight line (Widom, 1987). Rolling around a closed loop places defects within the loop. For certain types of loops, these defects maintain orientational order. These special loops bound the rhombic faces of the oblate and prolate rhombahedra of the three-dimensional Penrose pattern (Henley and Elser, 1986).

5.1 Unified Theory of Crystals, Glass, and Quasicrystals

The key to this unified theory of crystals, glass, and quasicrystals lies in the remarkable isomorphism between the structure of the icosahedral polytope {3,3,5} and the icosahedral symmetry group Y. In particular, $\pi/5$, the geodesic distance between nearest-neighbor atoms on the polytope of unit radius, is precisely one-half the smallest nonzero angle of rotation which leaves an icosahedron invariant. I shall show that the fact that this angle is not itself a symmetry of the icosahedron leads to metastable glass formation in general, but low-energy icosahedral structures exploit the fact that twice this angle is a symmetry of the icosahedron to propagate long-range orientational order.

Consider rolling the polytope along paths in $\mathbf{R}^3$ so as to fill up space with atoms. Every time an atom of the polytope touches $\mathbf{R}^3$, project that atom and as many of its icosahedral neighbors as can be deposited without conflicting with atoms already present in $\mathbf{R}^3$. Begin with a single icosahedral cluster. A natural direction to roll is from the center of that cluster to one of its 12 icosahedral neighbors. Then project the 12 neighbors of this atom into $\mathbf{R}^3$. Note, though, that 6 of the atoms fall on top of atoms already present; namely, the original atom and those 5 of its neighbors which are also neighbors of the atom we rolled to. Thus, do not project these 6 atoms. Rather, make whatever small adjustments are necessary to restore mechanical equilibrium in the new cluster of 19 atoms.

The resulting structure is precisely the pair of interpenetrating icosahedra observed in noble gas clusters (Harris *et al.*, 1984). Consider the orientations of these two interpenetrating icosahedra. Because they are related by rolling the polytope a distance d, the angular order parameter rotates an angle $\kappa d = \pi/5$. Since this angle is one-half the angle of rotation leaving an icosahedron invariant, the second icosahedron has its orientation exactly reversed with respect to the first. Of course, if we roll the polytope to the next atom going in the same direction, the third icosahedron restores the orientation of the first. As we continue to roll along a straight line, we create a string of interpenetrating icosahedra with alternating orientations (Fig. 24). Note that we can also join icosahedra on faces rotating $2 \times \pi/3$ each time or on edges rotating $2 \times \pi/2$ each time.

Obviously, this construction maintains long-range orientational and translational order. Is it physically plausible that nature should arrange its icosahedra in this manner? Examination of structures of Frank–Kasper phases answers this question with a definite yes. Many of the Frank–Kasper (1959) phases (in particular the laves and μ phases as well as the unit cells of β-W and other structures) are based on layers of Kagome nets (Fig. 25). Sachdev and Nelson (1985) point out that each atom on the net has icosahedral coordination, and the orientations of neighboring icosahedra are twisted by $\pi/5$. Thus each straight line of atoms in Fig. 25 represents a chain of interpenetrating icosahedra such as is shown in Fig. 24.

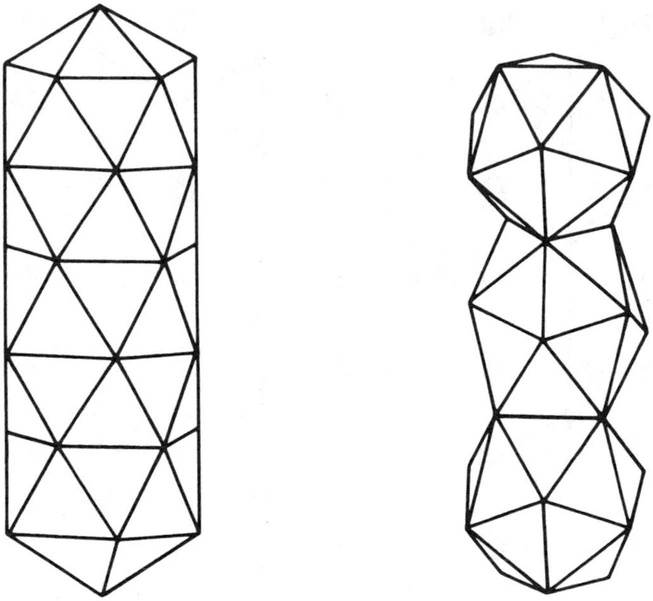

Figure 24. Rolling the polytope upwards along the z axis. (a) constructing the local icosahedron with a vertex aligned along the z axis at each point $(0,0,nd)$, where n is an integer. Note how the orientations of the icosahedra at odd values of n differ by $\pi/5$ from the orientations at even values of n. (b) icosahedra joined at their faces generated by rolling polytope along path through faces of icosahedral coordination shells.

Unless special care is taken, paths zig-zag back and forth through space combining many $\pi/5$ rotations, which are not in the icosahedral symmetry group, in random fashion (note, however, many approximate colineations extending over several atomic diameters in the Bernal structure in Fig. 11). Because rotations do not commute, the result of rolling the polytope along a sequence of paths pointing in different directions is to come arbitrarily close to any rotation and translation. The resulting structures have high energies, because rolling back close to the starting point of a loop often creates unresolvable conflicts in position and orientation. The resulting structures are, in fact, the glassy states of the free energy equation (4.13).

In order to guarantee orientational coherence when rolling the polytope,

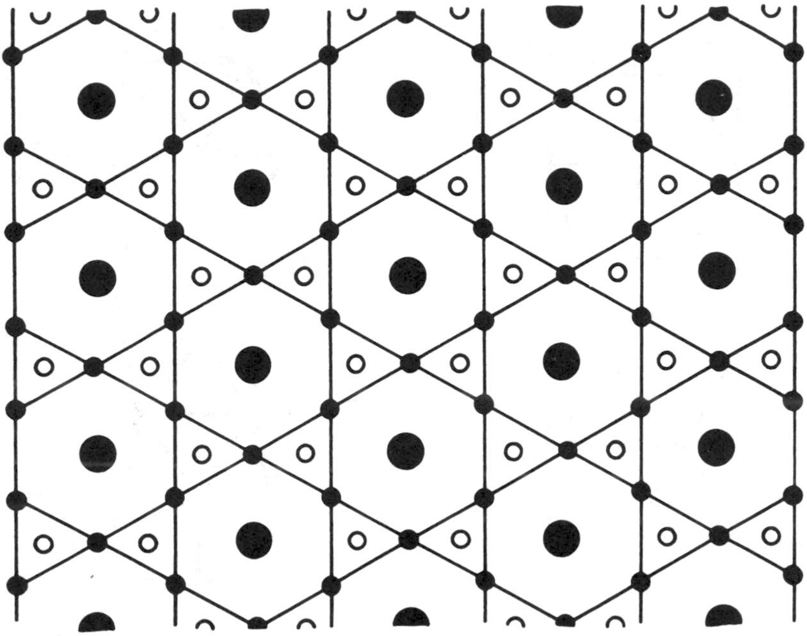

Figure 25. Kagome net decorated with atoms in μ phase structure. Small dots represent icosahedrally coordinated atoms on net. Large dots represent Z 14 atoms at $z = \pm 1/4$. Six-fold disclination lines thread hexagons. Small open circles represent icosahedrally coordinated atoms at $z = \pm 1/2$.

always roll an even number of atomic spacings in a single direction before turning a corner (Widom, 1987). This ensures that at each corner of the path, the orientation is identical to the starting orientation rotated by an amount corresponding to a symmetry of the icosahedron. Combining symmetry operations of an icosahedron at random always yields another symmetry because groups are closed under multiplication. Therefore orientational order is maintained by rolling along paths which continue in straight lines for an even number of atomic spacings.

The smallest such path is a rhombus with edge length $2d$ and angles 63.43° and 116.57°, where the angles correspond to the angles between vertices of an icosahedron. These loops combine to form the rhombic faces of the prolate and oblate rhombahedra (Fig. 26) of the three-dimensional Penrose pattern (Penrose, 1974, Mackay, 1982, and Levine and Steinhardt, 1984). These rhombahedra can pack to form a variety of space-filling structures. There appears to be no mechanism for enforcing the Penrose matching rules (Levine and Steinhardt, 1986), so the rhombuses

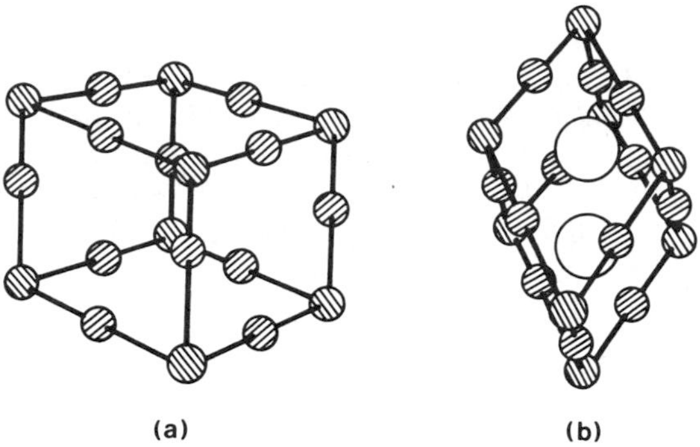

Figure 26. Oblate (a) and prolate (b) rhombahedra decorated by rolling polytope along rhombahedron edges. Six-fold disclination lines pass through each rhombic face.

need not pack into any particular local isomorphism class, or even form a quasi-periodic structure at all. In particular, many Frank–Kasper phases may be broken up into these rhombahedra (Henley and Elser, 1986).

In order to observe icosahedral symmetry in the diffraction patterns, all icosahedral bond angles must be equally likely. This can be accomplished by forming a perfect Penrose pattern from these rhombahedra. There are many other ways to pack the rhombahedra, however. The sharpness of the diffraction patterns is not necessarily lost when we construct these other objects. For instance, mild rearrangements of the building blocks alter the peak intensities and introduce a diffuse background but leave the delta-function peaks intact (Elser, 1985). Even a random packing of icosahedra joined at their vertices produces sharp (though not delta functions) peaks (Stephens and Goldman, 1986). Henley (1987) and Elser (1985) predict delta-function peaks in three dimensions but not in lower dimensionality.

Monte Carlo simulations of quasicrystal formation in two dimensions (Widom *et al.*, 1987) suggest that the correct model of quasicrystals is based on random packings of fundamental building blocks. When the system is rapidly quenched, structures such as Stephens and Goldman produce, with their obviously metastable defected regions, occur. More gentle cooling produces equilibrium quasicrystals which appear to lower their free energy below competing crystal structures by exploiting the entropy available by rearranging these building blocks at little or no energy cost.

Thus the free energy equation (4.13) provides a unified model of icosahedral order in crystals, glass, and quasicrystals. In all cases, there is a distinct local preference for local icosahedral order. Icosahedral frustration forces a minimum density of $-72°$ disclination line defects. Real atomic configurations in flat space $\mathbf{R}^3$ may be produced by rolling the perfect unfrustrated icosahedral polytope {3,3,5} along paths through space. Well-ordered crystalline networks of paths lead to low-energy Frank–Kasper phases. Random paths lead to metastable tangled defect line networks characteristic of glass. Finally, a class of paths creates structures with the local order and low energy of the crystal phases but possessing the long-range icosahedral symmetry and translational order of quasicrystals.

References

Allpress, J. G., and Sanders, J. V. (1970). The structure and stability of small clusters of atoms. *Aust. J. Phys.* 23, 23–36.

Andersen, H. C. (1987). Private communication.

Bander, M., and Itzykson, C. (1966). Group theory and the hydrogen atom (I). *Rev. Mod. Phys.* 38, 330–345.

Bennett, C. H. (1972). Serially deposited amorphous aggregates of hard spheres. *J. Appl. Phys.* 43, 2727–2734.

Bergman, G., Waugh, J. L. T., and Pauling, L. (1957). The crystal structure of the metallic phase $Mg_{32}(Al,Zn)_{49}$. *Acta Cryst.* 10, 254–259.

Bernal, J. D. (1964). The Bakerian lecture, 1962. The structure of liquids. *Proc. R. Soc. London Ser.* A280, 299–322.

Biedenharn, L. C. (1961). Wigner coefficients for the R^4 group and some applications. *J. Math. Phys.* 2, 433–441.

Brodsky, M. H., and DiVincenzo, D. P. (1983). Preservation of optical selection rules in an amorphous semiconductor. *Physica* 117B, 118B, 971–973.

Caflish, R. G., Levine, H., and Banavar, J. R. (1986). Glassy dynamics in icosahedral systems. *Phys. Rev. Lett.* 57, 2679–2682.

Cargill, G. S. (1970). Dense random packing of hard spheres as a structural model for noncrystalline metallic solids. *J. Appl. Phys.* 41, 2248–2250.

Cargill, G. S. (1975). Structure of metallic alloy glasses. In *Solid State Physics* 30 (Academic Press, New York).

Caspar, D. L. D., and Klug, A. (1962). Physical principles in the construction of regular viruses. In *Cold Spring Harbor Symposia on Quantitative Biology,* 32, 1–24.

Chaudhari, P., and Turnbull, D. (1978). Structure and properties of metallic glasses. *Science* 199, 11–21.

Cohen, M. H., and Turnbull, D. (1964). Metastability of amorphous structures. *Nature* 203, 964.

Coxeter, H. S. M. (1969). *Introduction to Geometry*. (Wiley, New York).

Coxeter, H. S. M. (1973). *Regular Polytopes* (Dover, New York).

Cyrot-Lackmann, F. (1967). On the electronic structure of liquid transition metals. *Adv. Phys.* 16, 393–400.

Darby, M. I., and Evans, G. R. (1978). Local fields in amorphous elements. *Phys. Stat. Sol.* B85 K63–K66.

Deng, D. P., and Widom, M. (1987). Vibrations and melting of an icosahedral polytope. Preprint.

DuVal, P. (1964). *Homographies, Quaternions and Rotations.* (Oxford, London).

Elser, V. (1985). Comment on "Quasicrystals: a new class of ordered structures." *Phys. Rev. Lett.* 54, 1730.

Elser, V. and Henley, C. L. (1985). Crystal and quasicrystal structures in Al-Mn-Si alloys. *Phys. Rev. Lett.* 55, 2883–2886.

Finney, J. L. (1970). Random packings and the structure of simple liquids. I. The geometry of random close packing. *Proc. R. Soc. London Ser.* A 319, 479–493; Random packings and the structure of simple liquids. II. The molecular geometry of simple liquids. *Proc. R. Soc. London Ser.* A 319, 495–507.

Frank, F. C. (1952). Supercooling of liquids. *Proc. R. Soc. London Ser.* A 215, 43–46.

Frank, F. C., and Kasper, J. S. (1958). Complex alloy structures regarded as sphere packings. I. Definitions and basic principles. *Acta Cryst.* 11, 184–190.

Frank, F. C., and Kasper, J. S. (1959). Complex alloy structures regarded as sphere packings. II. Analysis and classification of representative structures. *Acta Cryst.* 12, 483–499.

Gaspard, J. P. (1976). Density of states and cohesive energy of amorphous transition metals. In *Structure and Excitations of Amorphous Solids, A.I.P. Conference Proceedings* 31, 372–377 (A.I.P., New York).

Grest, G. S., Nagel, S. R., and Rahman, A. (1984). Zone boundaries on glasses. *Phys. Rev.* B29, 59-68-5971.

Guyot, P. and Audier, M. (1985). A quasicrystal structure model for Al-Mn. *Phil. Mag.* B52, L15–L19.

Hafner, J. (1983). Dispersion of collective excitations in a metallic glass. *J. Phys.* C16, 5773–5792.

Haggin, J. (1987). Catalytic promise sparks emerging discipline of cluster chemistry. *Chemical & Engineering News* 65, 9–15.

Harris, I. A., Kidwell, R. S., and Northby, J. A. (1984). Structure of charged argon clusters formed in a free jet expansion. *Phys. Rev. Lett.* 53, 2390–2393.

Henley, C. L. (1987). Quasicrystal order, its origins and its consequences: A survey of current models. *Comments in Condensed Matter Physics.*

Henley, C. L., and Elser, V. (1986). Quasicrystal structure of $(Al, Zn)_{49}Mg_{32}$. *Phil. Mag.* B53, L59–L66.

Hoard, J. L. and Hughes, R. E. (1967). Elemental boron and compounds of high boron content: structure, properties, and polymorphism. In *The Chemistry of Boron and Its Compounds* (Muetterties, E. L., ed.). Wiley, New York.

Hoare, M. (1976). Stability and local order in simple amorphous packings. *Ann. N.Y. Acad. Sci.* 279, 186–207.

Hornreich, R. M., Kugler, M., and Shtrikman, S. (1982). Localized instabilities and the order–disorder transition in cholesteric liquid crystals. *Phys. Rev. Lett.* 48, 1404–1407.

Honeycutt, J. D., and Andersen, H. C. (1987). Molecular dynamics study of melting and freezing of small Lennard–Jones clusters. *Preprint.*

Jarić, M. V. (1985). Long range icosahedral orientational order and quasicrystals. *Phys. Rev. Lett.* 55, 607–610.

Kleman, M. (1981). Crystallography of amorphous bodies. In *Continuum Models of Discrete Systems* 4, 287–296 (North-Holland).

Kleman, M. (1982). The geometrical nature of disorder and its elementary excitations. *J. Physique* 43, 1389–1396.

Kleman, M. (1985). Frustration in polymers. *J. Physique Lett.* 46, L723–L732.

Kleman, M. and Sadoc, J. F. (1979). A tentative description of the crystallography of amorphous solids. *J. Physique Lett.* 40, L569–L574.

Knight, W. D., Clemenger, K., de Heer, W. A., Saunders, W. A., Chou, M. Y., and Cohen, M. L. (1984). Electronic shell structure and abundances of sodium clusters. *Phys. Rev. Lett.* 52, 2141–2143.

Landau, L. D., and Lifshitz, E. M. (1965). *Quantum mechanics*. Pergamon, New York.

Levine, D., and Steinhardt, P. J. (1984). Quasicrystals: a new class of ordered structures. *Phys. Rev. Lett.* 53, 2477–2480.

Levine, D., and Steinhardt, P. J. (1986). Quasicrystals. I. Definition and Structure. *Phys. Rev.* B34, 596.

Ma, Y., Stern, E. A., and Gayle, F. W. (1987). Structure of icosahedral Al-Li-Cu. *Phys. Rev. Lett.* 58, 1956–1959.

Mackay, A. L. (1982). Crystallography and the Penrose pattern. *Physica* 114A, 609–613.

Mermin, N. D. (1979). The topological theory of defects in ordered media. *Rev. Mod. Phys.* 51, 591–648.

Michel, L. (1980). Symmetry defects and broken symmetry. Configurations hidden symmetry. *Rev. Mod. Phys.* 52, 617–651.

Mosseri, R., DiVincenzo, D. P., Sadoc, J. F., and Brodsky, M. H. (1985). Polytope model and the electronic and structural properties of amorphous semiconductors. *Phys. Rev.* B32, 3947–4000.

Mosseri, R., and Sadoc, J. F. (1984). Hierarchical structure of defects in non-crystalline sphere packings. *J. Physique Lett.* 45, L827–L832.

Nelson, D. R. (1983a). Liquids and glasses in spaces of incommensurate curvature. *Phys. Rev. Lett.* 53, 982–985.

Nelson, D. R. (1983b). Order, frustration, and defects in liquids and glasses. *Phys. Rev.* B28, 5515–5535.

Nelson, D. R., and Toner, J. (1981). Bond-orientational order, dislocation loops, and melting of solids and smectic-A liquid crystals. *Phys. Rev.* B24, 363–387.

Nelson, D. R., and Widom, M. (1984). Symmetry, Landau theory and polytope models of glass. *Nucl. Phys.* B240 [FS12], 113–139.

Nicolis, S., Mosseri, R., and Sadoc, J. F. (1986). Polytopes with tangled disclinations. *Europhys. Lett.* 1, 571–578.

Penrose, R. (1974). The role of aesthetics in pure and applied mathematical research. *B.I.M.A.* 10, 266–271.

Phillips, J. C. (1986). Chemical bonding, kenetics, and the approach to equilibrium structures of simple metallic, molecular, and network microclusters. *Chem. Rev.* 86, 619–634.

Rahman, A. (1966). Liquid structure and self-diffusion. *J. Chem. Phys.* 45, 2585–2592.

Rehr, J. J., and Alben, R. (1977). Vibrations and electronic states in a model amorphous metal. *Phys. Rev.* B16, 2400–2407.

Sachdev, S., and Nelson, D. R. (1984). Theory of the structure factor of metallic glasses. *Phys. Rev. Lett.* 53, 1947–1950.

Sachdev, S., and Nelson, D. R. (1985). Statistical mechanics of pentagonal and icosahedral order in dense liquids. *Phys. Rev.* B32, 1480–1502.

Sadoc, J. F., Dixmier, J., and Guinier, A. (1973). Theoretical calculation of dense random packings of equal and non-equal sized hard spheres. Applications to amorphous metallic alloys. *J. Non-Cryst. Solids* 12, 46–60.

Sadoc, J. F., and Mosseri, R. (1982). Order and disorder in amorphous, tetrahedrally coordinated semiconductors: A curved space description. *Phil. Mag.* B45, 467–483.

Sadoc, J. F., and Mosseri, R. (1985). Hierarchical interlaced networks of disclination lines in non-periodic structures. *J. Physique.* 46, 1809–1826.

Selinger, J. V., and Nelson, D. R. (1985). Polytope models of metallic glasses: S-band electronic properties. *Proceedings of the MRS Meeting, Boston, Massachusetts.*

Sethna, J. P. (1983). Frustration and curvature: glasses and the cholesteric blue phase. *Phys. Rev. Lett.* 51, 2198–2201.

Sethna, J. P. (1985). Frustration, curvature, and defect lines in metallic glasses and the cholesteric blue phase. *Phys. Rev.* B31, 6278–6297.

Shechtman, D., Blech, I., Gratias, D., and Cahn, J. W. (1984). Metallic phase with long range orientational order and no translational symmetry. *Phys. Rev. Lett.* 53, 1951–1953.

Singh, Y., Stoessel, J. P., and Wolynes, P. G. (1985). Hard-sphere glass and the density functional theory of aperiodic crystals. *Phys. Rev. Lett.* 54, 1059–1062.

Steinhardt, P. J., Nelson, D. R., and Ronchetti, M. (1983). Bond-orientational order in liquids and glasses. *Phys. Rev.* B28, 784–805.

Stephens, P. W., and Goldman, A. I. (1986). Sharp diffraction maxima from an icosahedral glass. *Phys. Rev. Lett.* 56, 1168–1171.

Straley, J. P. (1985). Crystal defects in curved three-dimensional space. *Materials Science Forum* 4, 93–98.

Straley, J. P. (1986). Effect of topological frustration on the freezing temperature. *Phys. Rev.* B34, 405–409.

Thompson, D. W. (1952). *Growth and Form.* Cambridge University Press, Cambridge, London.

Turnbull, D. (1952). Kenetics of solidification of supercooled liquid mercury droplets. *J. Chem. Phys.* 20, 411–424.

Von Heimendahl (1979). Structure and dynamics of a two-component metallic glass. *J. Phys.* F9, 161–169.

Warner, N. P. (1982). The application of Regge calculus to quantum gravity and quantum field theory in a curved background. *Proc. R. Soc. London Ser.* A383, 359–377.

Widom, M. (1984). Icosahedral symmetry of a polytope model of glass. In *XIIIth International Colloquium on Group Theoretical Methods in Physics* (World Scientific, Singapore), 348–355.

Widom, M. (1985). Icosahedral order in glass: Electronic properties. *Phys. Rev.* B31, 6456–6468.
Widom, M. (1986). Icosahedral order in glass: Acoustic properties. *Phys. Rev.* B34, 756–763.
Widom, M. (1987). Model of icosahedral order. In *Nonlinearity in Condensed Matter, Solid State Science Series* 69. Springer, Berlin.
Widom, M., Strandburg, K. J., and Swendsen, R. H. (1987). Quasicrystal equilibrium state. *Phys. Rev. Lett.* 58, 706–709.
Zallen, R. (1983). *Physics of Amorphous Solids*. Wiley, New York.

Chapter 3
Metallurgy of Quasicrystals

ROBERT J. SCHAEFER AND
LEONID A. BENDERSKY

Metallurgy Division,
National Bureau of Standards,
Gaithersburg, MD

Contents

1. Introduction

The discovery of quasicrystals has stimulated interactions between the communities of physicists, mathematicians, crystallographers, and me-

APERIODICITY AND ORDER
Introduction to
Quasicrystals

ISBN 0-12-04060-2

tallurgists. In this paper we shall discuss some aspects of quasicrystals that have been of particular interest to the metallurgical community. Specifically, we shall consider the conditions that lead to the formation of quasicrystals, and how quasicrystals are related to other phases. The intensive study of quasicrystals has not only lead to an understanding of how better quasicrystal samples can be produced for physical measurements, it has also provided information that is of general interest to metallurgists concerned with the nucleation and growth of stable or metastable intermetallic compounds.

After the icosahedral phase of Al–Mn was discovered (Shechtman *et al.*, 1984; Shechtman and Blech, 1985), it immediately became of interest to determine whether similar phases would form in other alloy systems. Many have now been found, in most cases as a result of work with systems in which there was a specific reason to expect quasicrystal formation. In the more intensively studied systems, we now have a fairly good understanding of how they fit into the alloy phase diagram.

The vast majority of quasicrystalline phases are metastable, and one of our goals is to understand the specific conditions that lead to the formation of these phases in place of more stable ones. The formation of quasicrystals stands in dramatic contrast to the formation of many metallic glasses, which are another type of metastable phase occurring in rapidly solidified alloys. As we shall discuss below, the contrast is accounted for by consideration of the nucleation and growth processes of quasicrystal phases.

The kinetics of a phase transformation and the microstructure that develops can be strongly affected if special crystallographic relationships exist between some of the alloy phases. Although orientation relationships between crystallographic phases are common, it might at first appear that they would be less likely in the case of aperiodic phases. In fact, several such crystallographic relationships have now been identified between quasicrystal phases or between quasicrystals and periodic phases. These relationships provide insight into the structure of the quasicrystals, as well as accounting for certain aspects of the observed microstructures.

An important question since the first discovery of quasicrystals has concerned their stability with respect to other phases. This has been investigated mainly by observing their transformation on heating. Some of these experiments have demonstrated very clearly the high sensitivity of quasicrystalline microstructures to the details of solidification conditions.

Several major aspects of quasicrystal metallurgy are thus very similar to phenomena that are important in any alloy system, in the ways they affect the origin, formation, and disappearance of the different phases. A challenge to the metallurgist is to understand how quasicrystal behavior may vary from that of other alloy phases, and to use this understanding to produce quasicrystal samples with specific desired characteristics.

2. Quasicrystal Alloy Systems

As more and more alloy systems forming quasicrystalline phases have been discovered, one common feature has become apparent: Almost all of the quasicrystalline phases can be associated with periodic crystalline phases containing icosahedrally packed groups of atoms. This correlation has been the basis for the discovery of many new icosahedral phases. In several cases it appears that the icosahedral phases form most readily at a composition slightly different from that of the associated periodic phases.

In alloys where no periodic phase related to the quasicrystal has been identified, the quasicrystalline phase may be a metastable relative to a crystalline phase that contains additional constituents not present in the quasicrystal. A good example of such an effect among periodic phases is the metastable orthorhombic Al_6Fe phase, which is isomorphous with the stable $Al_6(Fe_x,Mn_{1-x})$. In the ternary Al–Mn–Fe phase diagram, the $Al_6(Fe_x,Mn_{1-x})$ phase is stable only for $x < 0.6$ (Phillips, 1959). However, even at moderately rapid rates of solidification, the range of occurrence of this phase can be extended to $x = 1$ to make binary Al_6Fe (Hughes and Jones, 1976). Perepezko and Boettinger (1983) have discussed in some detail the ways in which rapid solidification can lead to this type of extension of phase fields. The occurrence of orthorhombic Al_6Fe would be unexpected if we did not know of the presence of the orthorhombic phase in the Al–Fe–Mn phase diagram, but is readily explained once we have seen it. Similarly, the icosahedral phase of Al–Mn appears to be most closely related to the ternary α phase of Al–Mn–Si, and other quasicrystals may be related to unknown periodic phases containing additional components.

2.1 Aluminum-Transition Metal Phases

Rapidly solidified alloys of aluminum with transition metals (TM) are of interest for structural applications, and it was the study of such alloys that lead to the initial discovery of the icosahedral phase in alloys of aluminum with Mn, Fe, and Cr (Shechtman *et al.*, 1984). Consequently, it was natural to seek icosahedral phases in alloys of aluminum with other transition metals, and several were identified by Bancel and Heiney (1986). Silicon, substituted for a few percent of the aluminum atoms, was found (Chen and Chen, 1986; Bendersky and Kaufman, 1986) to produce an icosahedral phase that is more perfect and more stable with respect to decomposition, at least in the Al–Mn system. Al–Mn and Al–Mn–Si have now been much more thoroughly investigated than any other of the metastable quasicrystal phases.

Icosahedral phases have been identified in binary alloys of Al with Mn, Fe, Cr (Shechtman *et al.*, 1984), Ru, V, W, Mo (Bancel and Heiney, 1986), and Ni (Dunlap and Dini, 1985). In some cases, such as that of Al–Fe, the amount of icosahedral phase formed appears to be small, such that while it is detected by transmission electron microscopy (TEM) (Shechtman *et al.*, 1984), it is not revealed by X-ray diffraction (Bancel and Heiney, 1986). The latter authors found that a feature common to many of the icosahedral phase-forming Al–TM binary systems is that the icosahedral phase is most readily formed in rather dilute alloys, where it is associated with a large amount of fcc Al. When the transition metal content is increased in an attempt to eliminate the free Al, new phases start to appear and replace the icosahedral phase. The concentration of TM at which the icosahedral phase first appears, as well as the concentration at which it starts to be replaced by other phases, are dependent upon the solidification rate of the alloy.

Aluminum–manganese appears to be the only Al–TM system in which it is possible to form all icosahedral phases, but this requires very high solidification rates. At more moderate solidification rates, the icosahedral phase is progressively replaced by the decagonal (Bendersky, 1985; Schaefer *et al.*, 1986a) phase as the Mn concentration increases. The addition of a few percent Si to the alloys completely eliminates the decagonal phase, making it much easier to form single-phase icosahedral samples by rapid solidification. As a result, rapid solidification of Al–Mn–Si alloys has been studied in considerable detail.

Elser and Henley (1985) and Guyot and Audier (1985) pointed out that the ternary α cubic phase of Al–Mn–Si (Cooper and Robinson, 1966) has characteristics that link it closely to the icosahedral phase. This phase contains, at the cube corners and body center, a 54-atom icosahedrally arranged group now commonly referred to as a Mackay Icosahedron (Mackay, 1962). Elser and Henley and Guyot and Audier proposed that this group could be used to build the icosahedral phase. Many experiments have now confirmed this concept.

Icosahedral groups of atoms occur in several other equilibrium Al–TM alloys: α Al–Fe–Si (Corby and Black, 1977), $Al_{10}V$ (Brown, 1957), and $Al_{12}Mn$ (Adam and Rich, 1954) are examples. The last was previously believed to be metastable, but it was recently shown to be an equilibrium phase (Schaefer *et al.*, 1986b). It has a body-centered cubic lattice with a Mn-centered icosahedron of aluminum atoms at each lattice point. There is, however, no evidence that this structure has any relationship to the icosahedral phase of Al–Mn (Bennnett *et al.*, 1987). Instead, icosahedral Al–Mn appears to be related to the Al–Mn–Si ternary α phase. Thus, it is not clear without detailed study if icosahedral phases observed in the various Al–TM alloys are related to binary phases such as $Al_{10}V$ or to other phases of more complex composition.

2.2 *Other Metastable Quasicrystal Phases*

Many quasicrystal phases have now been formed by identifying and rapidly solidifying alloys that have equilibrium phases containing icosahedrally arranged groups of atoms. The first of these was $Mg_{32}(Al, Zn)_{49}$ (Ramachandrarao and Sastry, 1985), and this was followed by several others, including Mg_4CuAl_6 (Sastry, Ramachandrarao, and Anantharaman, 1986), $Mg_{32}(Al, Zn, Cu)_{49}$ (Mukhopadhyay *et al.*, 1986), $(Ti_{1-x},V_x)_2Ni$ (for $x = 0$ to 0.3) (Zhang, Ye, and Kuo, 1985; Zhang and Kuo, 1986), Ti_2Fe (Dong *et al.*, 1986), Mn_3Ni_2Si (Kuo *et al.*, 1986), $V_{41}Ni_{36}Si_{23}$ (Kuo, Zhou, and Li, 1987), and Cd–Cu (Bendersky and Biancaniello, 1987). It is clear, however, that these are not all distinctly different phases. Thus Mukhopadhyay *et al.* (1986) pointed out that Mg_{32} (Al, Zn)$_{49}$, Mg_4CuAl_6, and Mg_{32} (Al, Zn, Cu)$_{49}$ are part of an extended phase field in the Mg–Al–Zn–Cu system. Similarly, we can speculate that Ti_2Fe is part of an extended system more generally represented as $(Ti_{1-x},V_x)_2(Ni_{1-y},Fe_y)$. The icosahedrally arranged group in the periodic phase of MgAlZn contains 117 atoms occupying several shells with icosahedral symmetry, whereas the crystalline phase of Mn_3Ni_2Si contains several different types of icosahedra with different orientations and with the different species of atoms packed in arrangements that do not have icosahedral symmetry. In the latter alloy, it is not clear what type of icosahedral units might be present in the icosahedral phase.

Villars, Phillips, and Chen (1986) have used quantum structural diagrams to predict new quasicrystal phases based on a structural mapping of ternary intermetallic compounds formed by a *p* element (Al, Ga, Ge, Sn), an *s* element (Li, Na, or Mg), and a near-noble metal (Ni, Cu, Zn, etc.). The success of these predictions has been demonstrated by Chen *et al.* (1987b) in the case of $AuLi_3Al_6$, $Zn_{17}Li_{32}Al_{51}$, $Ga_{16}Mg_{32}Zn_{52}$, and $Ag_{15}Mg_{35}Al_{15}$. Once again, we may speculate that some of these compounds represent limiting cases of quaternary phase fields such as $(Au_x,Cu_{1-x})Li_3Al_6$; therefore the number of nonisomorphous phases is limited.

Icosahedral Pd–U–Si (Poon, Drehman, and Lawless, 1985; Drehman, Pelton, and Noack, 1986; Shen, Poon, and Shiflet, 1986) is unusual in several respects; it forms only by divitrification of a metallic glass, it has a very narrow composition range, and it is not known to have a relationship to any crystalline phase.

2.3 Stable Quasicrystal Phases

Once Ball and Lloyd (1985) reported that precipitates with icosahedral symmetry formed in the Al–Li–Cu system, it was soon concluded that the phase Al_6Li_3Cu was a stable icosahedral phase. The existence of this phase had been known for many years, but its structure had remained undetermined (Hardy and Silcock, 1955/56). It was soon evident that it was closely related in structure to the cubic Al_5Li_3Cu phase (*R* phase), which is isostructural with $Mg_{32}(Al,Zn)_{49}$ (Bergmann, Waugh, and Pauling, 1957), containing large icosahedral groups. Thus, in this case, once again the quasicrystal and the related periodic phase have slightly different compositions, but here they are both stable.

Tsai *et al.* (1987) have now reported a thermodynamically stable icosahedral phase at the atomic composition $Al_{65}Cu_{20}Fe_{15}$. Recently, MgGaZn was also demonstrated to be a stable icosahedral phase (Ohashi and Spaepen, 1987). We may anticipate that additional stable icosahedral phases will be identified.

3. Formation of Quasicrystals

When quasicrystal phases are produced by solidification from the melt, rapid solidification with cooling rates of 10^3 to 10^4 K/sec is required to produce all except for the stable phases. It is the high rate of nucleation of quasicrystal phases compared to that of competing crystalline phases, rather than an especially high growth rate, that accounts for their occurrence in rapidly solidified melts.

3.1 Contrast to Metallic Glass Formation

The formation of quasicrystals from the melt contrasts dramatically with one of the other major phenomena produced by rapid solidification, namely the formation of classical metallic glasses in which nucleation and growth of crystal phases fails to occur. Such metallic glasses are most readily formed at compositions close to deep eutectics: These are compositions at which no crystalline phase is stable, so that at equilibrium the alloy must separate into two or more crystalline phases of different composition. Because this diffusion-controlled chemical separation is a slow process, nucleation and crystal growth are difficult, and in a rapidly cooled melt, conditions are favorable for the liquid to become a glass. For example, Boettinger (1982) has shown how in the Pd–Cu–Si system, a eutectic composed of two intermetallic phases can grow at low velocities, but if one attempts to accelerate the growth rate beyond 2.5 mm/sec, the eutectic simply ceases to grow, and the liquid transforms to a glass. The interface between the crystalline and glassy regions is sharp and distinct in optical or electron micrographs.

Contrastingly, no quasicrystal phases have been reported to form at compositions where the phase diagram shows a eutectic. Instead, the characteristic feature of phase diagrams where quasicrystals form is a peritectic or chain of peritectics, as in the Al-TM series. These phase diagram features are characteristic of systems where there are strong interactions between the different atomic species, and there is a strong tendency to form intermetallic compounds. Quasicrystals form in these systems by a nucleation and growth process, whereas the formation of a classical metallic glass requires the absence of nucleation and growth processes.

Nonetheless, icosahedral phases can be formed by crystallization of an amorphous phase in Pd–U–Si (Poon, Drehman, and Lawless, 1985) and Al–Cu–V (Garçon *et al.*, 1987), and an amorphous phase has been reported to form in Si-rich Al–Mn–Si alloys (Dunlap and Dini, 1986; Chen, Koskenmaki, and Chen, 1987a; Yamane *et al.*, 1987). In these cases, the amorphous material formed shows indications of strong chemical ordering, and it may represent a condition significantly different from the classical near-eutectic glass. Recently, Inoue *et al.* (1988) have demonstrated that at some compositions, the amorphous Al–Mn–Si alloys when heated transform first to the icosahedral phase and then to periodic phases. There is, in addition, a large body of literature describing the formation of amorphous materials at low temperatures and their subsequent transformation to quasicrystals by thermal or ion-beam activation (Follstaedt and Knapp, 1986; Lilienfeld *et al.*, 1985). This type of process is distinctly different from the formation of quasicrystals in rapidly cooled melts, and we shall not discuss it here except to note that the amorphous materials made by such processes also are not classical metallic glasses.

3.2 Aluminum–Manganese

The Al–Mn and Al–Mn–Si systems have been studied so thoroughly that they provide a rather complete picture of the quasicrystal formation process. We shall discuss them in detail, because we believe that the phenomena that explain their microstructures are common to many quasicrystal systems.

3.2.1 Composition of the Icosahedral Phase. Much of the early work with the icosahedral phase of Al–Mn was carried out with alloys containing about 14 at.% Mn, approximately equivalent to the Mn content of the orthorhombic Al_6Mn compound. Since rapidly solidified alloys of this composition contain icosahedral grains within a matrix of fcc Al, it was clear that the composition of the icosahedral phase itself must be greater than 14 at.% Mn. Several groups (Kimura *et al.*, 1985; Schaefer *et al.*, 1986a; Inoue *et al.*, 1986) have reported that as the Mn content of the alloy is increased, the X-ray diffraction peaks from the fcc Al decrease in intensity and disappear when the Mn concentration reaches 20–22 at.%. If the cooling rate is high enough to suppress the formation of the decagonal phase, single-phase icosahedral samples in this composition range may then be produced. Similarly, scanning transmission electron microscopy (STEM) measurements (Krishnan, Gronsky, and Tanner, 1986; Kimura *et al.*, 1985) of rapidly solidified Al-14 at.% Mn alloys show the icosahedral phase to contain slightly more than 20 at.% Mn.

While these observations indicate that the icosahedral phase composition tends to be closer to Al_4Mn than to Al_6Mn, there is also evidence that it can be formed with a range of compositions. By rapid solidification, single-phase fcc Al can be produced with up to 10 at.% Mn in solid solution, and the lattice parameter of these alloys decreases by 0.17% for each at.% of the smaller Mn atoms in solid solution. Similarly, as the alloy concentration increases from 10 to 26 at.%, the lattice parameter of the icosahedral phase decreases by $\sim$1% (Schaefer *et al.*, 1986a; Chen *et al.*, 1987a), indicating that the Mn content of this phase (and not just the volume fraction of the phase) changes with the alloy composition. By STEM measurements, Inoue *et al.* (1986) found the icosahedral phase to contain 18.2 at.% Mn in a 14.3 at.% Mn alloy, but 22.6 at.% Mn in a 22.5 at.% Mn alloy. There is at present no information on how the icosahedral phase structure accommodates changes in composition, i.e., by substitution of one type of atom for another or by introduction of additional structural disorder. All of these data indicate that in alloys containing less than about 21 at.% Mn, the icosahedral phase has a Mn content greater than that of the liquid. Solute diffusion will therefore play an important role in determining the solidification morphologies.

3.2.2 Phase Diagram. Figure 1 shows the Al–Mn phase diagram with several metastable phase boundaries. Metastable phase boundaries represent the equilibria which would occur when slow nucleation and growth rates suppress the formation of other, more stable phases. In the real world, we usually suppress these other phases by techniques such as rapid solidification. As a result, the metastable phases that we produce are not

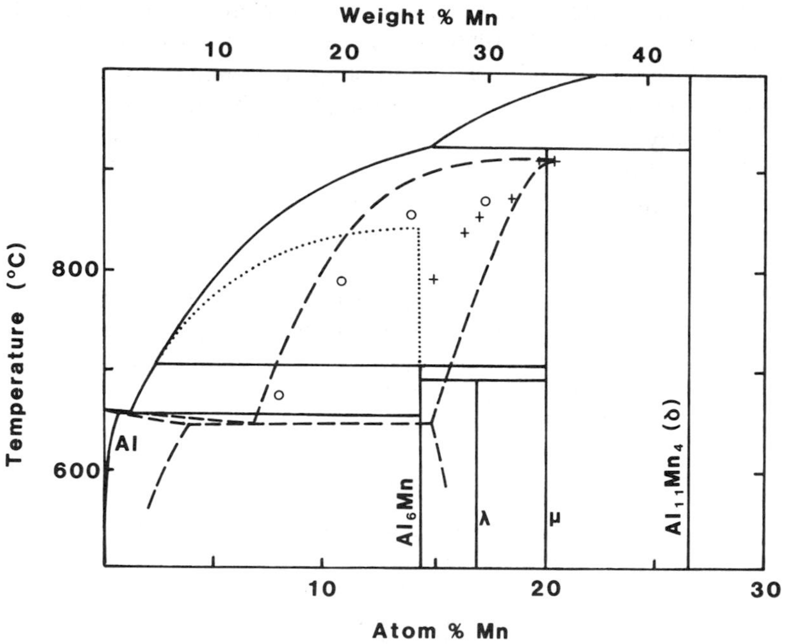

Figure 1. A portion of the Al-Mn phase diagram. The dotted lines show the metastable extension of the liquidus for Al_6Mn as calculated by Murray et al. (1987). Crosses are the T_o data of Knapp and Follstaedt (1987); circles are the nucleation temperatures of Mueller et al. (1987); dashed lines are the estimated metastable equilibria of the icosahedral phase and aluminum.

necessarily in a true state of metastable equilibrium, so that it is difficult to locate the boundaries exactly on the basis of measurements with rapidly solidified materials. Nonetheless, we can use such measurements to set limits on the positions of the true metastable equilibrium curves.

Knapp and Follstaedt (1987) determined the melting points of the icosahedral and decagonal phases at 14.8 to 20.5 at.% Mn by using high-rate heating experiments. They concluded that the temperatures they measured were close to the T_0 temperature, between the true metastable solidus and liquidus curves of the phases. Their measurements suggest a maximum melting temperature of 910 ± 20°C for the icosahedral phase at a composition of about 20 at.% Mn, in agreement with the tendency described above for the icosahedral phase to form at this composition. The solidus and liquidus curves shown on the phase diagram are thus drawn to encompass the melting point data of Knapp and Follstaedt.

At lower Mn contents, the positions of these curves are more difficult to locate. Yu-Zhang *et al.* (1986) observed a eutectic between Al and the icosahedral phase in rapidly solidified alloys containing 8.5 at.% Mn, but it is to be expected that the composition at which the eutectic forms will be significantly shifted from its equilibrium value toward higher Mn concentrations at rapid solidification rates, just as it is for the Al–Al_6Mn eutectic (Eady, Hogan and Davies, 1975; Juarez-Islas and Jones, 1987). The true metastable eutectic composition for Al/icosahedral phase is thus expected to be at < 8.5 at.% Mn.

Figure 1 also shows an extension of the liquidus temperature for the equilibrium Al_6Mn phase, as calculated by Murray *et al.* (1987), to a metastable congruent melting temperature of 875°C. Note that upon cooling a 14 at.% Mn alloy, the liquidus of the icosahedral phase is encountered before the liquidus of the Al_6Mn phase.

3.2.3 Nucleation. The icosahedral phase predominates in the microstructure of rapidly solidified Al–Mn alloys because of its high rate of nucleation: There is evidence that its growth rate is not especially high. The first reports of the phase were based on observations of melt-spun ribbons, where the icosahedral phase took the form of numerous branched crystallites that nucleated independently within the ribbons but never grew fast enough to become more than 1–2 μm in diameter (Fig. 2a).

The most dramatic evidence of the high nucleation rate of the icosahederal phase in supercooled melts comes from observations of crystallization of liquid droplets (Bendersky and Ridder, 1986). By using an electrohydrodynamic (EHD) atomization technique, Al–14 at.% Mn liquid was atomized into droplets less than 1 μm in diameter. It was found that the solidified droplets consisted of polycrystalline icosahedral phase with an extremely fine grain size (Fig. 2b). Moreover, the grain size was dependent upon the droplet size, with the finest (< 0.1μm diameter) particles having such a small grain size that the electron diffraction peaks were broadened so greatly as to be essentially indistinguishable from the diffraction from an amorphous structure. This structure was referred to as "microquasicrystalline," and appears to be the result of an extremely high nucleation rate, rather than the absence of nucleation that leads to the formation of a classical glass.

In a different type of droplet experiment, Mueller *et al.* (1987) studied the solidification of larger (50–100 μm) Al–Mn droplets containing up to 30 wt% Mn at cooling rates ranging from 0.5 to 500°C/sec. Thermal measurements enabled the determination of nucleation temperatures in alloys cooled at 25°C/sec or less. At the slowest cooling rate, it was found that

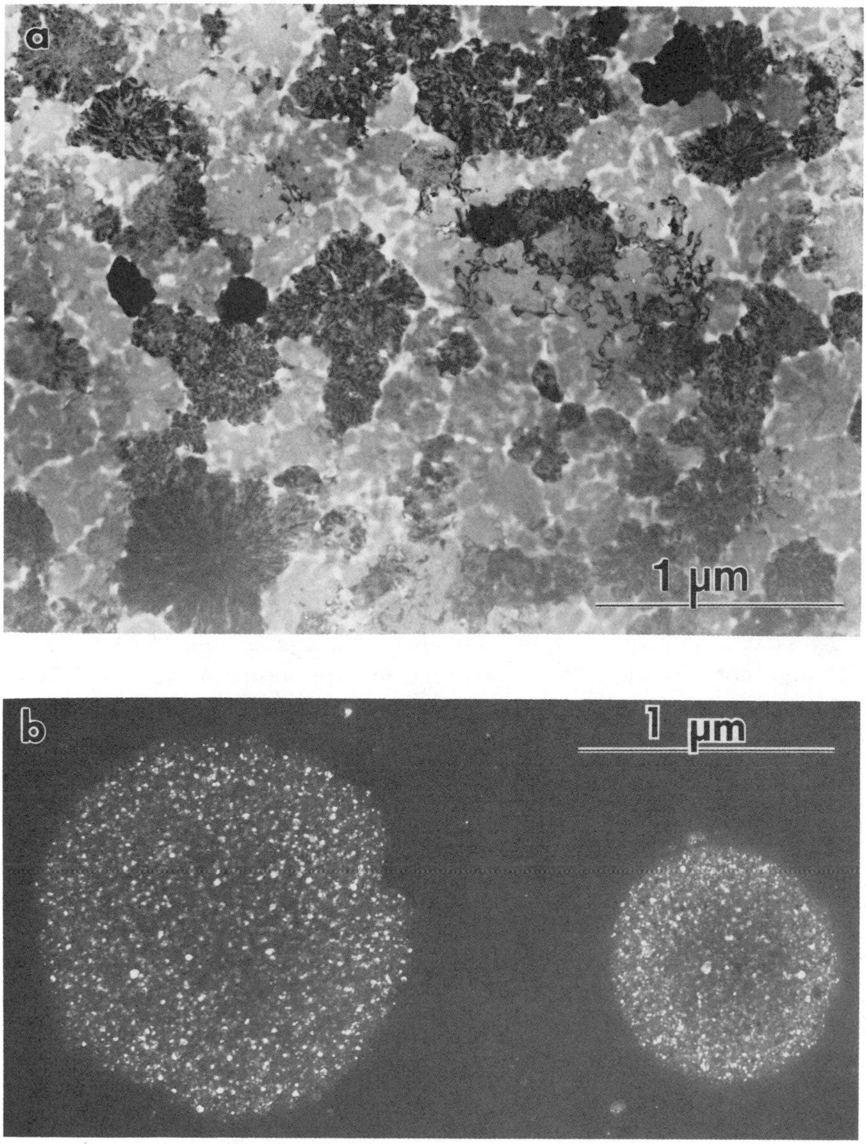

Figure 2. Microstructure of an Al-14 at % Mn alloy showing the dependence of icosahedral phase grain size on the conditions of solidification. (a) Melt-spun ribbon (bright field TEM micrograph). (b) Electrohydrodynamic atomized droplets (dark field TEM micrograph).

the solidified particles contained the hexagonal λ(Al_4Mn) phase plus fcc Al. Increasing the cooling rate to 25°C/sec resulted in the appearance of the decagonal phase, whereas a further increase of cooling rate to 500°C/sec resulted in the additional appearance of the icosahedral phase.

The nucleation temperatures measured in the droplets cooled at 25°C/sec are believed to represent nucleation of the icosahedral, rather than the decagonal phase, since it has been demonstrated that at moderate cooling rates, the decagonal phase is nucleated by the icosahedral phase and replaces it (Schaefer and Bendersky, 1986). In Fig. 1 the dots represent the nucleation data from the droplets cooled at 25°C/sec. They are seen to be displaced by a constant temperature interval (in this case about 40°C) below the proposed metastable liquidus curve. Such a constant displacement of the nucleation temperature from the liquidus for emulsified drops has been observed for stable phases in the Pb–Sn system (Richmond *et al.*, 1983). Reddy, Sekhar, and Rao (1987) cooled bulk samples of Al–14 at.% Mn at moderate rates by suddenly pressurizing the liquid and thereby increasing the thermal contact with the mold. They observed a thermal arrest at 875°C that they concluded resulted from the nucleation of the icosahedral phase. This point is in complete agreement with the droplet data of Mueller *et al.* (1987).

These data indicate that the icosahedral phase can nucleate at temperatures not far below the metastable melting temperature of Al_6Mn. The experiments also show that nucleation of Al_6Mn is extremely difficult. Although λ(Al_4Mn) appears to nucleate at higher temperatures, its growth rate is extremely slow (Schaefer *et al.*, 1986a). Consequently, even a moderately high rate of cooling (25°C/sec) allows insufficient time for formation of the Al_6Mn or λ phase, and the icosahedral phase can nucleate.

Although no such detailed studies of nucleation processes have been carried out in other quasicrystal-forming systems, the microstructures usually reported correspond to high nucleation rates and moderate growth rates (many small crystals growing from independent nucleation sites).

The icosahedral phase of Al–Mn serves as a powerful nucleant for the decagonal phase (Schaefer and Bendersky, 1986), with each icosahedral-phase crystal generating six equivalent orientation variants of the decagonal phase. In a series of experiments using a wide range of solidification rates, we were able to show that the decagonal-phase crystals were more stable than the icosahedral-phase crystal which nucleated them, and they therefore replaced it in slowly solidified samples. The icosahedral phase thus generated the seeds of its own destruction. As the solidification rate increased, however, this replacement was suppressed and the icosahedral crystals were preserved. In the slowly solidified samples, the replacement of the icosahedral phase was complete, but the presence of the full set of decagonal-phase orientation variants demonstrated that an icosahedral-phase crystal had initially been present. Thus, in these experiments, the prevalence of the icosahedral phase at high solidification rates and of the decagonal phase at lower solidification rates can be understood in terms of the high nucleation rate of the icosahedral phase in the liquid, followed by a high rate of epitaxial nucleation of the decagonal phase on the icosahedral phase.

The decagonal phase can also be seen as rather large dendritic crystals in chill-cast samples of 14 at.% Mn (Urban *et al.*, 1986). Because no sets of related orientational variants were seen in this case, it was not clear if the decagonal phase nucleated from the melt or on an icosahedral precursor.

3.2.4 Growth. Icosahedral crystals which nucleate in melt-spun ribbons of Al with 11–15 at.% Mn typically grow with a branched structure having little apparent tendency to propagate in any preferred direction. The maximum dimension of the crystals is typically 1–2 μm. From the dimensions of the branches and considerations of solute diffusion rates, Shechtman *et al.* (1984) estimated that the growth velocity was about 1 cm/sec.

Additional information about the growth rate of the icosahedral and other phases in the aluminum–manganese systems was obtained from experiments in which a travelling melt zone about 1mm in diameter passes across the surface of an alloy bar (Schaefer *et al.*, 1986a). The melt zone is produced by a beam of 25-kV electrons that is scanned across the sample surface at velocities from 10^{-3} to 2 m/s. The resolidified material is examined by optical metallography, TEM, and X-ray diffraction.

These experiments reveal that some of the equilibrium phases of the Al–Mn system, even when present in the substrate material, can propagate into the moving melt zone only at very slow scan velocities. Specifically, large blocky hexagonal crystals of the (Al_4Mn) λ phase in the substrate appeared unable to grow into the melt zone at all except when the scan rate was less than 1 mm/sec.

Figure 3 summarizes the phases that dominate the resolidified zone of the surface-melted alloys, as a function of composition and scan velocity. In some cases the domination is the result of high nucleation rates, whereas in other cases it is the result of a high growth rate of the dominant phase. At Mn contents of up to 7.5 at.%, rapidly solidified melts consist of fcc Al supersaturated with Mn (the equilibrium solubility limit is only 0.6 at.%). Metallographic examination reveals that, even at scan velocities of 1 m/sec or more, the Al crystals grow from the substrate to the top of the melt zone, with no evidence of nucleation of new crystals. The metastable extension of the solidus and liquidus for Al in Fig. 1 are very

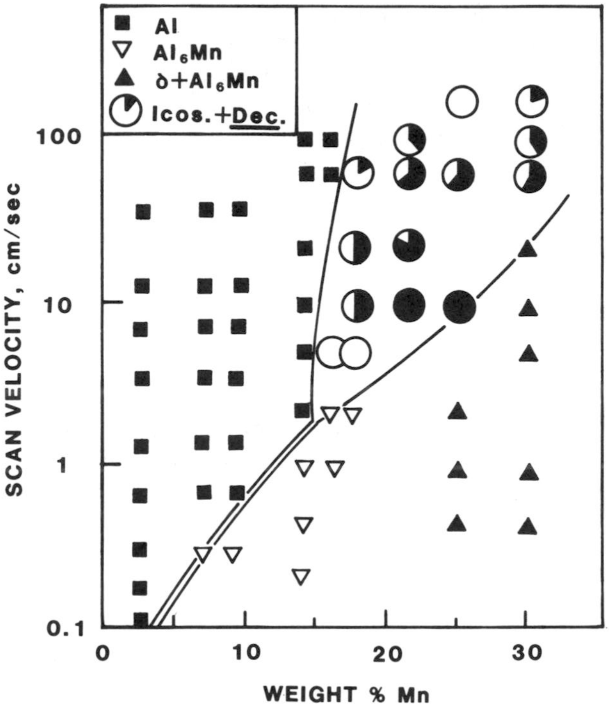

Figure 3. The phases dominating resolidified surface melts in Al-Mn alloys, as a function of scan velocity and Mn concentration.

close together, indicating that the fcc Al phase should be able to grow from the melt with relatively little change in composition. There is therefore nothing to prevent this phase from growing very rapidly from the melt, as long as the interface temperature is below the metastable extension of the liquidus.

As the alloy's Mn content increases, however, the liquidus temperature of the Al phase changes only slightly, whereas the liquidus temperatures of the various intermetallic phases (including the icosahedral phase) increase rapidly. Thus, the temperature at which the fcc phase can grow represents a progressively higher supercooling with respect to the intermetallic phases. At low velocities, this results in growth of the intermetallic phases from the substrate into the melt zone. At high velocities, it results in nucleation of the icosahedral phase within the melt zone.

In an alloy containing 10 at.% Mn, it was found that the orthorhombic Al_6Mn phase grew to the top of the melt zone at scan velocities of less than 2.5 cm/sec. but at this velocity or greater, the icosahedral phase nucleated within the melt zone and prevented further growth of the orthorhombic phase. The icosahedral crystals that nucleated within the melt zone grew to a maximum diameter of only 10–20 μm and showed little tendency to propagate in the direction of the melt scan, indicating that their velocity of growth was less than the scan velocity of 2.5 cm/sec. Therefore, they were able to replace the orthohombic phase in the melt zone because of their high nucleation rate, even though they did not grow as fast as the orthorhombic phase.

Icosahedral crystals growing in a 10 at.% Mn alloy grow with a composition considerably enriched in Mn with respect to the melt, and their growth rate is therefore limited by solute diffusion. In more concentrated alloys, the icosahedral phase can grow with less change of composition, and the growth velocity may be considerably higher. Thus, Robertson *et al.* (1986) found that in an Al–20 at.% Mn–6 at.% Si alloy, where the icosahedral phase forms with the same composition as the melt, the icosahedral phase grew as columnar grains across the thickness of melt-spun ribbons, suggesting that it had a high growth rate.

The important role of solute diffusion in limiting the growth rate of the icosahedral phase in alloys containing less than 20 at.% Mn is also indicated by the dendritic structure of the crystals, which was revealed with particular clarity in the surface-melted 10 at.% Mn alloys. At this composition, the icosahedral phase nuclei are sufficiently far apart so that the crystals can grow to sizes of 20 μm or more, and they develop a dendritic form with a conspicuous crystallographic symmetry. The dendritic structures seen by optical microscopy of polished sections of the surface-melted samples result from growth of primary stems along the threefold axes of the icosahedron, with branches growing in the directions of the adjacent threefold axes.

The preferred growth direction of dendritic structures can be correlated to anisotropy of solid/liquid surface energy or crystal growth kinetics. Although these anisotropies are extremely difficult to measure directly, their effects can sometimes be revealed by the tendency of equilibrated or growing phases to form facets. If the anisotropy is sufficiently large, facets will be formed on the planes that have the lowest interfacial energy or the lowest growth rate. In the case of Al–Mn–Si alloys, Robertson *et al.* (1986) have observed facets on the fivefold planes, resulting in the formation of pentagonal dodecahedra with apices along the threefold axes. With the apices being the most favorable sites for diffusion-limited growth, the preferred dendrite growth along the threefold axes would result.

In many cases the icosahedral phase of Al–Mn shows little indication of faceting, but only a very small anisotropy is required to produce growth in a preferred crystallographic direction. For example, studies of camphene, a body-centered cubic organic material, showed that it had a surface energy anisotropy of only about 2% (Blodgett, Schaefer, and Glicksman, 1974), but that this was sufficient to direct its dendritic growth in the [100] directions.

In melt-spun ribbons of alloys containing 6–8.5 at.% Mn, a eutectic between the icosahedral phase and Al can be observed (Yu-Zhang *et al.*, 1986). In most cases it appears that growth starts as a dendritic crystal of the icosahedral phase and then transforms to a coupled eutectic growth with the fcc phase. A great variety of morphologies are seen in these samples, including very fine icosahedral phase particles that appear to have formed by the break-up of eutectics. The eutectic often has an unusual morphology in which the two phases appear to have grown alternately to give a series of concentric rings, in place of the more normal side-by-side lamellar growth.

The decagonal phase grows with a distinctive morphology characterized by strong facetting on its tenfold plane but little if any tendency to facet in other directions. For small decagonal crystals, this results in a blocky cylindrical shape, but for larger crystals, it results in a dendritic form that spreads out in the directions normal to the tenfold axis.

3.2.5 Effects of Silicon Addition. The addition of a small amount of silicon to Al–Mn alloys eliminates the formation of the decagonal phase and results in formation of single-phase icosahedral material at a composition close to $Al_{73}Si_6Mn_{21}$. The unusually perfect icosahedral structure of alloys of this composition allowed Bendersky and Kaufman (1986) to use convergent beam electron diffraction to determine unambiguously that the crystals belonged to the $m\bar{3}\bar{5}$ point group. Elser and Henley (1985) and Guyot and Audier (1985) pointed out that the cubic α Al–Mn–Si phase could be described as either a bcc packing of Mackay icosahedra or a periodic packing of two types of decorated rhombohedral units; the icosahedral phase could be regarded as a quasiperiodic packing of these units. Dunlap and Dini (1986), Chen, Koskenmaki, and Chen (1987a), and Yamane *et al.* (1987) have determined the phases present in Al–Mn–Si alloys of many different compositions and found generally similar results. For example, in alloys containing 20 at.% Mn, the phases present in melt-spun ribbons change from icosahedral + decagonal (0% Si) to icosahedral (1–10% Si) to icosahedral + α (13–20% Si) to amorphous (30% Si). All investigators noted that the phases present are a function of solidification rate, with higher solidification rates favoring the icosahedral and amorphous phases at the expense of the α and decagonal phases.

These experiments all indicate that the most favorable composition for formation of the icosahedral phase is about $Al_{73}Si_6Mn_{21}$, compared to the α-phase composition of about $Al_{73}Si_{10}Mn_{17}$. Koskenmaki, Chen, and Rao (1986b) showed that melt-spun alloys of the latter composition contain nodules of the icosahedral phase surrounded by a layer of α-phase material, oriented in such a way that the symmetry axes of the Mackay icosahedra in the α phase are parallel to the corresponding symmetry axes of the icosahedral phase. Thus, the icosahedral phase nucleates first in this alloy, but it then nucleates the cubic phase before solidification is complete. In $Al_{73}Si_6Mn_{21}$, this secondary nucleation of the α phase does not occur, and single-phase icosahedral material is produced.

3.3 Stable Phases

The discovery that the "*T*2" phase of Al–Li–Cu was a stable icosahedral phase caused great excitement, because it indicated the possibility of growing large, high-quality single crystals by conventional crystal growth techniques. There are however, some practical difficulties involved in working with this system. The first of these is that the *T*2 phase does not freeze congruently, but forms by a peritectic reaction. To form it directly from the melt, it is necessary to employ a liquid which is aluminum-rich with respect to the *T*2 phase. This results in morphological instability of the solidifying crystals, unless very slow growth rates and high temperature gradients are used. Slow growth rates, however, aggrevate the effects of another difficult aspect of this system, which is the reactive and volatile nature of the lithium. Great care is required to prevent reaction with containers and loss of lithium during melting. A final difficulty is that the etching behavior of the *T*2 and cubic *R* phases is essentially identical, so that they are extremely difficult to distinguish by optical metallography; this makes evaluation of the samples more difficult.

Fortunately, a great deal can be learned from single crystals in the size range 0.1 to 1 mm, and many of these have been produced even by rather crude crystal growth techniques. The general type of process that has been successful is one in which the molten metal is cooled from all sides at a moderately rapid rate, such that pockets of liquid are trapped within the solid. Continuing solidification of the liquid within these pockets causes the formation of voids, in which the liquid drains away from some of the growing *T*2-phase crystals. The external form of these crystals is then revealed when the sample is later broken open.

Gayle (1987) and Dubost *et al*. (1986) have published dramatic scanning electron microscopy (SEM) pictures showing the triacontahedral form of crystals of the *T*2 phase revealed by this method. In some cases, the triacontahedra develop into dendrites that grow along the fivefold axes, again in the direction of the sharp apices of the triacontahedron. As growth continues, the liquid between the *T*2 crystals becomes progressively more enriched in Al, and finally, a eutectic between *T*2 and Al develops. The large size of the facets observed in this system indicates that they are kinetic in origin, since a very long time would be required to produce an equilibrium shape with such large dimensions.

The recently identified stable quasicrystals of $Al_{65}Cu_{20}Fe_{15}$ (Tsai et al., 1987) and MgGaZn (Ohashi and Spaepen, 1987) facet on the fivefold planes to form pentagonal dodecahedra. It should be possible to produce large single-crystal samples of these compounds by conventional crystal growth methods.

4. Orientation Relationships

The study of orientation relationships between different phases can give us additional information about their structural relation. Certain orientation relationships result from the requirement to have a coherent interface with minimum energy: These are frequently observed in precipitation. Another type of orientation relationship occurs between crystallographic variants of an ordered phase and a disordered one. All phases in this case share the same disordered lattice. Orientation relationships can also be favored between phases with a similar structural motif. In that case the lattices of the two phases are expected to be oriented so that their motif will have the same orientation. Phase intergrowth is frequently an example of this.

These orientation relationships indicate that one phase is serving as a heterogeneous nucleant for the other. Because of this, the formation of metastable phases is frequently possible kinetically. Several specific orientation relationships have been reported between aperiodic and periodic phases.

4.1 Icosahedral Phase and fcc Al

In several Al-based systems, the icosahedral phase frequently coexists with fcc Al. When Al forms as a secondary solidified phase between the arms of icosahedral dendrites or grains, no special orientation relationship was found. However, when the icosahedral phase precipitates from a supersaturated fcc solid solution, it has an exact orientation with the fcc matrix. Highly supersaturated Al prepared by implantation of Mn ions into the surface of an aluminum specimen and consequently annealed at low temperature (Budai and Aziz, 1986) shows precipitation of the Al–Mn icosahedral phase with the following orientation relationship:

$$[A2]_i \parallel [001]_{Al}$$
$$[A2]_i \parallel [\bar{1}10]_{Al}$$

where $[A2]_i$ is the orthogonal set of twofold axes of the icosahedral phase. Because of the strong orientation relationship, it was possible to form a large quantity of the Al–Mn icosahedral phase in a single orientation in large grains of Al, and thereby to obtain X-ray diffraction equivalent to that from a single crystal, unattainable for the Al–Mn system (Budai *et al.*, 1987).

The same orientation relationship was observed when precipitates formed from supersaturated Al, obtained by rapid quenching, in dilute Al–Mn (Yu-Zhang *et al.*, 1988a) and Al–Mn–Zr (Ohashi *et al.*, 1986) alloys. In the case of the Al–Mn alloy, the shape of the precipitated icosahedral phase particles was found to be dodecahedral. The dodecahedral shape of the icosahedral phase was also observed in earlier stages of solidification before the development of a dendritic morphology, controlled by slow growth on the fivefold planes. The later case probably corresponds to development of minimum surface energy, where the former can be explained by symmetry-dictated optimum shape.

For an Al–Li–Cu–Mg alloy, where the Al–Li–Cu icosahedral phase was precipitated from an Al matrix, another type of approximate orientation relationship was reported, different from that in Al–Mn (Loiseau and Lapasset, 1986, 1987): the twofold zone axis $(A2)_i$ of the icosahedral phase is parallel to the [110] of the Al matrix, where three different relationships are established by the following parallel directions in the zone axis plane: $(A2)_i \parallel [111]_{Al}$; $(A2)_i \parallel [100]_{Al}$; $(A5)_i \parallel [111]_{Al}$. The precipitates in that case have a needle-like form, with long axis along a $\langle 011 \rangle$ direction.

Until the atomic structure of the icosahedral phases has been determined and the structure of interfaces can be modeled, we can only speculate on the reason for the different orientation relationships. Nevertheless, it seems that the reasons are similar to these for formation of coherent crystalline precipitates. Overlapping of some spots of precipitates and matrix (e.g., 200_{Al} and 420000_i) can reflect a certain coherency between an average plane of the quasiperiodic lattice and the periodic planes of Al (Budai and Aziz, 1986).

4.2 *Icosahedral and Intermetallic Phases*

As was discussed before, the icosahedral phases grow in competition with different intermetallic phases, most of them with complex structures. The presence of icosahedral clusters as a major part of the motif for some of the phases with known structure seems to be a key for understanding the observed orientation relationships. In some Al–Mn–Si alloys (e.g., $Al_{74}Mn_{20}Si_{5.8}$), the icosahedral phase has been observed surrounded by the equilibrium cubic α phase, which is second to solidify (Guyot, Audier, and Lequette, 1986; Koskenmaki, Chen, and Rao, 1986b; Audier and Guyot, 1986). The orientation relationship is

$$[100]_\alpha \parallel [A2]_i$$
$$[111]_\alpha \parallel [A3]_i$$

which corresponds to the common orientation of fivefold axes of the icosahedral lattice and the icosahedral clusters of the cubic phase. In view of current modeling of the icosahedral phase structure, the orientation of the same icosahedral clusters is preserved across the interface between phases, where the packing of the clusters changes from quasiperiodic or random to periodic. Because the lower symmetry cubic phase nucleates and grows on top of the higher symmetry icosahedral phase, several crystallographic variants are expected. The α cubic phase formed by solid-state transformation of the icosahedral phase during annealing has the same orientation relationship (Fung and Zhou, 1986).

This type of orientation relationship seems to be quite general and has been observed for other types of icosahedral phases where an icosahedral motif of different chemistry and structure is expected. The Al–Li–Cu icosahedral phase can grow in conjunction with the cubic bcc *R* phase that is similar to the α(AlMnSi) phase in having bcc sites occupied by complex icosahedral clusters (however, of different structure). Several workers (Sainfort and Dubost, 1986; Audier, Sainfort, and Dubost, 1986) reported the same orientation relationship $[100]_R \parallel [A2]_i$ and $[111]_R \parallel [A3]_i$ observed in cast structures.

Two other icosahedral phases, the $FeTi_2$ and $(Ti_{1-x}V_x)_2Ni$, have been observed coexisting with fcc $FeTi_2$ and $NiTi_2$ isomorphous phases ($Fd\bar{3}m$, $a \approx 1.13$ nm) (Dong *et al.*, 1986; Dong, Chattopadhyay, and Kuo, 1987; Zhang, Ye, and Kuo, 1985; Zhang and Kuo, 1986, 1987). The common orientation relationship was observed: $[A5]_i \parallel [110]_{fcc}$; $[A2] \parallel [1\bar{1}0]_{fcc}$. The [110] diffraction pattern shows pseudo tenfold intensity modulation that probably indicates the presence of icosahedral clusters in the crystalline structure and suggests the same type of orientation relationship as between the icosahedral and bcc α phase. In fact, another phase that could be similar to the bcc α(AlMnSi) phases was found in $FeTi_2$ alloys by Dong, Chattopadhyay, and Kuo (1987), and again the phase is in the expected orientation relation with the icosahedral phase, which preserves orientation of the clusters.

4.3 Decagonal and Intermetallic Phases

In the Al–Fe system, the decagonal phase has been observed surrounded by crystalline twins of the monoclinic $Al_{13}Fe_4$ phase (Zou, Fung, and Kuo, 1987; Fung, Zou, and Yang, 1987). Tenfold twins frequently form in the

Al–Fe system (Louis, Mora, and Pastor, 1980) because of the pentagonal-triangle network of atoms in the (010) plane of the monoclinic phase. Therefore, the orientation relationship is natural and shows the tenfold axis of the decagonal phase parallel to the pseudo-tenfold [010] axis of the twins. The orientation relation is also supported by a good coincidence between the period of the decagonal and monoclinic phases along the decagonal axis: $c_{dec} \cong 1.6$ nm $= 2b_{Al_{13}Fe_4}$ ($b = 0.8083$ nm). A similar relationship was found in the Al–Fe–Ce alloy (Steeds *et al.*, 1986) where the transition between the aperiodic and crystalline phase appears to be gradual (according to high-resolution electron microscopy). This type of orientation relationship suggests modeling for the decagonal phase by using structural elements of crystalline phases to decorate a two-dimensional Penrose tile (Kumar, Sahoo, and Athithan, 1986). This approach will probably suggest a grouping of atoms (cluster) different from that in the icosahedral phase, and this seems to contradict results from diffraction experiments that support the similarity between the structural motif of icosahedral and decagonal phases (Bendersky, 1986; Dubois *et al.*, 1986).

A similar coexistence of the decagonal and decagonally twinned crystalline phase, different from the $Al_{13}Fe_4$-type, was also observed in the Al–Pd system (Thangaraj *et al.*, 1986; Sastry *et al.*, 1978), but crystallographic details of the crystalline phase are yet unclear. Another type of crystalline phase that seems to be closely related to the decagonal phase was observed in $Al_{60}Mn_{11}Ni_4$ (Van Tendeloo *et al.*, 1987). The phase is either already present as small domains inside the decagonal phase, or forms by its thermal decomposition. The observation of pseudo-fivefold axes in its diffraction space strongly suggests the presence of icosahedral clusters as the dominant feature. The phase also has a period comparable with that of the decagonal phase ($c = 1.24$ nm) and can be a good candidate as a structural model for the decagonal phase.

4.4 Icosahedral and Decagonal Phases in Al–Mn

The decagonal phase nucleates on the surface of the icosahedral phase with the tenfold axis of the former lying parallel to one of the fivefold axes of the later (Bendersky, 1985; Bendersky *et al.*, 1985). Each icosahedral grain thereby generates six orientational variants of the decagonal phase (Schaefer and Bendersky, 1986). This orientation relationship between two aperiodic phases and also intensity modulation in diffraction space suggests that the decagonal phase has icosahedral clusters similar to those in the icosahedral phase that are mirror related (Bendersky 1986). Observation of the icosahedral phase twinning in $Al_{86}Mn_7Fe_7$ alloy, where the twins are related by a mirror plane perpendicular to the fivefold axis, could reflect a tendency to form the decagonal phase instead of icosahedral when Mn atoms are replaced by Fe (Koskenmaki, Chen, and Rao, 1986a).

5. Transformations on Heating

Heating of the metastable quasicrystal phases in the electron microscope, differential thermal analysis (DTA) or differential scanning calorimetry (DSC) apparatus has revealed much about the way in which these alloys approach equilibrium. Some of the icosahedral phases survive to rather high temperatures, indicating that they are not far from being stable. In some cases, thermal measurements have provided a very sensitive indication of the dependence of the metastable microstructure on processing conditions. This sensitivity to processing conditions makes it difficult to interpret some of the thermal effects. Once again, the Al–Mn system has been studied more thoroughly than any other, and it will be discussed in detail.

The transformation of melt-spun Al–Mn ribbons upon heating has been studied by thermal measurements (Bagley and Chen, 1985; Chen *et al.*, 1985; Dunlap and Dini, 1985; Kelton and Wu, 1985; Kimura *et al.*, 1986; Luck *et al.*, 1986; McAlister *et al.*, 1987; Murray *et al.*, 1987; Quan and Giessen, 1986; Yu-Zhang *et al.*, 1988b), TEM (Kimura *et al.*, 1986; McAlister *et al.*, 1987; Samuel, de Lonckere, and Gerin, 1986; Samuel and Samuel, 1987), and real-time neutron diffraction (Pannetier *et al.*, 1987). It is difficult to make exact quantitative comparisons between the results of different workers because of the sensitivity of the results to the conditions under which the starting samples were made, but in a qualitative sense there is much consistency among the findings of various workers.

TEM observations of partially transformed samples (Samuel, de Lonckere, and Gerin, 1986; Samuel and Samuel, 1987; McAlister *et al.*, 1987) show that in alloys of approximately 14 at.% Mn, the orthorhombic Al_6Mn phase nucleates and grows at the interface between the icosahedral dendrites and the Al matrix. Since the orthorhombic phase is not known to deviate from its stoichiometric composition of 14.3 at.% Mn, and it forms from icosahedral material of 20–21 at.% Mn and aluminum with $<$ 5 at.% Mn, it is not surprising that Johnson–Mehl–Avrami analysis of thermal data (Bagley and Chen, 1985; Kelton and Wu, 1985; McAlister *et al.*, 1987) indicates kinetics limited by solute diffusion.

Whereas the orthorhombic phase is the normal reaction product in alloys containing 14 at.% or less Mn, in alloys containing about 20 at.% Mn one finds that the icosahedral phase transforms first to the decagonal phase and then, upon further heating, to the equilibrium λ phase (Kimura *et al.*, 1986; Murray *et al.*, 1987). The phases seen upon heating melt-spun samples of various compositions are summarized in Fig. 4, where one may

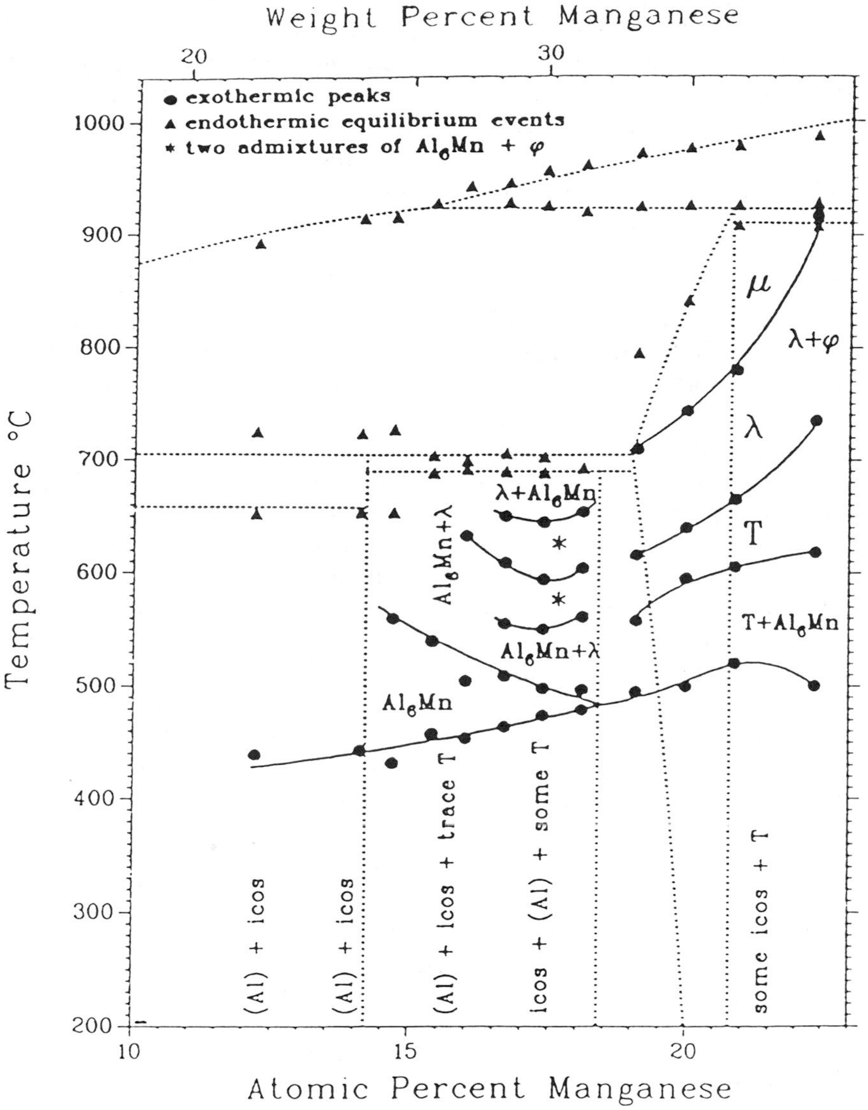

Figure 4. Phase transformations observed when melt-spun Al-Mn ribbons were heated in a differential thermal analysis instrument. The phases were identified by x-ray diffraction. From Murray et al. (1987).

also note that the temperature at which the icosahedral phase decomposes increases with increasing Mn content.

The sensitivity of the icosahedral phase decomposition temperature to the sample preparation technique was demonstrated by a comparison of splat-cooled and melt-spun samples (Luck *et al.*, 1986). The former are expected to have solidified more rapidly, and showed DSC curves that peaked at 713 K as compared to 735 K for the melt-spun samples.

Much more detailed studies of the sensitivity to preparation conditions have been carried out using melt-spun alloys with Mn concentrations near the lower limit for icosahedral phase formation (Yu-Zhang *et al.*, 1988b). Alloys in this composition range (6–10 at% Mn) typically contain a wide variety of microstructures (Yu-Zhang, 1987), apparently correlated to small variations in the thermal conditions within the sample as it solidified. Melt-spun ribbons can contain regions in which almost all of the Mn is in a highly supersaturated solid solution of fcc Al, other regions in which icosahedral phase dendrites in a matrix of Al are surrounded by a eutectic between the icosahedral phase and Al, and regions with icosahedral crystallites that appear to have formed by solid-state precipitation processes. The different regions can be expected to have different transformation kinetics, and the sensitivity of the transformation to processing conditions can be attributed to the different volume fractions of the various microstructures. Thus it is not necessary to conclude that icosahedral phase material formed at more rapid solidification rates has a more unstable atomic structure than that produced at lower rates, just because the more rapidly solidified samples transform at a lower temperature. A more simple explanation is that the more rapidly solidified samples have a microstructure that results in more rapid transformation kinetics.

6. Discussion

The nucleation, growth, and decomposition of quasicrystal phases as described above are at least qualitatively similar to the same processes in periodic crystal phases. They can thus be discussed without addressing the question of whether their structure is best described starting from the true quasicrystal (Penrose tiling) model of Levine and Steinhardt (1984) or the icosahedral glass model of Shechtman and Blech (1985) and Stephens and Goldman (1986). We have used the word "quasicrystal" as a generic term without intending to imply a specific structural model. Although there are many features of these phases that must eventually find their explanation in terms of their detailed structure (such as the TEM observation of a mottled contrast in Al–Mn and a radial growth pattern in Mg–Al–Zn), one can still discuss the phases simply as those having noncrystallographic point groups without reference to a specific structural model. The study of the metallurgical relationships between quasicrystals and similar periodic phases provides important clues about the atomic groups that occur within the quasicrystal structure but does not reveal much about the detailed arrangement of these clusters in the quasicrystal structure.

Relatively little attention has been devoted to the study of nucleation and growth of complex intermetallic phases: In most cases metallurgists are mainly interested in knowing how to prevent formation of such phases because of their extreme brittleness. The brittleness is a direct result of the difficulty of forming and moving dislocations in such structures, and quasicrystals are no exception to this rule. Indeed, one of the first indications of successful formation of a high-volume fraction of quasicrystals in a melt-spun ribbon is that the ribbon shatters into small pieces. The intense interest in the structure and properties of quasicrystals is not, however, based on their potential as structural materials but on their crystallographic structure, and it has thus provided an unusual opportunity for detailed metallurgical study of complex intermetallic phases.

Almost every quasicrystalline phase has now been correlated to a periodic phase of similar composition that contains icosahedrally-arranged groups of atoms. These groups are arranged in periodic arrays in the latter case but in aperiodic arrays in the quasicrystals. At least two types of large icosahedral groups appear capable of forming quasicrystals; the prototypes for these groups are found in α Al–Mn–Si and $Mg_{32}(Al, Zn)_{49}$. Additional atoms are present that may be described as links between adjoining icosahedral groups (Audier and Guyot, 1986) or as components of additional icosahedral shells present on some but not all of the basic groups (Fowler, Moser, and Sims, 1988). There is no clear evidence, however, that icosahedral phases can be formed from smallcr icosahedral groups such as the 13-atom motif of the $Al_{12}Mn$ phase.

One might hypothesize that the quasicrystal phases represent an imperfect packing of the periodic crystals, resulting from a rapid solidification process that allows insufficient time for the large atom groups to move into the proper crystallographic sites. This concept probably overemphasizes the physical role of the icosahedral groups, which are very useful conceptual entities for describing the structure but probably should not be regarded as molecular units that form in the liquid and become attached to the growing crystal. Instead, they appear to represent an energetically favorable structural motif that can occur in either of two distinct thermodynamic phases. This view is best illustrated by the case of the Al–Li–Cu quasicrystalline *T*2 and periodic *R* phases, which have slightly different compositions and can exist in equilibrium with each other. Such an equilibrium would be difficult to account for if the quasicrystal represented an imperfect packing of a periodic phase. The metastability of most quasicrystal phases does not imply that they represent an imperfect construction of the related crystal phase, it only indicates that their free energies are somewhat larger than those of other phases or combinations of phases.

The orientation relationships that have been identified between the quasicrystalline and periodic phases can in many cases be interpreted in terms of the parallel alignment of the same icosahedral motif in the two structures. This concept is particularly favored by the observation of several orientational variants of the lower symmetry phases, all having the same orientation of the motif, when they are nucleated by an icosahedral phase.

A particularly interesting phase relationship exists between the icosahedral and decagonal phases in the Al–Mn system. In this case it appears that the decagonal phase has icosahedral groups in two orientations, one of which is parallel to the groups in the icosahedral phase by which it is nucleated. Thus, in this case, it appears that a similar icosahedral group can exist with one orientation and aperiodic packing in the icosahedral phase, with two orientations and partly periodic packing in the decagonal phase (if silicon is absent), and with one orientation and periodic packing in the α phase (if about 6 at.% Si is present).

Although it is clear that the metastable quasicrystals can prevail in rapidly solidified melts because of their relatively high rates of nucleation, a quantitative explanation of their nucleation behavior has not been attempted. Indeed, it is difficult to predict nucleation rates even for simple crystal structures. Recent observations of Al–Fe–Si alloys (Bendersky, Biancaniello, and Schaefer, 1987) suggested that nucleation processes in systems with icosahedral atomic groups can be particularly complex. The microstructure of this alloy was interpreted to result from the nucleation and growth of a phase that contained icosahedral structural units similar to those in the crystalline α Al–Fe–Si alloy. In this case, however, the orientation relationship between the atomic groups was not preserved, so that the structure produced was neither a periodic crystal nor a quasicrystal, but an amorphous phase. If this interpretation is correct, it suggests that further study of quasicrystal formation may reveal still more types of processes by which solid phases can be formed with atomic structures outside the framework of classical crystallography.

References

Adam, J., and Rich, J. B. (1954). The crystal structure of WAl_{12}, $MoAl_{12}$, and $(Mn,Cr)Al_{12}$. *Acta Cryst.* 7, 813–816.

Audier, M., and Guyot, P. (1986). Al_4Mn quasicrystal atomic structure, diffraction data and Penrose tiling. *Phil. Mag. B* **53,** L43–L51.

Audier, M., Sainfort, P., and Dubost, B. (1986). A simple construction of the AlCuLi quasicrystalline structure related to the $(Al,Zn)_{49}Mg_{32}$ cubic structure type. *Phil. Mag. B* **54,** L105–L111.

Bagley, B. G., and Chen, H. S. (1985). Transformation kinetics for crystalline, quasicrystalline and glassy alloys in the binary $Al_{100-x}Mn_x$ and pseudobinary $Al_{84-y}Si_yMn_{16}$ systems. *Mat. Res. Soc. Symp. Proc.* **57** (Slade Cargill III, G., Spaepen, F., and Tu, K.-N., eds.). MRS, Pittsburgh, PA, 57, 451–464.

Ball, M. D., and Lloyd, D. J. (1985). Particles apparently exhibiting five-fold symmetry in Al–Li–Cu–Mg alloy. *Scr. Met.* **19,** 1065–1068.

Bancel, P. A., and Heiney, P. A. (1986). Icosahedral aluminum-transition metal alloys. *Phys. Rev. B* **33,** 7917–1922.

Bendersky, L. (1985). Quasicrystal with one-dimensional translational symmetry and a ten-fold rotation axis. *Phys. Rev. Lett.* **55,** 1461–1463.

Bendersky, L. A., (1986). Decagonal phase. *J. Phy. Coll.* C3, **47,** 457–464.

Bendersky, L. A., and Biancaniello, F. S. (1987). TEM observation of icosahedral, new crystalline, and glassy phases in rapidly quenched Cd–Cu alloys. *Scr. Met.* **21,** 531–536.

Bendersky, L. A., and Kaufman, M. J. (1986). Determination of the point group of the icosahedral phase in an Al–Mn–Si alloy using convergent beam electron diffraction. *Phil. Mag. B* **53,** L75–L80.

Bendersky, L. A., and Ridder, S. D. (1986). Nucleation behavior of Al–Mn icosahedral phase. *J. Mater. Res.* **1,** 405–414.

Bendersky, L., Schaefer, R. J., Biancaniello, F. S., Boettinger, W. J., Kaufman, M. J., and Shechtman, D. (1985). Icosahedral Al–Mn and related phases: resemblance in structure. *Scr. Met.* **19,** 909–914.

Bendersky, L. A., Biancaniello, F. S., and Schaefer, R. J., (1987). Amorphous phase formation in $Al_{70}Si_{17}Fe_{13}$ alloy. *J. Mater. Res.* **2,** 427–430.

Bennett, L. H., Cahn, J. W., Schaefer, R. J., Rubinstein, M., and Stauss, G. H. (1987). Icosahedral symmetry vs. local icosahedral environment in Al–Mn alloys from NMR *Nature* **326,** 372–373.

Bergmann, G., Waugh, J. L. T., and Pauling, L. (1957). The crystal structure of the metallic phase $Mg_{32}(Al,Zn)_{49}$. *Acta Cryst.* **10,** 254–257.

Blodgett, J. A., Schaefer, R. J., and Glicksman, M. E. (1974). A holographic system for crystal growth studies: design and applications. *Metallography* **7,** 453–504.

Boettinger, W. J. (1982). Growth kinetic limitations during rapid solidification. In *Rapidly Solidified Amorphous and Crystalline Alloys* (Kear, B. H., Giessen, B. C., and Cohen, M., eds.), 15–31. Elsevier, New York.

Brown, P. J. (1957). The structure of α(V–Al). *Acta Cryst.* **10,** 133–135.

Budai, J. D., and Aziz, M. J. (1986). Formation of icosahedral Al–Mn by ion implantation into oriented crystalline films. *Phys. Rev. B* **33,** 2876–2878.

Budai, J. D., Tischler, J. Z., Mabenschluss, A., Ice, G. E., and Elser, V. (1987). X-ray diffraction study of phason strain field in oriented icosahedral Al–Mn. *Phys. Rev. Lett.* **58,** 2304–2307.

Chen, C. H., and Chen, H. S. (1986). "Superlattices" in quenched Al–Si–Mn quasicrystals. *Phys. Rev. B* **33,** 2814–2816.

Chen, H. S., Chen, C. H., Inoue, A., and Krause, J. T. (1985). Density, Young's modulus, specific heat and stability of icosahedral $Al_{86}Mn_{14}$. *Phys. Rev. B* **32,** 1940–1944.

Chen, H. S., Koskenmaki, D., and Chen, C. H. (1987a). Formation and structure of icosahedral and glassy phases in melt-spun Al–Si–Mn. *Phys. Rev. B* **35,** 3715–3721.

Chen, H. S., Phillips, J. C., Villars, P., Kortan, A. R., and Inoue, A. (1987b). New quasicrystals of alloys containing *s, p* and *d* elements. *Phys. Rev. B* **35,** 9326–9329.

Cooper, M., and Robinson, K. (1966). The crystal structure of the ternary alloy alpha (AlMnSi). *Acta Cryst.* **20,** 614 –617.

Corby, R. N., and Black, P. J. (1977). The structure of alpha-(AlFeSi) by anomalous dispersion methods. *Acta Cryst. B* **33,** 3468–3475.

Dong, C., Hei, Z. K., Wang, L. B., Song, Q. H., Wu, Y. K., and Kuo, K. H. (1986). A new icosahedral quasicrystal in rapidly solidified $FeTi_2$. *Scr. Met.* **20,** 1155–1158.

Dong, C., Chattopadhyay, K., and Kuo, K. H. (1987). Quasicrystalline eutectic growth and metastable phase orientation relations in rapidly solidified Fe–Ti alloys. *Scr. Met.* **21,** 1307–1312.

Drehman, A. J., Pelton, A. R., and Noack, J. (1986). Nucleation and growth of quasicrystalline Pd–U–Si from the glassy state. *J. Mater. Res.* **1,** 741–745.

Dubois, J. -M., Janot, C., Pannetier, J., and Pianelli, A. (1986). Diffraction approach to the structure of decagonal quasicrystals. *Phys. Lett. A* **117,** 421–427.

Dubost, B., Lang, J. -M., Tanaka, M., Sainfort, P., and Audier, M. (1986). Large AlCuLi single quasicrystals with triacontahedral solidification morphology. *Nature* **324,** 48–50.

Dunlap, R. A., and Dini, K. (1985). Formation, structure and crystallization of metastable quasicrystalline Al-transition metal alloys prepared by rapid solidification. *Can. J. Phys.* **63,** 1267–1269.

Dunlap, R. A., and Dini, K. (1986). Amorphization of rapidly quenched quasicrystalline Al-transition metal alloys by the addition of Si. *J. Mater. Res.* **1,** 415–419.

Eady, J. A., Hogan, L. M., Davies, P. G. (1975). *J. Aust. Inst. Met.* **20,** 23–32.

Elser, V., and Henley, C. L. (1985). Crystal and quasi-crystal structures in Al–Mn–Si alloys. *Phys. Rev. Lett.* **55,** 2883–2886.

Follstaedt, D. M., and Knapp, J. A. (1986). Metastable Al–Mn phases formed by ion beam mixing. *J. Appl. Phys.* **59,** 1756–1758.

Fowler, H. E., Moser, B., and Sims, J. (1988). Triple-shell symmetry in α(Al,Si)–Mn. *Phys. Rev. B* **37,** 3906–3913.

Fung, K. K., and Zhou, Y. Q. (1986). Direct observation of the transformation of the icosahedral phase in $(Al_6Mn)_{1-x}Si_x$ into α(AlMnSi). *Phil. Mag. B* **54,** L27–L31.

Fung, K. K., Zou, X. D., and Yang, C. Y. (1987). Transmission electron microscopy study of $Al_{13}Fe_4$ ten-fold twins in rapidly cooled Al–Fe alloys. *Phil. Mag. Lett.* **55,** 27–32.

Garçon, S., Sainfort, P., Regazzoni, G., and Dubois, J. M. (1987). Formation of an icosahedral phase by crystallization of amorphous Al–Cu–V alloys. *Scr. Met.* **21,** 1493–1498.

Gayle, F. W. (1987). Free-surface solidification habit and point group symmetry of a faceted, icosahedral Al–Li–Cu phase. *J. Mater. Res.* **2,** 1–4.

Guyot, P. and Audier, M. (1985). A quasicrystal structure model for Al–Mn. *Phil. Mag. B* **52.** L15–L19.

Guyot, P., Audier, M., and Lequette, R. (1986). Quasi-crystal and crystal in AlMn and AlMnSi: model structure of the icosahedral phase. *J. Phys. Coll.,* C3, **47,** 389–404.

Hardy, H. K., and Silcock, J. M. (1955/56). The phase sections at 500°C and 350°C of aluminum-rich Al–Cu–Li alloys. *J. Inst. Met.* **24,** 423–428.

Hughes, I. R., and Jones, H. (1976). Coupled eutectic growth in Al–Fe alloys. *J. Mater. Sci.* **11,** 1781–1793.

Inoue, A., Arnberg, L., Lehtinen, B., Oguchi, M., and Masumoto, M. (1986). Compositional analysis of the icosahedral phase in rapidly quenched Al–Mn and Al–V alloys. *Met. Trans. A* **17,** 1657–1664.

Inoue, A., Bizen, Y., and Masumoto, T. (1988). Quasicrystalline phase in Al–Si–Mn system prepared by annealing of amorphous phase. *Met.Trans. A* **19,** 383–385.

Juarez-Islas, J. A. and Jones, H. (1987). Conditions for growth of extended Al-rich alloy solid solutions and Al–Al_6Mn eutectic during rapid solidification. *Acta Metall.* **35,** 499–507.

Kelton, K. F., and Wu, T. W. (1985). Density measurements, calorimetry, and transmission electron microscopy of icosahedral $Mn_{14}Al_{86}$. *Appl. Phys. Lett.* **46,** 1059–1060.

Kimura, K., Hashimoto, T., Suzuki, K., Nagayma, K., Ino, H., and Takeuchi, S. (1985). Stoichiometry of quasicrystalline Al–Mn. *J. Phys. Soc. Japan* **54,** 3217–3220.

Kimura, K., Hashimoto, T., Suzuki, K., Nagayma, K., Ino, H., and Takeuchi, S. (1986). Structure and stability of quasicrystalline Al–Mn alloys. *J. Phys. Soc. Japan* **55,** 534–543.

Knapp, J. H., and Follstaedt, D. M. (1987). Measurement of melting temperatures of quasicrystalline Al–Mn phases. *Phys. Rev. Lett.* **58,** 2454–2457.

Koskenmaki, D. C., Chen, H. S., and Rao, K. U. (1986a). Observation of mirror–related grains in icosahedral $Al_{86}Mn_7Fe_7$ and $Al_{86}Mn_{14}$. *Scr. Met.* **20,** 1631–1634.

Koskenmaki, D. C., Chen, H. S., and Rao, K. U. (1986b). Coherent orientation relationship between icosahedral and cubic α-phase in melt-spun Al–Si–Mn. *Phys. Rev. B* **33,** 5328–5332.

Krishnan, K. M., Gronsky, R., and Tanner, L. E. (1986). Determination of the composition of the icosahedral phase in rapidly solidified Al–Mn quasicrystals at high spatial resolution. *Scr. Met.* **20,** 239–242.

Kumar, V., Sahoo, D., and Athithan, G. (1986). Characterization and decoration of the two-dimensional Penrose lattice. Preprint.

Kuo, K. H., Dong, C., Zhou, D. S., Guo, Y. X., Hei, Z. K., and Li, D. X. (1986). A Friauf–Laues (Frank–Kasper) phase-related quasicrystal in rapidly solidified Mn_3Ni_2Si alloy. *Scr. Met.* **20,** 1695–1698.

Kuo, K. H., Zhou, D. S., and Li, D. X. (1987). Quasicrystalline and Frank–Kasper phases in a rapidly solidified $V_{41}Ni_{36}Si_{23}$ alloy. *Phil. Mag. Lett.* **55,** 33–39.

Levine, D., and Steinhardt, P. J. (1984). Quasicrystals: a new class of ordered structures. *Phys. Rev. Lett.* **53,** 2477–2480.

Lilienfeld, D. A., Nastasi, M., Johnson, H. H., Ast, D. G., and Mayer, J. W. (1985). Amorphous to quasicrystalline transformation in the solid state. *Phys. Rev. Lett.* **55,** 1587–1590.

Loiseau, A., and Lapasset, G. (1986). Crystalline and quasicrystalline structures in Al–Li–Cu–Mg alloy. *J. Phys. Coll. C3,* **47,** 331–340.

Loiseau, A., and Lapaset, G. (1987). Relations between quasicrystals and crystalline phases in Al–Li–Cu–Mg alloys: a new class of approximent structures. *Phil. Mag. Lett.* **56,** 165–171.

Louis, E., Mora, R., and Pastor, J. (1980). Nature of star-shaped clusters of $FeAl_3$ in aluminum–iron alloys. *Metal Science,* Dec. 1980, 591–593.

Luck, R., Haas, H., Sommer, F., and Predel, B. (1986). Relaxation and phase transformation of quasicrystalline Al–14% Mn investigated by differential scanning calorimetry. *Scr. Met.* **20,** 677–679.

McAlister, A. J., Bendersky, L. A., Schaefer, R. J., and Biancaniello, F. S. (1987). Crystallization of the icosahedral phase in rapidly quenched Al-rich Al–Mn alloys. *Scr. Met.* **21,** 103–106.

Mackay, A. L. (1962). A dense non-crystallographic packing of equal spheres. *Acta Cryst.* **15,** 916–918.

Mueller, B. A., Schaefer, R. J., and Perepezko, J. H. (1987). The solidification of Al–Mn powder. *J. Mater. Res.* **2,** 809–817.

Mukhopadhyay, N. K., Subbanna, G. N., Ranganathan, S., and Chattopadhyay, K. (1986). An electron microscopic study of quasicrystals in a quaternary alloy: $Mg_{32}(Al,Zn,Cu)_{49}$. *Scr. Met.* **20,** 525–528.

Murray, J. L., McAlister, A. J., Schaefer, R. J., Bendersky, L. A., Biancaniello, F. S., and Moffat, D. L. (1987). Stable and metastable phase equilibria in the Al–Mn system. *Met. Trans. A* **18,** 385–390.

Ohashi, T., Dai, L., Fukatsu, N., and Miwa, K. (1986). Precipitation of quasicrystalline phase in rapidly solidified Al–Mn–Zr alloys. *Scr. Met.* **20,** 1241–1244.

Ohashi, W., and Spaepen, F. (1987). Stable Ga–Mg–Zn quasi-periodic crystals with pentagonal dodecahedral solidification morphology. *Nature* **330,** 555–556.

Pannetier, J., Dubois, J. M., Janot, C., and Bilde, A. (1987). Thermal transformation of icosahedral quasiperiodic crystals of the Al–Mn system. *Phil. Mag. B* **55,** 435–457.

Perepezko, J. H., and Boettinger, W. J. (1983). Use of metastable phase diagrams in rapid solidification. In *Mat. Res. Soc. Symp. Proc.* **19** (Bennett, L. H., Massalski, T. B., and Giessen, B. C., eds.), 223–240.

Phillips, H. W. L. (1959). *Annotated Equilibrium Phase Diagrams of Some Al Alloy Systems.* The Institute of Metals, London.

Poon, S. J., Drehman, A. J., and Lawless, K. R. (1985). Amorphous to icosahedral phase transformation in Pd–U–Si alloys. *Phys. Rev. Lett.* **55,** 2324–2327.

Quan, M. X., and Giessen, B. C. (1986). Formation and thermal stability of quasicrystalline and other metastable Al-rich Al–Mn phases. In *Rapidly Solidified Alloys and Their Mechanical and Magnetic Properties* (Giessen, B., C., Polk, D. E., and Taub, A. I., eds.), 241–248. MRS, Pittsburgh, PA.

Ramachandrarao, P., and Sastry, G. V. S. (1985). A basis for the synthesis of quasicrystals. *Pramana* **24,** L225–L230.

Reddy, G. S., Sekhar, J. A., and Rao, P. V. (1987). The effect of moderate cooling rates on the formation temperatures of icosahedral and *T* phases in the Al–14 at.% Mn alloy. *Scripta Met.* **21,** 13–18.

Richmond, J. J., Perepezko, J. H., LeBeau, S. E., and Cooper, K. P. (1983). Solidification microstructures of undercooled droplets. In *Rapid Solidification Processing: Principles and Technologies, III* (Mehrabian, R., ed.), 90–95. NBS, Washington, DC.

Robertson, J. L., Misenheimer, M. E., Moss, S. C., and Bendersky, L. (1986). X-ray and electron metallographic study of quasicrystalline Al–Mn–Si alloys. *Acta Met.* **34,** 2177–2189.

Sainfort, P., and Dubost, B. (1986). The *T*2 compound: a stable quasi-crystal in the system Al–Li–Cu–(Mg)? *J. Phys. Coll.* C3, **47,** 321–330.

Samuel, A. M., de Lonckere, A., and Gerin, F. (1986). On the microstructure of rapidly solidified Al–15 wt% Mn ribbons: effect of annealing in the temperature range 300 to 600°C. *Met. Trans. A* 17, 1671–1683.

Samuel, F. H., and Samuel, A. M. (1987). Crystallization of the icosahedral structure in Al–15% Mn melt-spun ribbons on continuous annealing: heating rate 10 K/min and 2 K/min. *Scr. Met.* **21,** 411–414.

Sastry, G.V.S., Suryanarayana, C., Van Sande, M., and Van Tendeloo, G. (1978). A new ordered phase in the Al–Pd system. *Mat. Res. Bull.* **13,** 1065–1070.

Sastry, G. V. S., Ramachandrarao, P., and Anantharaman, T. R. (1986). A new quasi-crystalline phase in rapidly solidified Mg_4CuAl_6. *Scr. Met.* **20,** 191–193.

Schaefer, R. J., and Bendersky, L. A. (1986). Replacement of icosahedral Al–Mn by decagonal phase. *Scr. Met.* **20,** 745–750.

Schaefer, R. J., Bendersky, L. A., Shechtman, D., Boettinger, W. J., and Biancaniello, F. S. (1986a). Icosahedral and decagonal phase formation in Al–Mn alloys. *Met. Trans. A* **17,** 2117–2125.

Schaefer, R. J., Biancaniello, F. S., and Cahn, J. W. (1986b). Formation and stability range of the *G* phase in the Al–Mn system. *Scr. Met.* **20,** 1439–1444.

Shechtman, D., and Blech, I. (1985). The microstructure of rapidly solidified Al_6Mn. *Met. Trans. A* **16,** 1005–1012.

Shechtman, D., Blech, I., Gratias, D., and Cahn, J. W. (1984). Metallic phase with long-range orientational order and no translational symmetry. *Phys. Rev. Lett.* **53,** 1951–1953.

Shen, Y., Poon, S. J., and Shiflet, G. J. (1986). Crystallization of icosahedral phase from glassy Pd–U–Si alloys. *Phys. Rev. B.* **34,** 3516–3519.

Steeds, J. W., Ayer, R., Lin, Y. P., and Vincent, R. (1986). Electron diffraction and microscopy of incommensurate phases and quasi-crystals. *J. Phys. Coll.* C3, **47,** 437–446.

Stephens, P. W., and Goldman, A. I. (1986). Sharp diffraction maxima from an icosahedral glass. *Phys. Rev. Lett.* **56,** 1168–1171.

Thangaraj, N., Chattopadhyay, K., Ranganathan, S., Small, C., and Davies, H. A. (1986). Quasicrystals and ordered twins in rapidly solidified Al–Pd alloys. *Proc. XIth Inter. Cong. on Electron Microscopy* (Imura, I., *et al.*, eds.) *Jap. Soc. Elec. Microsc., Tokyo,* Vol. 1, 1525–1526.

Tsai, A.-P., Inoue, A., and Masumoto, T. (1987). A stable quasicrystal in Al–Cu–Fe system. Japan. *J. Appl. Phys.* **26,** L1505–L1507.

Urban, K., Mayer, M., Rapp, M., Wilkens, M., Csanady, A., and Fidler, J. (1986). Studies of aperiodic crystals in Al–Mn and Al–V alloys by means of transmission electron microscopy. *J. Phys. Coll. C3,* **47,** 465–476.

Van Tendeloo, G., Van Landuy, J., Amelinckx, S., and Ranganathan, S. (1987). Quasi-crystals and their crystalline homologues in the $Al_{60}Mn_{11}Ni_{21}$ ternary alloy. *J. of Microscopy-Oxford,* **149,** 1–19.

Villars, P., Phillips, J. C., and Chen, H. S. (1986). Icosahedral quasicrystals and quantum structural diagrams. *Phys. Rev. Lett.* **57,** 3085–3088.

Yamane, A., Kimura, K., Shibuya, T., and Takeuchi, S. (1987). Production of Al-based ternary and quaternary icosahedral phases. *Technical Report of ISSP,* Ser. A, No. 1848. The University of Tokyo.

Yu-Zhang, K. (1987). Icosahedral phase formation in low Mn-content A1–Mn alloys. To be published in *Proc. of Inter. Workshop on Quasicrystals,* Beijing.

Yu-Zhang, K., Bigot, J., Martin, G., Portier, R., and Gratias, D. (1986). Microstructures in rapidly quenched (A1-Mn) alloys. *Proc. XIth Cong. on Electron Microscopy, Kyoto,* 167–168.

Yu-Zhang, K., Bigot, J., Chevalier, J.-P., Gratias, D., Martin, G., and Portier, R. (1988a). Dodecahedral shaped quasicrystalline precipitates in dilute Al–Mn solid solutions. *Phil. Mag.* (to appear).

Yu-Zhang, K., Harmelin, M., Quivy, A., Calvayrac, Y., Bigot, J., and Portier, R. (1988b). Thermal stability of the icosahedral phase in A1–Mn alloys prepared by planar flow casting (Mn = 4–15 at%). To be published in *J. Mat. Sci. Eng.*

Zhang, Z., and Kuo, K. H. (1986). Orientation relationship between the icosahedral and crystalline phases in $(Ti_{1-x}V_x)_2Ni$ alloys. *Phil. Mag. B* **54,** L83–L87.

Zhang, Z., and Kuo, K. H. (1987). Local translational order in the $NiTi_2$ icosahedral quasicrystal. *J. of Microscopy* **146,** 313–321.

Zhang, Z., Ye, H. Q., and Kuo, K. H. (1985). A new icosahedral phase with the *m*35 symmetry. *Phil. Mag. A* **52,** L49–L52.

Zou, X. D., Fung, K. K., and Kuo, K. H. (1987). Orientation relationship of decagonal quasicrystal and ten-fold twins in rapidly cooled A1–Fe alloys. *Phys. Rev. B* **35,** 4526–4528.

Chapter 4

Quasicrystallography

PER BAK AND ALAN I. GOLDMAN

Department of Physics
Brookhaven National Laboratory
Upton, New York

Contents

1. Introduction

The discovery of the icosahedral phase (IP) alloys[1] has stimulated a flurry of both experimental and theoretical activity. Despite great efforts over the last two years, an unambiguous structural determination for any of the IP systems has yet to be made. In fact, there is still unresolved disagreement concerning the basic *structural character* of these novel materials; are they best described as conventional 3D crystals in disguise,[2] icosahedral crystals[3–6] (to be defined below), "liquid crystals with orientational order",[7] or intrinsically disordered networks of icosahedral packing units?[8,9] We wish to concentrate on the crystallography of ideal icosahedral crystals, and leave the discussion of the applicability to specific alloys to others. We also note that there are several planned review ar-

APERIODICITY AND ORDER
Introduction to
Quasicrystals

ISBN 0-12-040601-2

ticles,[10–12] as well as a conference proceeding[13] and reprint collection[14] which are or will be available to the interested readers.

We define an icosahedral crystal simply as an arrangement of atoms which, in a diffraction experiment, produce infinitely sharp (δ-function) Bragg peaks in a pattern which exhibits overall icosahedral symmetry. We shall not engage in detailed atomic modelling, but rather discuss general arrangements and their symmetries. We shall see how the concept of space groups can be generalized to icosahedral crystals, allowing a systematic crystallographic description. Regardless of whether the real IP alloys may properly be called icosahedral crystals, we hope that this chapter provides an adequately detailed introduction to the art of *quasicrystallography*.

No three-dimensional periodic lattice can have icosahedral symmetry or any other point group symmetry which include fivefold or sevenfold (and higher) rotations.[15] Much of the original excitement surrounding the IP alloys may be attributed to the common (but erroneous) belief that only crystals with a regular (periodic) lattice can exhibit sharp peaks in their diffraction pattern, and consequently it should be possible to label all crystals by one of the 230 three-dimensional space group symmetries.

There is a trivial way of obtaining diffraction patterns with icosahedral symmetry, such as that shown in Fig. 1 for a rapidly quenched Al-Mn alloy. Pauling[2] has suggested that these diffraction patterns arise from twinned cubic crystals with several hundred atoms in the unit cell. The dimensions of the cubic cell vary from description to description, a point that emphasizes the danger of structural determinations from only powder diffraction data—*any set of lines in a diffraction pattern can be indexed using a sufficiently large cubic unit cell.* It is instructive, however, to consider the minimum-size cubic cell required for consistency with the scale of the experimental patterns. For any periodic structure, the d-spacings of any two collinear spots must be in an integer ratio. Of course, the symmetries of various diffraction spots are not evident in a powder pattern, they must be deduced from single-grain electron or X-ray diffraction measurements. The two lowest-order diffraction spots observed[16] along a twofold direction in IP Al–Mn have d-spacings of 8.58(7)Å and 5.42(3)Å, respectively, and their ratio is 1.633. The lowest rational approximate, within experimental error, to this number is $\frac{13}{8}$ (which is also one of the lowest rational approximates to the golden mean, $\tau = \frac{\sqrt{5}+1}{2}$). This then implies a minimum edge length for the cubic cell of over 70Å, significantly larger than those proposed by any proponents of the twinning arguments. Such a large cell would contain well over 10,000 atoms! Con-

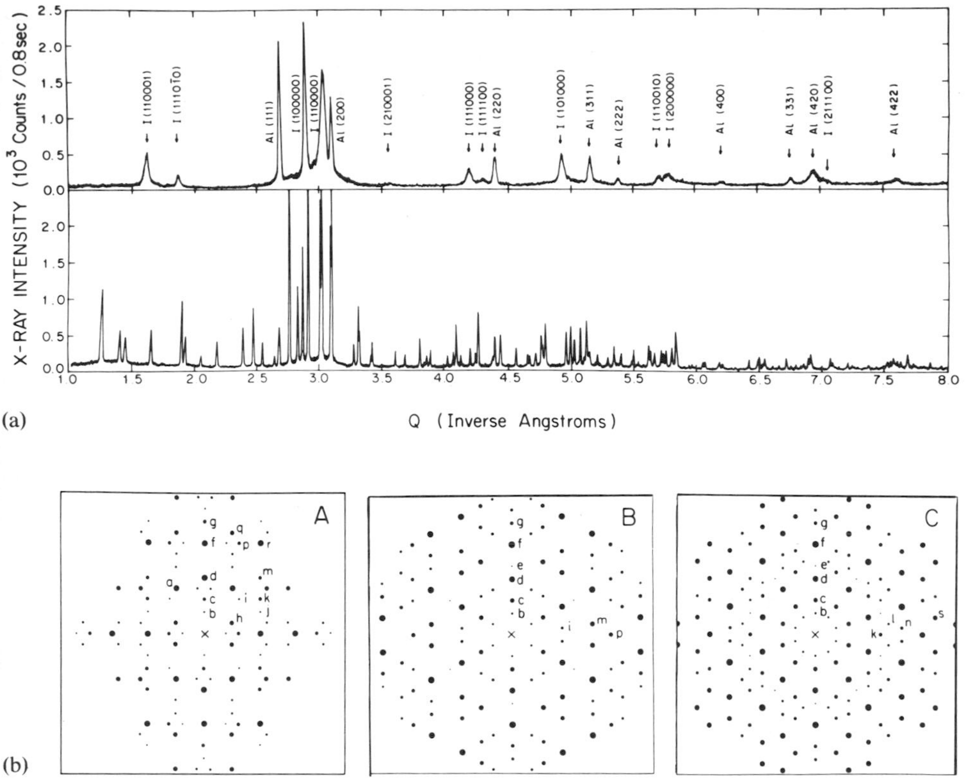

(a)

(b)

Figure 1. (a) Top Panel: High-resolution powder diffraction pattern from a rapidly quenched IP Al–Mn alloy (after Ref. 22). The IP peaks are labeled by six Miller indices (and letters which correspond to the spots in Fig. 1b). The indexing scheme corresponds to a different choice of vertex vectors as the basis set from that of Fig. 3. FCC aluminum impurity peaks are also labeled. Bottom Panel: Diffraction pattern from the Al-Mn alloy after annealing. All peaks can be indexed to the orthorhombic Al_6Mn structure. (b) Schematic of the electron diffraction pattern from the Al-Mn IP alloy in a two-fold, three-fold and five-fold plane.

verging beam electron diffraction measurements,[1] with beams focussed to less than 70Å, reveal no evidence of twinning on this scale. Along with this observation, there now seems to be abundant experimental evidence[17] which contradicts this traditional view.

It is now clear that we should not be so restrictive in our requirement of *periodic* translational order for the occurrence of perfectly sharp diffraction peaks. Incommensurate crystals such as α-quartz or $NaNO_3$ are examples of structures which yield perfectly sharp diffraction spots, but which nevertheless cannot be described by the periodic stacking of identical unit cells. In fact, we shall see that the crystallography of icosahedral crystals has much in common with the crystallography of incommensurate structures[18,19]: in both cases, the proper crystallographic description involves regular lattices in some higher dimension.

Icosahedral crystals can be naturally described in terms of regular crystals in six dimensions.[5] For instance, the space group symmetry of all of the known IP alloys is consistent with P352 or P$\bar{3}\bar{5}$2, where 3, 5, and 2 refer to the threefold, fivefold, and twofold rotational symmetries, respectively. As we shall point out, fivefold symmetry is compatible with a regular lattice in six dimensions. The real crystal is defined by the intersection of the six-dimensional crystal with an imbedded three-dimensional hyperplane. As we shall discuss below, the icosahedral symmetry of the real crystal does place constraints on the orientation of the real world plane. We also point out that the *cut-and projection* technique, commonly employed to describe the IP alloys, place severe constraints on the description of the atomic basis in six dimensions.

One can make certain restrictive assumptions which allow us to describe the icosahedral crystal in terms of stacking structural units, or tilings, in three dimensions. Penrose[20] has demonstrated that it is possible to construct a two-dimensional crystal with fivefold rotational symmetry from a set of two rhombic tiles. An example of such a 2D Penrose tiling is shown in Fig. 2. This tiling can be generalized[20] to three dimensions by using rhombohedral bricks to build a structure with icosahedral symmetry. However, the requirement of icosahedral symmetry *does not* impose a tiling structure, although these particular tilings must be included as a special case: they correspond to choosing a particular atomic motif in the 6D unit cell. This does not mean that such tiling schemes are not a useful starting point for a more rigorous structural determination,[21] but in principle, one must perform a complete six-dimensional crystal analysis in order to determine the three-dimensional atomic arrangement.

We wish to emphasize that thinking of the icosahedral crystal as a periodic structure in six dimensions is not merely an amusing mathematical abstraction. Consider, for example, the indexing scheme required to characterize the diffraction patterns of the IP alloys. The diffraction patterns from crystal structures based on any of the 3D space groups may be indexed to a reciprocal lattice with three basic vectors, $\mathbf{q}_i$, which are either orthogonal, or are rational combinations of three orthogonal vectors. Dif-

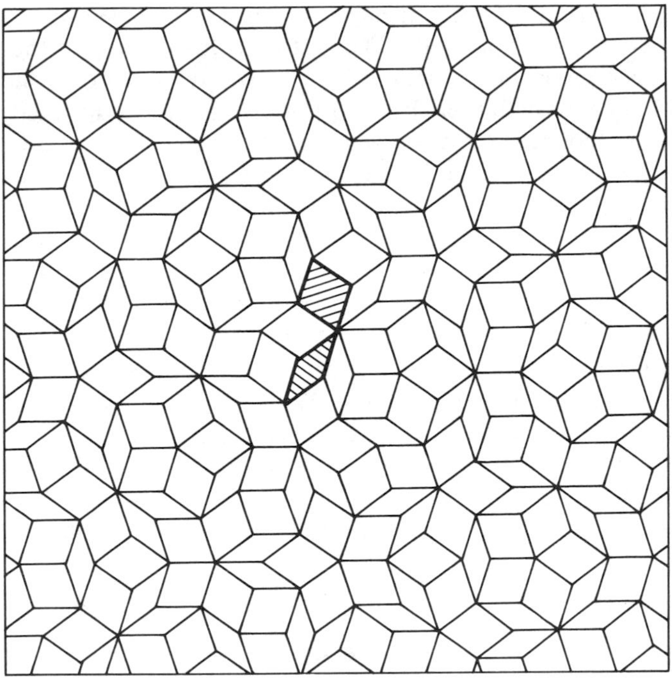

Figure 2. A two-dimensional Penrose tiling of the plane generated by the appropriate stacking of "fat" and "skinny" rhombuses (shaded).

fraction peaks may be labelled by an integer combination of these reciprocal lattice basis vectors.:

$$\mathbf{Q} = \sum_{i=1}^{3} n_i \mathbf{q}_i$$

In the course of analyzing diffraction data from the IP Al–Mn alloy,[22] it was found that all X-ray and electron diffraction peaks, shown in Fig. 1, could be successfully indexed (with integer n_i), by using six independent vectors pointing to the vertices of an icosahedron as a basis set (see Fig. 3):

$$\mathbf{Q} = \frac{|\mathbf{Q}_0^*|}{\sqrt{1 + \tau^2}} \sum_{i=1}^{6} n_i \mathbf{q}_i;$$

$$\mathbf{q}_1 = (1,\tau,0);\ \mathbf{q}_2 = (\tau,0,1);\ \mathbf{q}_3 = (\tau,0,-1); \qquad (1.1)$$

$$\mathbf{q}_4 = (0,1,-\tau);\ \mathbf{q}_5 = (-1,\tau,0);\ \mathbf{q}_6 = (0,1,\tau);$$

The $\mathbf{q}_i$ are then our reciprocal lattice basis vectors. The choice of the magnitude of the fundamental wavevector corresponding to the (100000) reflection is somewhat ambiguous due to the invariance of the pattern under inflations or deflations by τ^3. In addition, one could easily choose the set of six independent basis vectors as combinations of other vertex vectors.[23] Finally, one could choose to label the peaks using only the three orthogonal vectors ($\mathbf{q}_x$,$\mathbf{q}_y$,$\mathbf{q}_z$), also shown in Fig. 3. However, this choice would result in irrational values for the n_i. Therefore, we see that the six-dimensional character of an icosahedral crystal arises naturally from the symmetries evident in the diffraction patterns.

Our review is organized as follows. For pedagogical reasons, we shall first review the theory of crystallography for 1D incommensurate crystals. The theory was first developed by deWolff,[18] and Janner and Janssen.[19] We shall introduce a slightly more transparent version than the one originally put forward. Our discussion serves to familiarize the reader with the idea that crystals can be perfectly well ordered even if the structure is not periodic, and to demonstrate the usefulness of working in a higher-

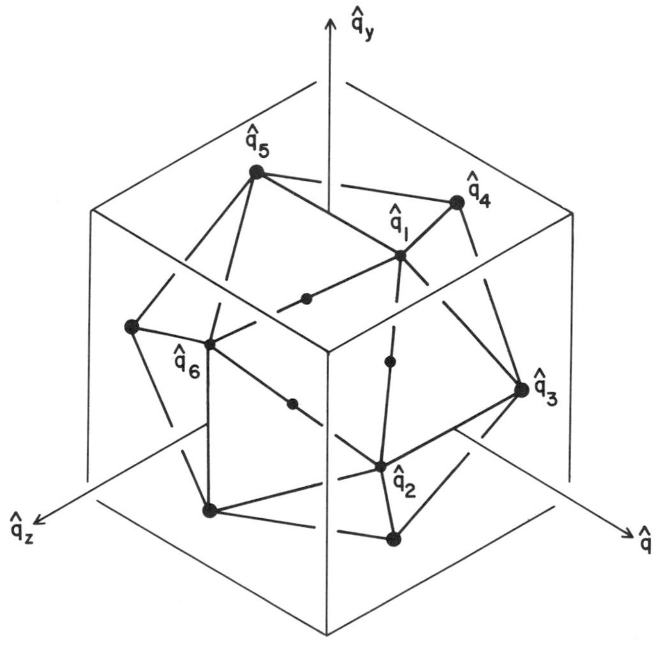

Figure 3. Fundamental vertex reciprocal lattice vectors for an icosahedron as given in Eq. (1.1).

dimensional space. For incommensurate structures in d dimensions with one modulation wavevector, the description is in $d+1$ dimensions. The theory has been useful not only in explaining the diffraction patterns from incommensurate systems, but also in discussing elastic properties, excitations, and phase transitions in a variety of systems such as $NaNO_3$,[24] the mercury chain compounds $Hg_{3-\delta}AsF_6$,[25] and quasi one-dimensional conductors such as TTF–TCNQ.[26] Armed with the insight from the simple incommensurate case, we shall proceed directly to the description of icosahedral crystals. The six-dimensional crystallography will be constructed, and the nature of the long-wavelength hydrodynamic modes will be derived, since the symmetry of these modes can be derived directly from the 6D space group symmetries. The thermodynamics of the hydrodynamic modes are important since at nonzero temperatures, lattice vibrations will reduce the intensities of Bragg peaks with an effective Debye–Waller factor. It has also been suggested[27] that phonons, or *phasons*[5] as some of these modes are called, are frozen-in in real materials, so that the diffraction peaks may be broadened or even displaced from their symmetric positions.

2. Quasicrystallography

2.1 Crystallography of Incommensurate Systems

The density, $\rho(\mathbf{r})$ of a regular periodic three-dimensional crystal can be expanded in terms of three fundamental reciprocal lattice vectors $\mathbf{q}_1$, $\mathbf{q}_2$, and $\mathbf{q}_3$:

$$\rho(\mathbf{r}) = \sum A_{n_1, n_2, n_3} \cos(n_1\mathbf{q}_1\cdot\mathbf{r} + n_2\mathbf{q}_2\cdot\mathbf{r} + n_3\mathbf{q}_3\cdot\mathbf{r} + \phi_{n_1, n_2, n_3}), \quad (2.1)$$

and the diffraction spots are at $n_1\mathbf{q}_1 + n_2\mathbf{q}_2 + n_3\mathbf{q}_3$ with intensities A^2, where A is the Fourier transform of the contents of the unit cell and thus depends upon the basis which is associated with the lattice points. In (2.1), ϕ_{n_1, n_2, n_3} describes the relative phases of the density waves. The above function is periodic in its arguments with period 2π. An incommensurate structure has (at least) one additional independent vector $\mathbf{q}_4$ which can not be written as a simple sum of rational fractions of the original three vectors, and so the most general expansion for the density of such a structure is

$$\rho(\mathbf{r}) = \sum A_{n_1, n_2, n_3, n_4} \cos(n_1\mathbf{q}_1\cdot\mathbf{r} + n_2\mathbf{q}_2\cdot\mathbf{r} + n_3\mathbf{q}_3\cdot\mathbf{r} + n_4\mathbf{q}_4\cdot\mathbf{r} + \phi_{n_1, n_2, n_3, n_4}).$$

This function is periodic in the *four* variables

$$(\theta_1,\theta_2,\theta_3,\theta_4) = (\mathbf{q}_1\cdot\mathbf{r},\mathbf{q}_2\cdot\mathbf{r},\mathbf{q}_3\cdot\mathbf{r},\mathbf{q}_4\cdot\mathbf{r}). \tag{2.2}$$

A function which is periodic in four variables can be viewed as a four-dimensional crystal, so in general, incommensurate crystals can be thought of as higher-dimensional crystals. If the function $\rho(\theta_1,\theta_2,\theta_3,\theta_4)$ is specified, the real space structure can be thought of as the three-dimensional surface in four dimensions which is traced out when **r** traverses the 3D real space. Note that there is a one-to-one correspondence between the incommensurate structures and periodic functions in four dimensions; a four-dimensional experimentalist performing a diffraction measurement on such a system would observe diffraction spots at the periodic reciprocal lattice vectors with intensities $A^2_{n_1,n_2,n_3,n_4}$. Hence, real space experimentalists should think of their reflections as Bragg spots on a 4D lattice, find the contents of the 4D unit cell by a 4D structural analysis, and construct the real 3D structure by inserting the appropriate 3D surface. Of course, this is easier said than done, since the most general description of the atomic basis in four dimensions will correspond to 1D curves rather than points. To illustrate these points, consider how one can most generally describe a one-dimensional incommensurate system. Some examples are shown in Fig. 4.

2.1.1 One-Dimensional Incommensurate Crystals. Following the discussion above, we may write the density in 2D as $\rho(\theta_1,\theta_2) = \rho(q_1x,q_2x)$, and so the incommensurate structure is defined by the density along a line, E, with slope $q = \frac{q_2}{q_1}$ in a regular 2D crystal. This slope could be, for example, the golden mean τ, but this is not a necessity and is not imposed by any symmetry. If the slope of the line is irrational, at one place or another, the line traverses all points in the 2D unit cell; hence, all the information in two dimensions is required to form the 1D incommensurate crystal. All of the possible incommensurate structures differ only with respect to the motif associated with the 2D unit cell, or the orientation of the real space line with respect to the 2D lattice. In periodic crystals, the motif can be assigned as a collection of pointlike atoms, but this will not work in this scheme, since for a given distribution of pointlike objects in the 2D unit cell, a line with irrational slope will rarely intersect any given point; real space would be pretty empty. Therefore, one must choose the motif in two dimensions to be a 1D curve, or curve segment, or a number of curves or curve segments. One of the simplest choices corresponds to a set of straight line segments perpendicular to the real world line E, as shown in Fig. 4b. This yields a structure where the distance between successive atoms is either a short distance s, which is the projection of one lattice vector on E, or a long distance l, which is the projection of the other lattice vector. The position m_i of the ith long segment in a stacking sequence may be specified by a simple formula:

$$m_i = Int(iq), \quad i \in integer.$$

If q is the golden mean, τ, the long segments take up the positions $m_i = 1,3,4,6,\ldots$, with the short segments placed at $m_i = 2,5,\ldots$.

The structure that we obtain from this particular choice for the basis, with τ for the slope of the real world line, is the well-known one-dimensional Fibonacci sequence,[28] whose two- and three-dimensional analogs have been dubbed *quasilattices* by Mackay[3] or *quasicrystals* by Levine and Steinhardt.[4] Note that the Fibonacci sequence is constructed from the stacking of two different tiles (l and s) as opposed to one for a conventional periodic 1D crystal. Of course, we might have chosen the line segments to lie along some other direction, and this would have led to a different ratio between the two tile lengths, as shown in Fig. 4c. Alternatively, we could have chosen the slope of the real world line through the 2D lattice to have some other value, irrational or rational, as shown in Fig. 4d; and this would have altered the sequence of short and long segments. In particular, if the slope of the line is a rational number, the density along the line will be periodic, albeit perhaps with a very long period.

Although no naturally occurring physical system corresponds to the limit of Fig. 4b, one can construct such 1D systems artificially by molecular beam epitaxy. Figure 5 shows the Fibonacci sequence of layers of GaAs/AlAs studied by Merlin *et al.*,[29] and Todd *et al.*[30] using X-ray diffraction. The resulting high-resolution diffraction pattern is also reproduced here, and must be indexed by a set of two integers (m,n) such that

$$Q = \frac{2\pi}{\tau l + s} * (m + n\tau).$$

How can we calculate the diffraction intensities from this 1D *quasicrystal*? According to the discussion above, the amplitude of the scattering is simply the Fourier transform of the contents of the 2D unit cell,

$$A_{m,n} = \frac{1}{(2\pi)^2} \int_{unit\ cell} d\boldsymbol{\theta}_1 d\boldsymbol{\theta}_2\, \rho(\boldsymbol{\theta}_1,\boldsymbol{\theta}_2) e^{i\mathbf{q}\cdot\boldsymbol{\theta}},$$

where $\mathbf{q} = m\hat{\mathbf{e}}_1 + n\hat{\mathbf{e}}_2$, and $\hat{\mathbf{e}}_1$ and $\hat{\mathbf{e}}_2$ are unit vectors in the 2D lattice. However, it is simpler to use a coordinate system where the axes are

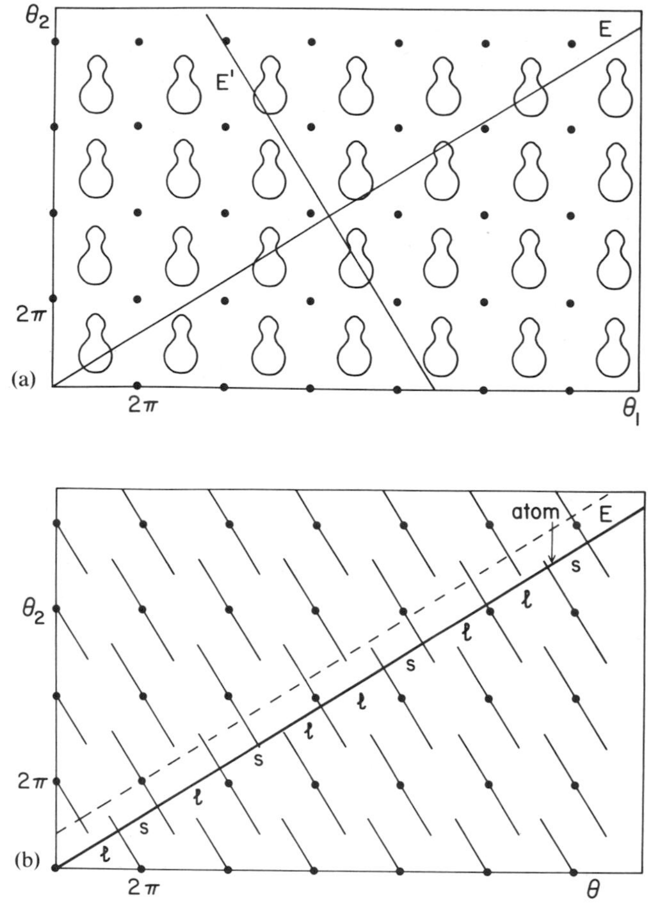

Figure 4. One-dimensional incommensurate ("quasiperiodic") structures represented as cuts along a line *E* in two-dimensional space. (a) A general structure. (b) The basis is a line segment perpendicular to *E*, and the resulting 1D structure is a Fibonacci lattice. The ratio of *l* to *s* is the golden mean τ. (c) The basis is a line segment in a skew direction, yielding a Fibonacci lattice where the ratio of the two lengths is different from τ. (d) The slope of *E* is rational so the sequence of *l* and *s* segments is periodic. (e) The basis is a curve segment, and the resulting 1D structure is a simple incommensurate modulation. The broken lines in these figures represent a shift of the real world *E* along the perpendicular direction *E'*. In (b), the effect is a rearrangement of tiles; in (e), the effect is a continuous displacement of atoms.

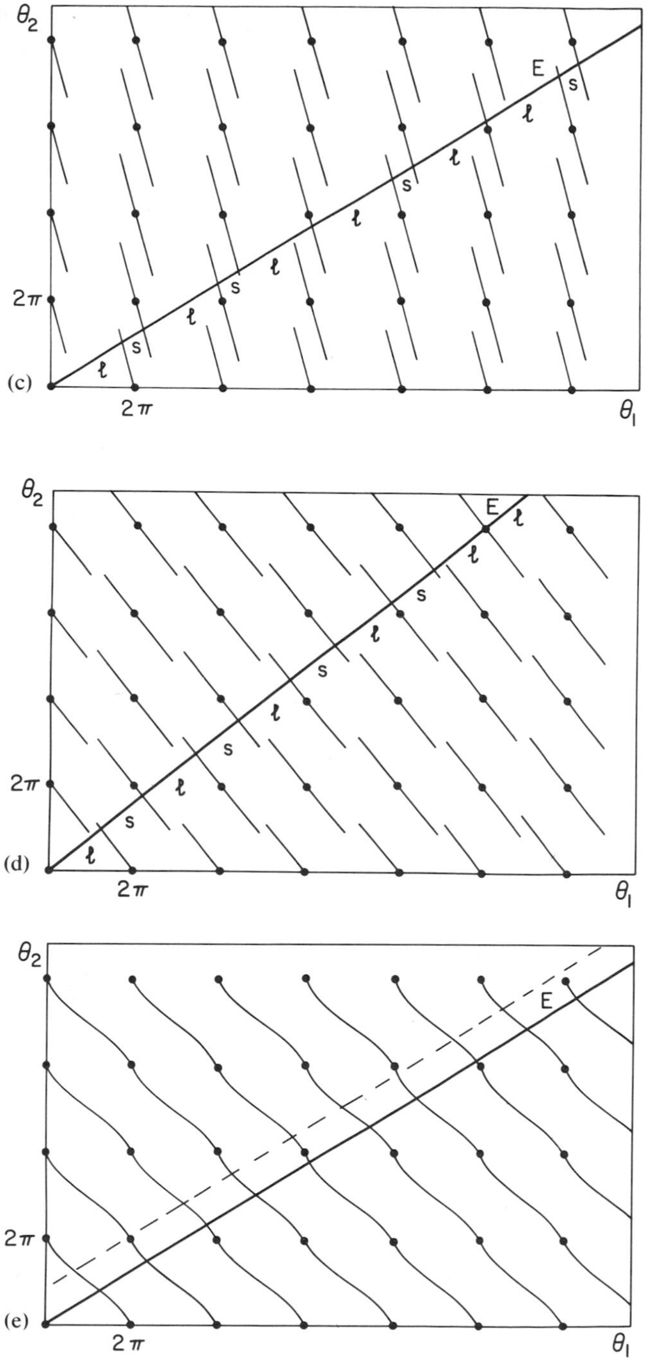

Figure 4. (Continued).

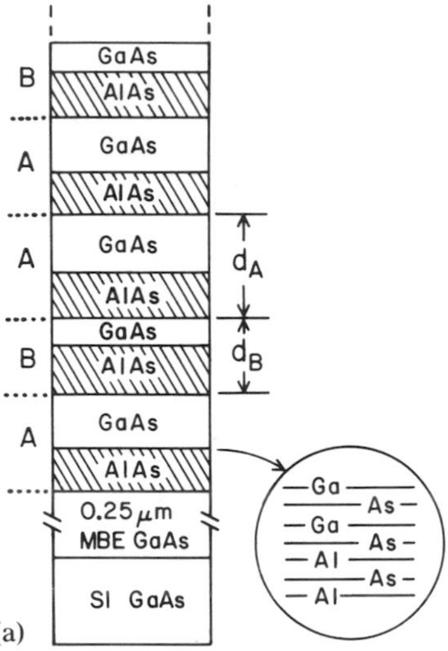

(a)

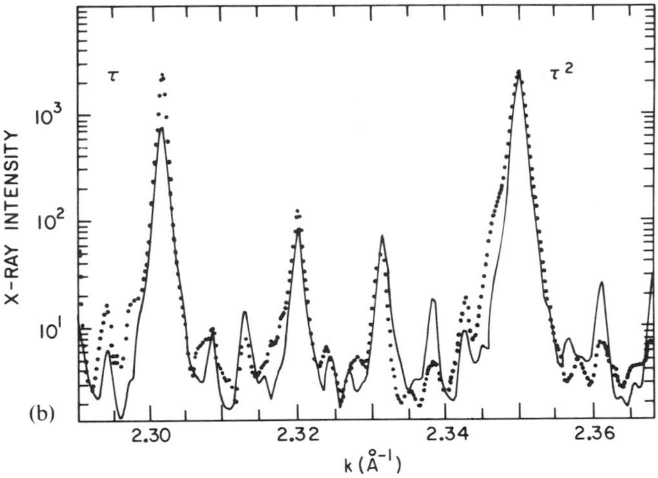

(b)

Figure 5. (a) Schematic arrangement of GaAs and AlAs layers deposited in a Fibonacci sequence with $d_a \approx$ 59Å, $d_b \approx$ 37Å, and $\frac{d_a}{d_b} \approx \tau$ (after Ref. 30.). (b) Diffraction profile of the GaAs/AlAs Fibonacci superlattice. Solid dots represent the high-resolution synchrotron data, and the solid line is a fit using the model described in Ref. 30. Inset: Low-resolution scan showing the superlattice peaks and their scaling as powers of τ. Shaded region indicates the range of the high-resolution scan.

parallel and perpendicular to the real world line, since we can then write the density as $\delta(e_{\parallel})$. Here we follow the generally accepted convention and denote the real and reciprocal space axes parallel to the real world line as $e_{\parallel}$ and $q_{\parallel}$, respectively, and the coordinate axes perpendicular to the real world line as $e_{\perp}$ and $q_{\perp}$, so that

$$A_{m,n} = \int de_{\parallel}\,\delta(e_{\parallel}) \int_{-L/2}^{L/2} de_{\perp} e^{iq_{\perp}e_{\perp}}$$
$$= \frac{\sin\left(\frac{q_{\perp}L}{2}\right)}{\frac{q_{\perp}L}{2}}.$$

The Fourier transform of Fig. 4b is illustrated in Fig. 6 and consists of a series of δ-functions along $q_{\parallel}$. The intensity of the scattering is determined by the distance between the 2D reciprocal lattice point and the $q_{\parallel}$ axis. Thus, in this simple case, the intensity depends only on the $q_{\perp}$ component of the 2D wavevector.

The fact that the intensity depends only on this seemingly unphysical perpendicular component is a consequence of the special basis that was

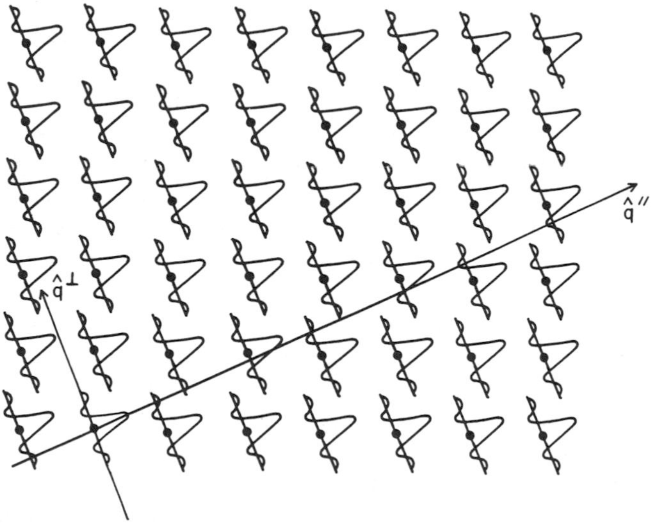

Figure 6. Fourier transform of Fig. 4b. Here, $\hat{\mathbf{q}}_{\parallel}$ is parallel to E and $\hat{\mathbf{q}}_{\perp}$ is perpendicular to E (after Duneau and Katz, Ref. 35).

chosen to describe the contents of the 2D unit cell. In general, the basis will not be a simple line segment, and incommensurate structures generally do not yield diffraction peaks which depend on only one component of the wavevector. Figure 4e shows the more general case, where the basis is described by a wavy line, and the lines in different cells connect to form an infinite curve. The resultant incommensurate crystal is a modulated structure where the positions of the atoms oscillate around regularly spaced average sites.

Sometimes the situation is even more complicated. There exist simple model systems for incommensurate structures where the basis is not a simple continuous curve, but a fractal curve which is broken into an infinity of pieces. The Frenkel–Kontorowa model,[31] an array of particles connected with harmonic strings and subjected to a periodic potential has this property when the potential is sufficiently strong. The message here is that one should not expect simplicity from real physical systems.

The two-dimensional representation is also useful for describing the continuous degrees of freedom and hydrodynamic modes associated with the breaking of these symmetries. A periodic 1D crystal has continuous translational symmetry and one acoustic phonon branch. Since the energy of the system is invariant under a simple translation of the entire structure. Likewise, the energy of the incommensurate system is invariant under any displacement of the 2D crystal, since any displacement simply corresponds to a shift of the origin. A shift along the real world line $e_{\parallel}$ describes a uniform spatial displacement of the crystal. A shift along the $e_{\perp}$ axis (dotted line) describes some internal rearrangement of the atoms. For the case, shown in Fig. 4e, of a sinusoidal modulation along $e_{\perp}$, a shift of the real space axis along $e_{\perp}$ involves an apparent shift in the phase of the modulation, and so the corresponding hydrodynamic mode may be called a *phason mode*. For a general discussion of the hydrodynamic modes in incommensurate systems, see the review by Axe and Bak.[25] While for the simple modulation shown in Fig. 4e, a continuous displacement along $e_{\perp}$ will result in a continuously varying motion of the atoms about their average position, there are discontinuous changes in the density along $e_{\parallel}$, for the Fibonacci case (Fig. 4b), as E is shifted continuously along the $e_{\perp}$ direction (dotted line). For instance, a sequence of tiles *ls* will be replaced by the sequence *ll*, or vice versa. Hence, we have a peculiar situation where a continuous degree of freedom represents discontinuous jumps of atoms. The physical interpretation of this is that the phase degree of freedom must be pinned, and there are no well-defined long-wavelength gapless modes in this case. We shall see that these arguments also carry through for icosahedral crystals.

In regular crystals, the intensity of the Bragg spots is reduced by the

Debye–Waller factor $W = \exp(-<\mathbf{q}\cdot\mathbf{u}>^2)$, where $\mathbf{u}$ is the displacement vector from an average position, and $<>$ denotes a thermal average. In the $d+1$ dimensional description of incommensurate crystals, displacement fluctuations in the direction parallel to the real world line contribute $W_{\parallel} = \exp(-q_{\parallel}^2 < u^2 >)$, while fluctuations in the displacement perpendicular to E (phase fluctuations) yield $W_{\perp} = \exp(-q_{\perp}^2 < \delta\phi^2 >)$. Note that the phase fluctuations most strongly affect the peaks with a large $q_{\perp}$ component.

2.2 *Icosahedral Crystals*

In the spirit of Eq. (2.1), we can describe any icosahedral structure in terms of a set of mass-density waves:

$$\rho(\mathbf{r}) = A \sum cos(\vec{\mathbf{q}}_i \cdot \vec{\mathbf{r}}_i + \phi_i) + \textit{higher harmonics}, \qquad (2.2.1)$$

where the $\mathbf{q}_i$ might, for example, correspond to the reciprocal lattice basis vectors pointing to the vertices of an icosahedron, as in Fig. 3. The form of expression (2.2.1) should not be taken literally to mean that the actual mass distribution looks anything like a smooth sinusoidal function, since the higher harmonics may be significant; the expansion can perfectly well describe a collection of point particles. Nevertheless, expansions like (2.2.1) can be used to describe the symmetries of the structure. In Landau's theory for phase transitions, the ordered phase is described this way, and the coefficient A is the order parameter for the transition. The various structures (e.g., bcc, fcc crystals) arise from the minimization of a phenomenological free energy. Alexander and McTague[32] have, with some success, used Landau's theory to describe freezing transitions into cubic crystals, and even noted that there should be some possibility of having icosahedral structures. Rather than predicting specific numerical coefficients in the expansion, the theory is most useful for specifying the symmetries of the structure.

In Landau's theory, the symmetry of the crystal is defined by the lowest order term in the density wave expansion, since higher order terms will generally not lower the symmetry. In addition, one can show (as we shall demonstrate later) that the energy cannot depend upon the phases, ϕ_i, of the fundamental density waves, since a phase shift simply corresponds to a shift of the origin. The phases of the higher harmonics are not independent, but coupled to the phases of the basic vectors through higher order terms in the Landau expansion. For $\phi_i = 0$ in (2.1.1), the density has the full icosahedral symmetry, just like the diffraction pattern. Of course, one could also choose the ten independent vectors pointing towards the triangular faces of an icosahedron, or the fifteen edge vectors, as the basis set. We shall also describe the kinds of structures which result from these choices.

2.2.1 The Simple Icosahedral Crystal. By choosing the set of basis vectors given by expression (1.1), the density expansion can be written,

$$\begin{aligned}\rho(\mathbf{r}) &= \sum A_{n_1,n_2,\ldots,n_6} \cos\left(\sum n_i \mathbf{q}_i \cdot \mathbf{r} + \phi_{n_i}\right) \\ &= \rho(\mathbf{q}_1\cdot\mathbf{r},\mathbf{q}_2\cdot\mathbf{r},\ldots,\mathbf{q}_6\cdot\mathbf{r}) \\ &= \rho(\theta_1,\theta_2,\theta_3,\theta_4,\theta_5,\theta_6).\end{aligned} \tag{2.2.2}$$

The requirement that ρ has icosahedral symmetry imposes restrictions on the amplitudes and phases. For instance, the amplitudes A, corresponding to components with wavevectors related by icosahedral symmetry, have to be identical; Since ρ can be expressed in terms of a periodic function of six variables (with period 2π), this function can be viewed as a 6D crystal. In analogy with the discussion above for 1D incommensurate crystals, the density at some point $\mathbf{r}$ in real space is given by the density in 6D space at the point

$$(\theta_1,\theta_2,\theta_3,\theta_4,\theta_5,\theta_6) = (\mathbf{q}_1\cdot\mathbf{r},\mathbf{q}_2\cdot\mathbf{r},\mathbf{q}_3\cdot\mathbf{r},\mathbf{q}_4\cdot\mathbf{r},\mathbf{q}_5\cdot\mathbf{r},\mathbf{q}_6\cdot\mathbf{r}). \tag{2.2.3}$$

As $\mathbf{r}$ runs through the 3D Euclidean space, a three-dimensional plane is traced out in 6D space. *Therefore, the density of a 3D icosahedral crystal can be simply thought of as the density along a three-plane in a periodic 6D crystal.* The orientation of this three-plane is defined uniquely in terms of the six reciprocal lattice basis vectors $\mathbf{q}_1 \ldots \mathbf{q}_6$, through (2.2.3), so it is constrained by the requirement of icosahedral symmetry in real space. Note that the golden mean, τ, plays a fundamental role here, since the relative directions of the six basis vectors are expressed in terms of τ (eg., $\mathbf{q}_1 \cdot \mathbf{q}_2 = \tau$). In contrast, for the simple 1D incommensurate case, the slope of the real world line, E, is not fixed by any symmetry to a specific value, like τ. Whereas the Fibonacci sequence shares some interesting properties with the higher-dimension Penrose tilings, there is no real physics associated with τ in one dimension.

A general vector in 6D has three components, $\mathbf{e}^i_{\parallel}$, along the real space, which we denote E, and three components, $\mathbf{e}^i_{\perp}$, along the complementary space, E'. Although E is a 3D plane, it samples all points in the 6D unit cell, because it approaches any point in the 6D unit cell arbitrarily closely due to the irrational slope defined by (2.2.3). One can also construct a reciprocal space in 6D, so that a general vector in the reciprocal space has three real space components, $\mathbf{q}_{\parallel}^{i}$, which define the physical momentum transfer measured in a diffraction experiment, and the complementary space partners $\mathbf{q}_{\perp}^{i}$.

In addition to the six-dimensional translational symmetries, the icosahedral crystal has rotational point group symmetries. The assumption that the density has full icosahedral symmetry about a specific point translates into symmetry operations involving planes in 6D space. This is completely analagous to a threefold rotational symmetry in a cubic crystal about a line, or a reflection through a specific plane. For instance, the fivefold axis along, say, $\mathbf{q}_1$ requires that the density in 3D space is invariant under the transformation

$$\mathbf{q}_1 \rightarrow \mathbf{q}_1, \; \mathbf{q}_2 \rightarrow \mathbf{q}_3 \rightarrow \mathbf{q}_4 \rightarrow \mathbf{q}_5 \rightarrow \mathbf{q}_6 \rightarrow \mathbf{q}_2,$$

which in turn implies that the density in the six-dimensional space is invariant under the transformation

$$\theta_1 \rightarrow \theta_1, \quad \theta_2 \rightarrow \theta_3 \rightarrow \theta_4 \rightarrow \theta_5 \rightarrow \theta_6 \rightarrow \theta_2.$$

This is a fivefold rotation in 6D which leaves the plane spanned by the vectors (1,0,0,0,0,0) and (0,1,1,1,1,1) invariant. Hence, the symmetry is a fivefold plane spanned by these vectors and is shown in Figure 7. Note that it has nonzero components both parallel and perpendicular to the real world. The real space component yields the density of the icosahedral crystal for $\mathbf{r}$ along the $\mathbf{q}_1$ direction. Thus a fivefold rotation in six dimensions is compatible with discrete translational symmetry in contrast to three dimensions. This fivefold plane in six dimensions is responsible for the fivefold symmetry of the diffraction pattern.

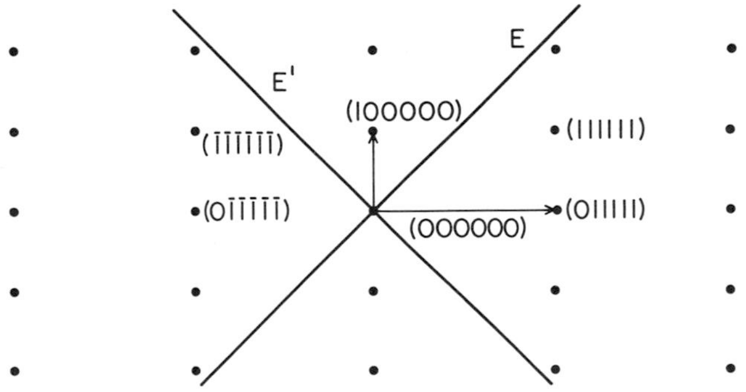

Figure 7. Fivefold plane of the six-dimensional crystal representing an icosahedral crystal. The real space is along E, and the perpendicular space is along E'.

Similarly, the threefold symmetry,

$$\mathbf{q}_1 \rightarrow \mathbf{q}_2 \rightarrow \mathbf{q}_3 \rightarrow \mathbf{q}_1, \quad \mathbf{q}_4 \rightarrow \mathbf{q}_5 \rightarrow \mathbf{q}_6 \rightarrow \mathbf{q}_4$$

requires threefold symmetry of the structure in six-dimensions, so that ρ must be invariant under the transformation

$$\theta_1 \rightarrow \theta_2 \rightarrow \theta_3 \rightarrow \theta_1, \quad \theta_4 \rightarrow \theta_5 \rightarrow \theta_6 \rightarrow \theta_4.$$

This operation leaves the plane spanned by (1,1,1,0,0,0) and $0,0,0,\bar{1},1,1)$, which also has components along *E* and *E*′ (Fig. 8), invariant. Finally, there are twofold "mirror" planes spanned by vectors such as (1,1,0,0,0,0) and (0,0,1,1,1,1).

In summary, for each rotation which transforms the vectors $\mathbf{q}_1 \ldots \mathbf{q}_6$ among themselves, there is a corresponding symmetry operation in six dimensions. Therefore, the point group of the six-dimensional crystal is isomorphous with the icosahedral group. Vectors along *E* and *E*′, respectively, transform among themselves under all symmetry operations (*E* and *E*′ are invariant subspaces). It turns out that vectors within *E* and *E*′ transform as two different 3D irreducible representations, *E* and *E*′, of the icosahedral group.

All the symmetry operations of the point group have now been derived. A general space group operation combines a six-dimensional translation with one of the above point group operations. In analogy with the usual notation for three-dimensional crystals, we may denote the space group P352, and term the structure *simple icosahedral* or si, in analogy with the *simple cubic* group in three dimensions. Such a structure will produce

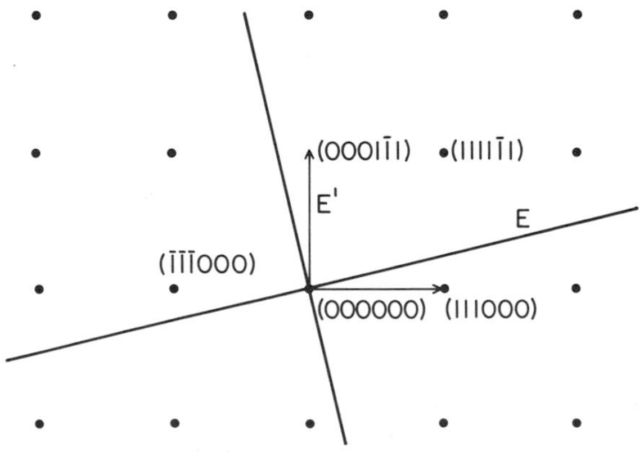

Figure 8. A threefold plane of the six-dimensional crystal.

diffraction spots at all positions $\mathbf{Q} = \Sigma\, n_i \mathbf{q}_i$ with no systematic extinctions. All missing reflections result from "accidental" extinction due to the contents of the 6D unit cell.

Another possibility is a structure where the inversion symmetry has been replaced by a twofold inversion axis, $P5\bar{2}\bar{3}/m$, where the threefold inversion planes define operations such as

$$\theta_1 \to -\theta_2 \to \theta_3 \to -\theta_1, \quad \theta_4 \to \theta_5 \to \theta_6 \to -\theta_4.$$

2.2.2 The Body-Centered Icosahedral Crystal. Another structure which can arise from the Landau expansion is formed by the fifteen pairs of edge vectors, $\boldsymbol{\varepsilon}_\alpha$, of the regular icosahedron:

$$\rho(\mathbf{r}) = \sum A_{n_1\, n_2, \ldots, n_{15}} \left(\sum n_\alpha \boldsymbol{\varepsilon}_\alpha \cdot \mathbf{r} + \phi_{\{n\}} \right).$$

However, an edge vector can always be written as the sum or difference of two vertex vectors, $\varepsilon_\alpha = \pm \mathbf{q}_{\alpha,1} \pm \mathbf{q}_{\alpha,2}$. The 6D reciprocal lattice basis vectors are directed toward 30 of the face centers of a 6D hypercube, so the reciprocal lattice is *face-centered icosahedral,* or fci. Note that the faces are not equivalent, since $\mathbf{q}_1 - \mathbf{q}_2$ and $\mathbf{q}_1 + \mathbf{q}_2$ are vectors of different lengths.

We can expand the density again in terms of the vertex vectors,

$$\rho(\mathbf{r}) = \sum A'_{n1,\, n_2, \ldots, n_6} cos \left(\sum n_i \mathbf{q}_i \cdot \mathbf{r} + \phi'_{\{n\}} \right),$$

where now the Σn_i must be an even number since the $\boldsymbol{\varepsilon}$-vectors introduce the $\mathbf{q}$-vectors in a pairwise manner. Hence the edge structure has systematic extinction rules which allow one to distinguish it from the si lattice described above: it produces Bragg scattering at $(n_1,n_2,n_3,n_4,n_5,n_6)$ with Σn_i even.

The density in real space can also be written as a periodic function of six variables as in expression (2.2.2). However, since the $\mathbf{q}$-vectors enter pairwise, the density is a function only of the sums or differences of the θ_i,

$$\rho = \rho(\theta_i \pm \theta_j).$$

Therefore, ρ is invariant under the noninteger translations,

$$(\theta_1,\theta_2,\theta_3,\theta_4,\theta_5,\theta_6) \to (\theta_1,\theta_2,\theta_3,\theta_4,\theta_5,\theta_6)$$
$$+\ 2\pi\ (\pm 1/2, \pm 1/2, \pm 1/2, \pm 1/2, \pm 1/2, \pm 1/2).$$

So, while the 6D reciprocal lattice is fci, the 6D direct space lattice is *body-centered icosahedral,* bci, and the space group is naturally denoted C235, in a direct analogy to the 3D body-centered cubic lattice and its face-centered 3D reciprocal lattice. It is unfortunate that nature has not yet provided any examples of the 6D bci structure. All present icosahedral alloys are consistent with the si structure.

2.2.3 The Face-Centered Icosahedral Crystal. From our experience with cubic crystals in 3D, we expect that in addition to the si and bci lattices defined in Ref. 5, and discussed above, there is also a face-centered icosahedral crystal, fci, whose reciprocal lattice is bci. Indeed, this is the case, but the structure is more subtle and probably less relevant than the structures above.

Let us first consider a structure which is formed by waves with wave-vectors spanned by ten pairs of vectors in the face directions of a regular icosahedron:

$$\boldsymbol{\gamma}_1 = (1111\bar{1}1),\ \boldsymbol{\gamma}_2 = (11111\bar{1}),$$

$$\boldsymbol{\gamma}_3 = (1\bar{1}1111),\ \boldsymbol{\gamma}_4 = (11\bar{1}111),$$

$$\boldsymbol{\gamma}_5 = (111\bar{1}11),\ \boldsymbol{\gamma}_6 = (111\bar{1}\bar{1}\bar{1}).$$

The projection of the γ_i vectors onto 3D corresponds to vectors along the face directions of a regular icosahedron. The diffraction pattern, which is formed by projecting linear combinations of these vectors onto three dimensions, has peaks with all even or all odd indices whose sum is divisible by 4 when expressed in terms of the vertex basis (1.1). Thus it appears that the pattern is different from those of the si or bci structures described above. One could also expand in terms of six vectors having 3D projections along the vertex directions,

$$\boldsymbol{\gamma}'_1 = (\bar{1}11111),\ \boldsymbol{\gamma}'_2 = (1\bar{1}1\bar{1}\bar{1}1),$$

$$\boldsymbol{\gamma}'_3 = (11\bar{1}1\bar{1}\bar{1}),\ \boldsymbol{\gamma}'_4 = (1\bar{1}1\bar{1}1\bar{1}),$$

$$\boldsymbol{\gamma}'_5 = (1\bar{1}\bar{1}1\bar{1}1),\ \boldsymbol{\gamma}'_6 = (11\bar{1}\bar{1}1\bar{1}).$$

Note here that the lengths of these projected vectors are $\gamma'_i = \frac{2}{\tau}|\mathbf{q}_i|$, where the $\mathbf{q}_i$ are the vertex vectors defined in (1.1). The vectors $\boldsymbol{\gamma}_i$ can be expressed in terms of the $\boldsymbol{\gamma}'_i$ and vice versa, for instance, $\boldsymbol{\gamma}_1 = \boldsymbol{\gamma}'_1 + \boldsymbol{\gamma}'_2 + \boldsymbol{\gamma}'_3$ and $\boldsymbol{\gamma}'_6 = \boldsymbol{\gamma}_4 - \boldsymbol{\gamma}_1 + \boldsymbol{\gamma}_6$. Hence, the space which is spanned by the vectors $\boldsymbol{\gamma}_i$ is also spanned by the vertex vectors $\boldsymbol{\gamma}'_i$, so the diffraction pattern from the lattice is the same, within a scale factor of $2/\tau$, as the si structure. If one removes all spots of mixed parity from the diffraction pattern of an si structure and rescales the pattern by $2/\tau$, one obtains the original pattern. Hence, while in six dimensions these vectors form a different Bravais class, the cut along E through the 6D lattice has the same symmetry as the cut through the simple lattice.

We now add *another* set of vectors in the face direction,

$$\gamma_7 = (111\bar{1}1\bar{1}),\ \gamma_8 = (1\bar{1}11\bar{1}1),$$

$$\gamma_9 = (11\bar{1}11\bar{1}),\ \gamma_{10} = (1\bar{1}1\bar{1}11),$$

$$\gamma_{11} = (11\bar{1}1\bar{1}1),\ \gamma_{12} = (\bar{1}1111\bar{1}1).$$

These vectors are a factor τ shorter than the vectors above. They define another cubic lattice. If the two lattices are combined, one obtains a complete body-centered six-dimensional lattice. This is the reciprocal lattice of the fci structure. Its projection in three dimensions cannot be formed by integral superpositions of vectors of the same lengths, but requires two lengths. Therefore, the structure does not emerge from a simple Landau expansion involving a single star of wavevectors.

The density in real space can be written as a periodic function of six variables. However, since the indices are obviously all odd or all even when the density is expanded in terms of the vertex vectors, the density is invariant under the noninteger translation

$$(\theta_1,\ldots,\theta_6) \rightarrow (\theta_1,\ldots,\theta_6) + 2\pi\left(\frac{1}{2},\frac{1}{2},0000\right)$$

and the equivalent face translation in 6D space. This justifies calling the structure face-centered icosahedral. The space group is naturally denoted F235, in analogy with the notation for 3D cubic crystals.

2.2.4 Where Are the Atoms? Now that we have discussed the structure in six dimensions, we must try to understand how to think of the crystal in our three-dimensional world. There is one crucial difference between the description of a conventional 3D crystal in terms of a unit cell plus an atomic basis, and icosahedral crystals. In a regular crystal, one can always describe the structure as a stacking of "unit cells," all of which are decorated with atoms in an identical manner. The icosahedral crystal in six dimensions can be envisioned in the same way. That is, we associate a basis with each unit cell in six dimensions, and repeat this motif for all cells. However, this stacking arrangement in six dimensions does not translate to a tiling in three dimensions: we can not, in general, reduce our crystallographic analysis to a determination of a repeated atomic basis. This is not a completely unusual situation, since incommensurate systems cannot, in general, be visualized as a stacking of unique building blocks as discussed in Section 2.1. Of course, it would thus seem that any attempt at a structural refinement in three dimensions is doomed to failure. What is the most appropriate way to describe the atomic motif? We consider this in several steps, beginning with some more or less reasonable assumptions which allow specific solutions, and then by proceeding towards a more general description (which might lead one to conclude that the problem is intractable).

Let us first assume that the structure in three dimensions consists of a set of pointlike atoms. The motif in six dimensions then cannot consist of pointlike objects, since the real world three-plane, E, will rarely intersect these points. The atomic motif in six dimensions must consist of one or more 3D surface segments, which intersect E at the positions of the atoms. This is perfectly analagous to the 1D incommensurate case where the basis was a 1D curve or line intersecting E at the atomic positions.

In the case of the simple 1D incommensurate structure, there are no symmetry constraints on the motif. This is not generally true for incommensurate structures in higher dimensions, and is certainly not the case for the icosahedral structures. Just as the basis associated with a cubic unit cell in three dimensions must have cubic symmetry, the basis associated with the 6D unit cell must have the appropriate point group symmetry. For instance, in a regular 3D cubic system, if an atom is placed at some general position (x, y, z), then 23 additional atoms must be placed at positions in the unit cell which are connected to (x, y, z) through the cubic symmetry operations. Similarly, if a three-surface is imbedded at some general position in the 6D hypercubic unit cell, the icosahedral point group symmetry demands that we place 119 additional three-surfaces at symmetry-equivalent positions, resulting in a complicated mess, impossible to analyze, but fortunately very unphysical, since these surfaces will intersect each other and lead to structures with arbitrarily small interatomic distances. This violates the "Delauny condition," which requires the distance between neighboring atoms to exceed a minimum value.

One can considerably simplify the situation by restricting the motif to special positions in the unit cell. For instance, one could choose to place the three-plane at a high symmetry position which is invariant under all point group operations, just as one could choose to place a single atom at the center of a 3D cubic unit cell. One particular decoration (although not the only one by far) is to place a three-plane segment along E', perpendicular to E. Since the three-plane E' transforms into itself under all symmetry operations, there is no need for additional planes to satisfy the symmetry conditions. Since it is not possible to show the geometry in six dimensions, we must be content with displaying 2D cuts or projections. Figure 9 shows the fivefold plane of Fig. 7 decorated with hyperplane segments. The positions of the atoms are at the points where the plane segments along E' intersect E. This construction is equivalent to the projection of a set of 6D lattice points onto the real world line; and so the *cut-and-projection method,* introduced by several authors,[35] thus arises as a special case of the general method described here. If the three-plane segment is chosen to be a regular tricontahedron of appropriate size, the resulting atomic arrangement is precisely the 3D Penrose tiling, consisting of a "stacking" of two rhombohedral tiles with atoms at the vertices.

A significant effort[36] has been directed toward further decorating these individual 3D tiles with a motif of atoms. This, of course, is not the most general way of describing the structure and is, in fact, quite restrictive. A more systematic method involves placing additional planes in the 6D unit cell, allowing for many more structures beyond the simple Penrose tiles. It is actually quite difficult to speculate about how to choose a specific motif corresponding to the real IP alloys. For the bimetallic alloys such as Al–Mn, one must introduce at least two types of planes corresponding to Al and Mn, respectively. One possibility would be to place Mn planes centered at the vertices of the 6D hypercube, (0,0,0,0,0,0), and Al planes through the six fivefold directions (1/2,0,0,0,0,0).[37] If the extent of the Al planes along E' is chosen appropriately, one ends up with a structure which is a decoration of Penrose tiles with Mn atoms at the corner and Al atoms at the edge positions. By inspection of Fig. 4b and Fig. 9, one can see that the density of the structure, defined by the intersection of the planes along E' with the real world E, can be altered simply by an expansion or contraction of the length of the planes along E'.

The symmetries do not really place any restriction on the extent of the Mn or Al planes, so that in principle, one could "remove" a Mn atom by having a hole in the Mn plane. For instance, the environment around a fraction of the corner sites may resemble a MacKay icosahedron.[4] Since the center of the MacKay icosahedron is vacant, the Mn plane in six dimensions must have a "hole" in it. We can avoid the problems associated with the low density of structures which result from the simple stacking of structural units. Denser structures may be formed by extending the Al planes in six dimensions beyond the lengths required for a Penrose tiling. This would result in additional atoms situated at sites other than between Mn atoms.

For the ternary alloys such as Al–Mn–Si, U–Pd–Si, and Al–Li–Cu, a specification of the atomic motif is correspondingly more difficult. At a minimum, each new constituent requires one additional three-plane which may or may not be described by a mixture of, for example, Al and Si. Since the addition of Si to the Al–Mn alloy stabilizes the icosahedral phase,[38] the correct description of the three-surface in six dimensions is probably of some importance. Similarly, alloys such as Al–Ru–Mn–Si and Al–Cr–Ru–Si require additional surfaces in six dimensions to describe the Cr or Ru positions, since there is some experimental evidence[39] that indicates that the sites of these transition metals are not equivalent to those of the Mn.

There are even more degrees of freedom associated with the decoration of the 6D lattice. Symmetry does not constrain the atomic motif to be a set of three-plane segments, but rather three-surface segments. In general, the position of the surface can be specified by a three-dimensional vector function, $\mathbf{f}(\mathbf{e}_\perp^i)$, of the perpendicular coordinate $\mathbf{e}_\perp^i$, which describes the displacement of the surface relative to the E' plane [i.e., the surface includes the point $(\mathbf{e}_\parallel^i, \mathbf{e}_\perp^i) = (\mathbf{f}(\mathbf{e}_\perp^i), \mathbf{e}_\perp^i)$]. Icosahedral point symmetry does place some restrictions on the function. For instance, if $\mathbf{e}_\perp^i$ is directed along a fivefold direction, $\mathbf{f}(\mathbf{e}_\perp^i)$ must be directed along an orthogonal fivefold direction in real space so that the three-surface intersects the fivefold plane along a 1D curve. Thus, there will be arrays of atoms along symmetry directions in 3D space. Figure 9 shows the intersection of a more general surface with the fivefold plane.

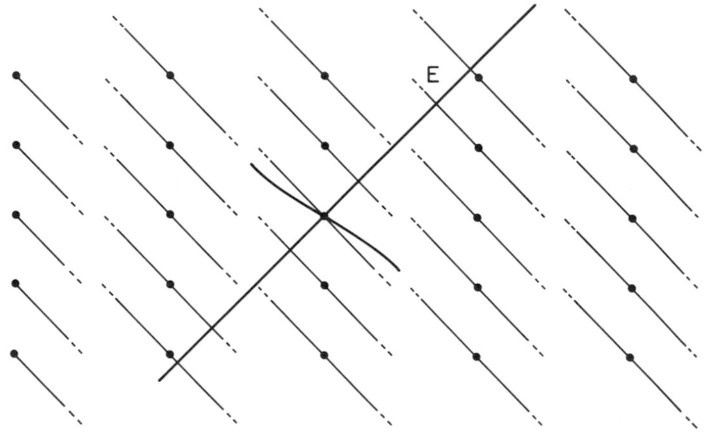

Figure 9. The fivefold plane of Fig. 7, decorated with hyperplane segments. The positions of the atoms are at the intersection of the hyperplane segments and E. The broken lines represent extensions of the planes leading to denser structures without violating "Delauney" condition. The wavy line represents a more general three-surface intersecting the fivefold plane.

Rather than specifying the motif as a set of three-surface segments, it might be possible to connect the surfaces in adjacent 6D unit cells to form extended surfaces in analogy with the situation shown in Fig. 4e. Indeed, it is possible to connect the surfaces along some directions, but it is incompatible with icosahedral symmetry to have continuous surfaces with the topology of infinite 3D planes.[40,41] This is analogous to the situation for regular cubic crystals where one cannot have an infinite invariant plane, but rather infinite surfaces with a different topology, such as the extended Fermi surface of copper (which has "holes" in it). This point is important with respect to the nature of the hydrodynamic modes.

The diffraction pattern of the icosahedral crystal can be obtained by Fourier transforming the contents of the 6D unit cell. It is again convenient to choose a reciprocal space coordinate system with three reciprocal lattice basis vectors along the real world plane E,

$$\mathbf{q}_{\parallel}^{i}; \quad i = 1,2,3,$$

and three vectors along the perpendicular directions E',

$$\mathbf{q}_{\perp}^{i}; \quad i = 1,2,3,$$

such that for the 6D hypercubic structure, $\hat{\mathbf{q}}_{\parallel}^{i} \cdot \hat{\mathbf{e}}_{\parallel}^{j} = \delta_{ij}$, and $\hat{\mathbf{q}}_{\perp}^{i} \cdot \hat{\mathbf{e}}_{\perp}^{j} = \delta_{ij}$. Therefore, the 6D reciprocal lattice can be decomposed along the parallel and perpendicular directions:

$$\mathbf{q}_{6D} = \mathbf{q}_{\parallel} + \mathbf{q}_{\perp}.$$

For the case where the atomic motif is but a single three-plane segment, the diffracted intensity depends only on the perpendicular component, $\mathbf{q}_{\perp}$, since we integrate over only the three-plane segment which lies along E' (i.e., over a rhombic tricontahedron in the case of a simple Penrose tiling). In the more realistic case where there is more than a single plane, or when the three-surface is not a plane, the diffraction intensity will also depend upon the parallel component, $\mathbf{q}_{\parallel}$, of the 6D reciprocal lattice vector.

2.2.5 Lattice Dynamics. Finally, let us discuss the nature of the hydrodynamic modes. As is true in a regular 3D crystal, the energy is invariant under any displacement of the 6D crystal. A shift along E simply represents a uniform translation of the entire system; a continuous motion of the individual atoms. In contrast, a *phase* shift of E along E' represents some rearrangement of the atoms. However, the energy remains invariant under this phase shift. Since E traverses the 6D unit cell ergodically, it passes infinitely close to all points at some time or another. A shift of E along E' merely displaces the origin some distance along E, although it may now be quite far from its original site. There are three continuous degrees of freedom corresponding to the three $\mathbf{e}_\perp^i$ components of E', and consequently, there are three hydrodynamic *phason* modes associated with these degrees of freedom, in addition to the acoustic modes associated with a uniform translation along E. In general, the phason and phonon modes will be coupled. For a thorough discussion of the elastic properties and thermodynamic modes, see Ref. 5.

In the case where the atomic motif in six dimensions corresponds to a number of three-planes (as for a decorated Penrose tiling), the phase translation involves a discontinuous shift of a fraction of the atoms, while the remaining atoms are unmolested. This corresponds to a rearrangement of the Penrose tiles. Since this involves significant discontinuous motion, or atomic diffusion, these phason modes must be pinned. One might speculate that it might be possible to choose a three-surface such that the phase shifts involve continuous motions only. However, this requires the surfaces to have the topology of a continuous three-plane, not just a three-plane segment, and so it is not compatible with icosahedral symmetry. The surface must have holes in it, which implies that at least a fraction of the atoms are pinned. Hence, icosahedral crystals cannot have unpinned *phason* modes.

The Debye–Waller factor will have contributions from both displacement fluctuations, u, and phase fluctuations, $\delta\phi$, just as for the 1D case. While the displacement fluctuations will strongly affect those peaks with a large $q_\parallel$ component, phase fluctuation most strongly affect those peaks with a sizeable $q_\perp$ component. At low temperatures, thermal phase fluctuations are relatively unimportant because the phason modes are pinned, but there may be a significant contribution from frozen-in fluctuations, *phase disorder*. Such a random rearrangement of the Penrose tiles can be represented by a random displacement of the three-plane segments along the E' direction. Unlike the displacement fluctuations, these phase fluctuations are not restricted to small deviations from some average position, so they may give rise not only to a reduction in the intensity of the diffraction peaks, but also to a broadening of the peaks (signalling the breakdown of long-range order), or a shift in the actual position of the peak if the phase fluctuation has a nonzero average gradient along some direction.[27]

3. Conclusions

Clearly, there is no simple way to describe an icosahedral crystal. In this review, we have attempted to outline the kinds of general structures, with icosahedral symmetry, that can arise. The real difficulty in applying these ideas to real systems is in how one attempts to specify the atomic positions.

Unlike the case for a regular crystal in three dimensions, the icosahedral crystal, in its most general form, cannot be described as a "stacking" of one or more unit cells. Rather than attempting to solve the structure in three dimensions, it is probably more appropriate to first perform a 6D structural refinement, since diffraction measurements directly yield the structure factor associated with a 6D lattice point. Once the nontrivial job of describing the atomic basis in six dimensions is completed, a well-defined 3D plane defined by (2.2.2) can be embedded in the 6D lattice, and the 3D density be determined. If nature is kind, the 3D density of a particular system may be adequately described by a unique decoration of Penrose tiles. Regardless, the existence of the IP alloys has reminded us that nature is not constrained by the fourteen 3D Bravais lattices.

References

1. D. Schechtman, I. Blech, D. Gratias, and J. W. Cahn, *Phys. Rev. Lett.* **53,** 1951 (1984).
2. L. Pauling, *Nature* **317,** 512 (1985); *Powder Diffraction* **1,** 14 (1986); *Phys. Rev. Lett.* **58,** 365 (1987).
3. A. L. MacKay, *Physica* **114A,** 609 (1982).
4. D. Levine and P. J. Steinhardt, *Phys. Rev. Lett.* **53,** 2477 (1984); *Phys. Rev. B* **34,** 596 (1986).
5. P. Bak, *Phys. Rev. Lett.* **54,** 1517 (1985); *Phys. Rev. B* **32,** 5764 (1985).
6. N. D. Mermin and S. M. Troian, *Phys. Rev. Lett.* **54,** 1524 (1985).
7. M. V. Jarić, *Phys. Rev. Lett.* **55,** 607 (1985).
8. D. Shechtman and I. Blech, *Metall. Trans.* **16A,** 1005 (1985).
9. P. W. Stephens and A. I. Goldman, *Phys. Rev. Lett.* **56,** 1168 (1986); *Phys. Rev. Lett.* **57,** 2770 (1986).
10. P. Heiney and P. Bancel, *Advances in Physics* (in preparation).
11. A. L. MacKay, *Rep. Prog. Phys.* (in preparation).
12. V. Elser and C. L. Henley, *Rev. Mod. Phys.* (in preparation); C. L. Henley, *Comments in Condensed Matter Physics,* **23,** 59 (1987).
13. *Proc. Workshop on Aperiodic Crystals, J. Physique Colloq.* **47-C3,** (1986).
14. P. J. Steinhardt and S. Ostlund, *Selected reprints on Quasicrystals* (Plenum, New York, 1987).
15. See for example, C. Kittel, *Introduction to Solid State Physics* (Wiley, New York, 1976).
16. P. Bancel, P. Heiney, P. W. Stephens, and A. I. Goldman, *Nature* **319,** 104 (1986).
17. See for example, *Nature* **319,** 105–104 (1986).
18. P. M. deWolff, *Acta Crystallog. A* **30,** 777 (1974).
19. A. Janner and T. Janssen, *Phys. Rev. B* **15,** 643 (1977).
20. R. Penrose, *Bull. Inst. Math. Appl.* **10,** 266 (1974); see also, M. Gardner, *Sci Am.* **236,** 110 (1977); three-dimensional tilings with icosahedral symmetry were first constructed by R. Amman.

21. P. Guyot and M. Audier, *Philos. Mag. B* **53,** L15 (1985); V. Elser and C. L. Henley, *Phys. Rev. Lett* **55,** 2883 (1985); M. Kuriyama, G. G. Long, and L. Bendersky, *Phys. Rev. Lett.* **55,** 849 (1985); Y. Shen, S. J. Poon, W. Dmowski, T. Egami and G. J. Shiflet, *Phys. Rev. Lett.* **58,** 1440 (1987).
22. P. A. Bancel, P. A. Heiney, P. W. Stephens, A. I. Goldman, and P. M. Horn, *Phys. Rev. Lett.* **54,** 2422 (1985).
23. Note here that we have chosen our set of reciprocal lattice vectors as a different (but equivalent) set of vertex vectors than Ref. 22.
24. A. Janner and T. Janssen, *Acta Crystallog. A* **36,** 399 (1980).
25. J. D. Axe and P. Bak, *Phys. Rev. B* **26,** 4963 (1982).
26. P. Bak and T. Janssen, *Phys. Rev. B* **17,** 436 (1979).
27. T. C. Lubensky, J. E. S. Socolar, P. J. Steinhardt, P. A. Bancel, and P. A. Heiney, *Phys. Rev. Lett* **57,** 1440 (1986).
28. See for example, J. Riordan, in *Applied Combinatorial Mathematics,* E. Beckenbach, ed. (Wiley, New York, 1964).
29. R. Merlin, K. Bajemi, R. Clarke, F. Y. Juang, and P. K. Bhattacharya, *Phys. Rev. Lett.* **55,** 1768 (1985).
30. J. Todd, R. Merlin, R. Clarke, K. M. Mohanty, and J. D. Axe, *Phys. Rev. Lett.* **57,** 1157 (1986).
31. P. Bak, *Rep. Prog. Phys.* **45,** 587 (1981).
32. S. Alexander and J. P. McTague, *Phys. Rev. Lett.* **41,** 702 (1978).
33. T. Janssen, *Acta Crystallog. A* **42,** 261 (1986).
34. D. S. Rokhsar, N. D. Mermin, and D. C. Wright, *Phys. Rev. B.* **35,** 5487 (1987).
35. P. A. Kalugin, A. Y. Kitaev, and L. C. Levitov, *JETP Lett.* **41,** 145 (1985); V. Elser, *Phys. Rev. B* **32,** 4892 (1985); M. Duneau and A. Katz, *Phys. Rev. Lett.* **54,** 2688 (1985).
36. See for example, M. V. Jaric, *Phys. Rev. B* **34,** 4685 (1986).
37. P. Bak, in *Scaling Phenomena in Disordered Systems,* R. Pynn and A. Skjeltorp, eds. (Plenum, New York, 1986).
38. P. A. Bancel and P. A. Heiney, p. 341 in Ref. 13.
39. P. A. Heiney, P. Bancel, A. I. Goldman, and P. W. Stephens, *Phys. Rev. B* **34,** 6746 (1986).
40. P. Bak, *Phys. Rev. Lett.* **56,** 861 (1986).
41. D. M. Frenkel, C. L. Henley, and E. D. Siggia, *Phys. Rev. B* **34,** 3649 (1986).

Chapter 5

Stability and Deformations in Quasicrystalline Solids

OFER BIHAM

Department of Physics
The Weizmann Institute of Science, Rehovot, Israel

DAVID MUKAMEL

IBM T. J. Watson Research Center, Yorktown Heights, New York 10598
and
Department of Physics
The Weizmann Institute of Science, Rehovot, Israel

S. SHTRIKMAN

Department of Electronics
The Weizmann Institute of Science, Rehovot, Israel

Contents

APERIODICITY AND ORDER
Introduction to
Quasicrystals

ISBN 0-12-040601-2

1. Introduction

The surprising discovery of Shechtman *et al.* (1984) that certain Mn-Al alloys exhibit a phase with noncrystallographic icosahedral symmetry has stimulated extensive experimental (Bancel *et al.*, 1985; Poon *et al.*, 1985; Bendersky, 1985; Bendersky *et al.*, 1985; Ishimasa *et al.*, 1985; Urban *et al.*, 1986; Fung *et al.*, 1986; Loiseau and Lapasset, 1986; Horn *et al.*, 1986) and theoretical (Bak, 1985a,b; Levine *et al.*, 1985; Mermin and Troian, 1985; Kalugin *et al.*, 1985; Jarić, 1985; Elser and Henley, 1985; Stephens and Goldman, 1986; Ho, 1986; Biham *et al.*, 1986a,b) work in this field. Since icosahedral symmetry is not compatible with periodic structures, it has been suggested that these systems may be described by an incommensurate, or aperiodic density function $\rho(\mathbf{r})$. It has subsequently been observed that other aperiodic structures with noncrystallographic point symmetries exist as well. One such phase, which has been studied extensively during the last year (1986) is the decagonal phase, which is also known as the T-phase (Bendersky, 1985; Fung *et al.*, 1986; Urban *et al.*, 1986). It consists of a periodic stacking of layers, each exhibiting a two-dimensional incommensurate structure with tenfold symmetry. Other aperiodic uniaxial phases which have been found are the T'' phase, which exhibits a structure with sixfold symmetry (Bendersky *et al.*, 1985), and a phase with a twelvefold symmetry, occurring in certain Ni-Cr alloys (Ishimasa *et al.*, 1985). Electron microscopy measurements on Mn-Al alloys (Shechtman *et al.*, 1984) have indicated that the electron diffraction patterns consist of sharp spots with icosahedral symmetry. These results suggest that these alloys may have both positional and orientational long-range order (LRO). However, more recent high-resolution X-ray scattering measurements (Bancel *et al.*, 1985; Horn *et al.*, 1986, 1987) indicate that although the Bragg peaks are relatively sharp, they have a finite width. It has therefore been suggested that while these systems have long-range orientational order, they seem to possess only short-range positional order with a typical correlation length of about 100 Å. Similar results (Horn *et al.*, 1987) have recently been obtained on relatively large (a fraction of a millimeter) single quasicrystals of Al_6CuLi_3.

Two main theoretical approaches have been applied to study these systems. The first is a geometrical approach, in which the properties of two-dimensional aperiodic tiling of the plane (Penrose, 1974; Gardner, 1977) have been extended to three-dimensional space-filling structures (Levine and Steinhardt, 1984, 1985; Socolar and Steinhardt, 1986; Socolar *et al.*, 1986; Kramer and Neri, 1984). It has been demonstrated that the diffraction patterns from such structures exhibit sharp Bragg peaks and show remarkable agreement with the experimental diffraction patterns. To account for the fact that the experimentally observed diffraction peaks have a finite width, Stephens and Goldman (1986) have modified these space-filling structures and considered a model of densely packed random assemblies of identically oriented vertex-sharing icosahedra. It has been shown that within this model, the diffraction peaks are expected to develop an intrinsic finite width. Recent Monte Carlo studies (Widom *et al.*, 1987) of two-component Lennard–Jones systems in two dimensions also demonstrated that these systems may exhibit quasicrystalline decagonal structure with long-range orientational order but only short-range positional order.

The second approach is based on the Landau theory (Bak, 1985a,b; Mermin and Troian, 1985; Kalugin *et al.*, 1985; Jarić, 1985; Biham *et al*, 1986a,b). This approach is phenomenological in nature and cannot yield information regarding the microscopic structure of these aperiodic systems. However, it provides a powerful tool for studying questions such as the symmetry and stability of these phases. When considering a phase transition from an isotropic liquid to a solid phase, one may use the Landau theory to study the possible symmetries of the resulting solid and, in particular, to study the question of whether or not this solid may be a quasicrystal.

In this chapter, we review the studies of stability of quasicrystalline structures. These studies, which are mainly based on Landau and mean-field theories, indicate that quasicrystals may exist as locally and globally stable phases with long-range (LR) positional and orientational order. We then examine the possible instabilities of icosahedral structures. These instabilities may be of two types: (a) ordinary elastic deformations, in which the point symmetry is reduced, and (b) lock-in distortions, which are peculiar to aperiodic structures. The possible symmetries which may arise when such deformations take place are studied. The chapter is organized as follows. In Section 2 we consider the question of global stability of quasicrystalline solids. Here the stability of quasicrystals relative to the conventional crystalline structures is examined. We first review in Section 2.1 the phenomenological Landau theory of formation and stability of crystals and quasicrystals. We then discuss in Section 2.2 the density functional theory, which provides a scheme for actually calculating the Landau free energy functional. The results of numerical simulations, which indicate that quasicrystals may indeed form an equilibrium state, are described in Section 2.3. Local stability and possible deformations of icosahedral structures are discussed in Section 3. The Landau theory is used in Section 3.1 to demonstrate that quasicrystals may be stable to small deformations. Possible elastic instabilities and lock-in deformations are considered in Sections 3.2 and 3.3, respectively. The possible symmetries which may result of these deformations, and the phase transitions which may take place between the various phases are discussed. A summary of the main results is presented in Section 4.

2. Global Stability of Quasicrystals

2.1 Landau Theory

In this section, we review the studies of global stability of crystalline and quasicrystalline solids using the Landau theory. In this approach, one starts by identifying the local order parameter associated with the phase transition from the isotropic, translationally invariant liquid to the low-symmetry condensed phase. In the case of a transition to a crystalline or quasicrystalline phase, the order parameter is the local density $\rho(\mathbf{r})$. For convenience, the average density of the liquid phase is subtracted from the density function so that $\rho(\mathbf{r}) = 0$ in the liquid phase. In the condensed phase, $\rho(\mathbf{r})$ is either a periodic or quasiperiodic function of $\mathbf{r}$, for crystalline or quasicrystalline phases, respectively. The next step is to consider the free energy density $f\{\rho(\mathbf{r})\}$. As usual, one assumes that $f\{\rho(\mathbf{r})\}$ may be expanded in powers of $\rho(\mathbf{r})$ and its gradient. This expansion takes the form:

$$f\{\rho(\mathbf{r})\} = -\frac{1}{2}k_0^2(\nabla\rho)^2 + \frac{1}{4}(\nabla^2\rho)^2 + \sum_{n=2}^{\infty} a_n\rho^n, \qquad (2.1)$$

where k_0 and a_n are parameters which depend on temperature and the other thermodynamic variables which define the system. For simplicity, higher order terms in the gradient operator ∇ have been neglected. The Landau free energy functional $F\{\rho(\mathbf{r})\}$ is given by:

$$F\{\rho(\mathbf{r})\} = \int f\{\rho(\mathbf{r})\}d\mathbf{r}. \qquad (2.2)$$

The density function $\rho(\mathbf{r})$ of the solid phase is obtained by minimizing the free energy. This is a formidable task and cannot be carried out in general. However, one may use the expansion (2.1) to calculate the free energy of given structures in order to find out which one of them is the most stable. This is usually done by truncating the expansion (2.1) and keeping only low-order terms in ρ. The results of such analyses clearly depend on the phenomenological parameters which appear in the free energy expansion and on the truncation scheme which is adopted. However, they may provide a general picture of the possible crystalline and quasicrystalline structures which may be obtained in the condensed phase. In the following, we review the results of these studies which have been carried out in recent years.

The Landau theory approach has first been applied by Baym *et al.* (1971) and later by Alexander and McTague (1978) to study the possible crystallographic structures of the solid below the transition. This analysis suggests that if the transition is almost second-order, such that the expansion (2.1) may be truncated at the lowest nontrivial order, the expected crystalline structure is body centered cubic (bcc). To carry out this analysis, one first rewrites the free energy (2.2) in terms of the Fourier components ρ_k of the local density $\rho(\mathbf{r})$. The components ρ_k which minimize the second-order term in (2.2) are those for which the magnitude of the vector $\mathbf{k}$ satisfies $|\mathbf{k}| = k_0$. It has therefore been assumed that the order parameter $p(\mathbf{r})$ may be written in terms of Fourier components ρ_k with $|\mathbf{k}| = k_0$, and that higher harmonics may be neglected. In order to determine the actual structure below the transition, one has to consider higher order terms in the expansion. Within this single shell scheme, one finds that to third order in $\rho(\mathbf{r})$ the free energy (2.2) becomes:

$$F\{\rho(\mathbf{r})\} = A_2 \sum_k \rho_k \rho_{-k} + a_3 \sum_{k_i} \rho_{k_1}\rho_{k_2}\rho_{k_3}\delta(\mathbf{k}_1 + \mathbf{k}_2 + \mathbf{k}_3) + O(\rho^4), \tag{2.3}$$

where $A_2 = a_2 - \frac{1}{4}k_0^4$, and the sums are restricted to wavevectors satisfying $|\mathbf{k}_i| = k_0$. One may now consider density functions $\rho(\mathbf{r})$ with various crystallographic symmetries such as simple cubic (sc), bcc, face centered cubic (fcc), or hexagonal close-packed (hcp) and calculate their free energy by using (2.3). For example, for sc structures, one takes:

$$\rho(\mathbf{r}) = \sum_{l=1}^{3} \rho l e^{i\mathbf{k}_l \cdot \mathbf{r}} + c.c., \tag{2.4}$$

where the sum runs over the three pairs of wavevectors $\pm\mathbf{k}_l$, forming the edges of a cube [Fig. 1(a)]. Here we have denoted ρ_{k_l} by ρ_l. Similarly for a bcc structure, $\rho(\mathbf{r})$ is given by a similar expression where the sum runs over the six pairs of wavevectors, forming the edges of an octahedron [Fig. 1(b)]. One proceeds by inserting these density functions into (2.3) and by minimizing the free energy. When truncated at the third-order term, the free energy (2.3) is clearly unstable. To avoid this instability, one minimizes (2.3) subject to the constraint $\Sigma\ \rho p_k \rho_{-k} =$ const. This approach

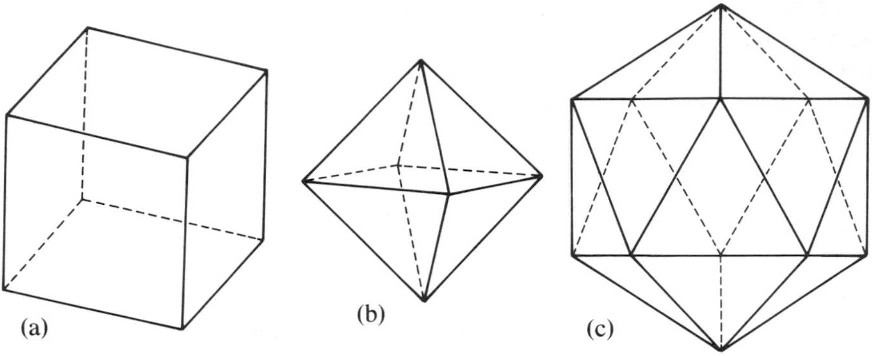

Figure 1. Regular polyhedra formed by the wavevectors in reciprocal space associated with the sc (a), bcc (b), and icosahedral (c) structures.

is equivalent to minimizing the free energy (2.3) but with a quartic term which is proportional to the square of the quadratic local bulk term per unit volume, namely $(\Sigma\ \rho_k \rho_{-k})^2$ (see Jarić, 1985; Troian, 1986). Such quartic term is nonlocal. It corresponds to an infinite range interaction of the form

$$\frac{1}{V} \int d\mathbf{r} d\mathbf{r}'\ \rho(\mathbf{r})^2 \rho(\mathbf{r}')^2, \tag{2.5}$$

where V is the volume of the system. Carrying out this minimization, one finds that the sc and fcc structures yield zero contributions to the third-order term in (2.3). The reason is that one cannot form equilateral triangles out of the wavevectors $\mathbf{k}_l$ which define these structures. On the other hand, bcc and hexagonal structures do yield nonzero contributions and therefore have lower free energy. The structure which minimizes the free energy (2.3) was found to be the bcc. This conclusion is expected to be valid only if the liquid-to-solid transition is almost second-order, such that higher order terms in (2.3) may be neglected near the transition. However, far from the transition, or in the case that the transition is strongly first-order, higher order terms may become important, leading to crystallographic structures which are not bcc. This result may also change when the analysis is carried out by using the full local quartic term rather than the nonlocal one (2.5). It has been found (Mermin and Troian, 1985) that at high temperatures, in the interval $-0.055 < A_2 < 0.089$, just below the liquid-solid transition, the bcc phase is indeed the most stable one. However, at lower temperatures, other structures become more stable. For $-2.15 < A_2 < -0.055$, the hexagonal phase wins, while for $A_2 < -2.15$ a one-dimensional, smecticlike structure becomes more stable.

This approach has recently been extended (Bak, 1985a,b) to consider the stability of quasicrystalline structures with an icosahedral symmetry. Here one expresses the density $\rho(\mathbf{r})$ by Fourier components associated with the fifteen pairs of wavevectors which form the edges of a regular icosahedron [Fig. 1(c)]. Inserting this density into (2.3), one finds that although it yields a nonzero contribution to the third-order terms, its free energy is higher than that of the bcc phase. However, when higher order terms are taken into account, and in particular for a sufficiently large fifth-order term,

$$a_5 \sum_{k_i} \rho_{k_1}\rho_{k_2}\rho_{k_3}\rho_{k_4}\rho_{k_5}\delta(k_1 + k_2 + k_3 + k_4 + k_5), \tag{2.6}$$

the icosahedral structure may become favorable. This is due to the many regular pentagons which can be formed from the set of **k**-vectors of the icosahedral structure.

A slightly different approach for studying the stability of the icosahedral phase has been taken by Kalugin *et al.* (1985), who argued that the effect of higher order harmonics in the density function may be important in determining the free energy of quasicrystalline structures. The point is that unlike in crystalline structures, the length of the wavevectors of some higher harmonic components may be sufficiently close to k_0. These harmonics may therefore yield a substantial contribution to the free energy. They considered a density function which is composed of two classes of wavevectors:

$$\rho(\mathbf{r}) = \rho_1 \sum_{l=1}^{6} e^{i\mathbf{k}_l\cdot\mathbf{r}} + \rho_2 \sum_{mn} e^{i\mathbf{k}_{mn}\cdot\mathbf{r}} + c.c., \tag{2.7}$$

where the first sum runs over the six pairs of wavevectors $\pm\mathbf{k}_l$ which connect the center to the twelve corners of a regular icosahedron. The second sum is over the fifteen pairs $\mathbf{k}_{mn} \equiv \mathbf{k}_m - \mathbf{k}_n$ of adjacent corners of the icosahedron. The length of the edge vectors $\boldsymbol{k}_{mn}$ is only about 5% larger than that of the vectors $\mathbf{k}_l$. Kalugin *et al.* (1985) have calculated the free energy associated with this density to third order in $\rho(\mathbf{r})$, subject to the constraint $\Sigma\, \rho_k\rho_{-k}$ = const. The free energy is of the same form as (2.3), except that the sums are over the two sets of wavevectors $\mathbf{k}_l$ and $\mathbf{k}_{mn}$. Moreover, since $|\mathbf{k}_{mn}| \neq |\mathbf{k}_l| \equiv k_0$, one has to consider the dependence of A_2 on the magnitude of the wavevectors. To second order in $|\mathbf{k} - \mathbf{k}_0|$, one has:

$$A_2 = a_2 - \frac{1}{4}k_0^4 + k_0^4\left(\frac{k}{k_0} - 1\right)^2. \tag{2.8}$$

By minimizing this free energy with respect to ρ_1 and ρ_2, it was found that the icosahedral structure is the most stable one for $k_0^4/\bar{a}_2 < 69$, where $\bar{a}_2$

$= a_2 - \frac{1}{4}k_0^4$. For $k_0^4/\bar{a}_2 > 69$, the bcc phase becomes more favorable.

This analysis has been extended to include the full local fourth-order term (Biham *et al.*, 1986a; Troian, 1986). It has been found that in this case the bcc structure has a lower free energy for any value of k_0. More recently, Gronlund and Mermin (1987) have further extended this analysis by allowing the amplitudes ρ_k of wavevectors belonging to the same shell to be different from each other. They found that if such amplitude variations are allowed and when a local fourth-order term is considered, the icosahedral structure is not even a local minimum of the free energy.

A different, although closely related approach to that of Kalugin *et al.* (1985) has been adopted by Mermin and Troian (1985). They, too, consider a density function $\rho(\mathbf{r})$ which can be written in terms of Fourier components belonging to two shells in reciprocal space, with radii k and q, respectively. However, unlike the work of Kalugin *et al.*, they left the ratio q/k as a free parameter. In this approach, it is assumed that the k-components of the density function serve as a primary order parameter. The q-components on the other hand are secondary order parameters, and they are induced by the primary ones. This requires a coupling of the form $\rho_q\rho_{k_1} \ldots \rho_{k_n}$, where $|\mathbf{q}| = q$, $|\mathbf{k}_i| = k$, $i = 1, \ldots n$, $\mathbf{q} + \mathbf{k}_1 + \ldots + \mathbf{k}_n = 0$, and $n > 1$. If one considers, for example, the case $n = 2$, such a coupling may contribute to the free energy, provided $0 < q < 2k$. The ratio q/k such that it corresponds to the ratio of the length of the higher harmonics of the basic icosahedral star of vertex vectors to the length of these basic vertex vectors. Possible choices for q/k are 1.0515, as in the case of Kalugin *et al.*, 1.7013, and many others. It has been argued by Mermin and Troian (1985) that in this way, one may stabilize an icosahedral structure. One may study the effect of the secondary order parameter by minimizing the free energy with respect to it. Since this order parameter is induced by the primary one, it is sufficient to truncate the free energy, keeping terms up to quadratic order in the secondary order parameter. In this way, one may express the secondary order parameter in terms of the primary one. Inserting this expression in the free energy density, one obtains an effective single-shell free energy. In this model, however, the quartic term is not purely local, as in the starting Landau model. It has the form:

$$\sum_{k_i} a_4(\mathbf{k}_1,\mathbf{k}_2,\mathbf{k}_3,\mathbf{k}_4),\rho_{k_1}\rho_{k_2}\rho_{k_3}\rho_{k_4}\delta(\mathbf{k}_1 + \mathbf{k}_2 + \mathbf{k}_3 + \mathbf{k}_4), \tag{2.9}$$

where a_4 is not a constant, but rather a function of the $\mathbf{k}$ vectors. In fact, due to rotational invariance, a_4 may depend only on the scalar products $\mathbf{k}_1 \cdot \mathbf{k}_2$ and $\mathbf{k}_2 \cdot \mathbf{k}_3$ (see Jarić, 1985) since these scalar products uniquely define the relative orientations of these four vectors. It is this dependence of a_4 on the **k** vectors which makes the icosahedral structure more favorable for appropriate choices of q/k.

Another possibility which has been studied within the Landau theory is that the icosahedral phase exhibits LR orientational order but not LR translational order close to the transition to the liquid phase. This approach (Jarić, 1985) was motivated by the experimental observation that many supercooled liquids and metallic glasses show short-range icosahedral bond-orientational order. Moreover, it fits nicely with the X-ray diffraction data which indicate that the translational order in the systems studied is only short-range. The order parameter for the orientational transition is $Q(\hat{n})$, which gives the density of bonds in the direction $\hat{n}$. This order parameter may be expanded in terms of the spherical harmonics $Y_{LM}(\mathbf{n})$ with expansion coefficients Q_{LM}. As usual one considers an order parameter associated with a single irreducible representation L of the rotational group. The lowest L which may accomodate icosahedral symmetry is $L = 6$ (Jarić *et al.*, 1984), and therefore the order parameter which is taken to describe the transition is Q_{6M}. Expanding the Landau free energy in powers of this 13-component order parameter, it was found that the model has one third-order and three fourth-order terms. Minimizing this free energy is a very complex problem. This minimization was carried out using group-theoretical and numerical techniques. It was found that there exists a region in the three-dimensional parameter space which defines the model in which the icosahedral phase is stable. This structure is obtained as a result of a first-order transition from the isotropic liquid phase.

An interesting possibility which may be examined within this approach is the onset of LR translational order at some temperature below the transition to the orientationally ordered phase. In this case, one has to consider a transition from a translationally invariant icosahedral phase to a phase with the same point symmetry but with quasicrystalline LR order. The order parameter of such transition is the density function $\rho(\mathbf{r})$. The directions of the **k**-vectors associated with the primary order parameter will now be determined by the anisotropy of the high-temperature phase. By considering the quadratic term in $\rho(\mathbf{r})$ and by expanding it to sixth order in **k**, it has been found (Jarić, 1985) that the fundamental star of wave-vectors is either the icosahedral vertex or the face vectors, pointing from the origin to the center of the twenty faces. The edge vectors are never favored in this scheme.

2.2 Density Functional Theory

The Landau theory presented in the previous section is phenomenological in nature. It provides a relatively simple tool for studying the possible crystalline or quasicrystalline structures which may occur below the liquid–solid transition. In this theory, the free energy depends on many parameters whose signs and magnitudes are determined by *ad hoc* assumptions. It therefore cannot predict the structure of a given system but rather makes a statement about the possible structures in general. On the other hand, the mean-field density functional theory, introduced by Ramakrishnan and Yussouff (1979), provides a scheme for determining the coefficients in the Landau expansion to all orders. It is then a matter of extensive numerical calculations to minimize the free energy and to determine the resulting structure. In the following, we briefly describe this approach and the results obtained by applying it to quasicrystalline systems.

The physical idea behind the density functional theory is the fact that at the (first-order) transition temperature, the translational correlation length is of the order of a few atomic spacings. One can therefore divide the liquid into cells with dimensions of the order of the correlation length, with a local density $\rho(\mathbf{r})$ associated with the cell at $\mathbf{r}$. Note that here $\rho(\mathbf{r})$ describes the full density, and that the density of the liquid, ρ_0, has not been subtracted. Treating all correlations at distances greater than the correlation length within mean-field approximation, the free energy associated with $\rho(\mathbf{r})$ is found to be

$$\begin{aligned} \frac{1}{K_B T} F\{\rho\} = & \int d\mathbf{r}\rho(\mathbf{r}) \left[ln\left(\frac{\rho(\mathbf{r})}{\rho_0}\right) - 1 \right] \\ & - \frac{1}{2} \int d\mathbf{r}_1 d\mathbf{r}_2 C(\mathbf{r}_1 - \mathbf{r}_2)[\rho(\mathbf{r}_1) - \rho_0][\rho(\mathbf{r}_2) - \rho_0], \end{aligned} \tag{2.10}$$

where T is the temperature. The first term in this free energy is the entropy associated with distributing the particles into the various cells, while the second term is the leading term in the expansion of the energy associated with the variations of the density. The direct pair-correlation function $C(\mathbf{r}_1 - \mathbf{r}_2)$ is determined by fitting the observed static structure factor $\langle \rho_k \rho_{-k} \rangle$. Here C_k is the Fourier component of $C(\mathbf{r}_1 - \mathbf{r}_2)$.

This approach has been applied by Sachdev and Nelson (1985) to study the possible structures which may be obtained in the solid phase. In this study, the free energy was obtained by using the experimental structure factor of amorphous cobalt and that obtained from a relaxed dense random packing model. One then assumes a structure with a given symmetry (like bcc, fcc, or icosahedral structure with either vertex, edge, or face fundamental set of wavevectors), and calculates the free energy associated with this structure. It turns out that one has to keep a large number of shells in $\mathbf{k}$-space (around 100 shells for the icosahedral structures) in order to obtain accurate free energies. These studies, therefore, involve extensive numerical calculations. It has been found that although the icosahedral phase has a lower free energy than the undercooled liquid, the bcc and fcc structures have even lower free energies, with the fcc being the most stable one.

2.3 *Numerical Simulations*

The studies described so far do not provide a microscopic picture of the atomic arrangement in space. To gain insight into the structure of quasicrystals and to study their stability (both local and global), numerical simulations have recently been carried out (Widom *et al.*, 1987). The model considered in this study is a simple two-component Lennard–Jones system in two dimensions. The potential was taken to be of the form

$$V_{\alpha\beta}(r) = E_{\alpha\beta}\left[\left(\frac{\sigma_{\alpha\beta}}{r}\right)^{12} - 2\left(\frac{\sigma_{\alpha\beta}}{r}\right)^{6}\right], \tag{2.11}$$

where $\alpha = L$ or S denotes large or small atoms, $\sigma_{\alpha\beta}$ is the bond length, and $E_{\alpha\beta}$ is the bond strength. The bond lengths are chosen to promote local decagonal order and are taken to be

$$\begin{aligned} \sigma_{LS} &= 1.0, \\ \sigma_{LL} &= 2sin36° = 1.176, \\ \sigma_{SS} &= 2sin18° = 0.618. \end{aligned} \tag{2.12}$$

In addition, to avoid phase separation into single-species triangular lattices, the interaction E_{LS} between unlike species was chosen such that it dominates the interactions E_{LL} and E_{SS} between like species. It was found that at low temperatures the system spontancously forms a quasicrystalline state with a decagonal symmetry. This state, which seems to be in a thermodynamic equilibrium, is characterized by long-ranged decagonal bond-orientational order, but with short-range positional order.

3. Local Stability and Deformations in Quasicrystals

3.1 *Local Stability*

The studies described so far (with the exception of the numerical simulations) are concerned with the question of *global* stability of icosahedral phases relative to ordinary crystalline structures. However, the Landau theory may also be used to study the fundamental question of *local* stability of these phases. In particular, the local stability to deformations which make quasicrystals commensurate (and thus periodic) in one or more directions may be analyzed. It has recently been shown (Biham *et al.*, 1986a,b) that, indeed, icosahedral structures are generically stable to such deformations and may thus exist as thermodynamically stable phases with both positional and orientational LRO. Let us introduce some of the considerations which are involved in this analysis. We consider for simplicity an icosahedral density function $\rho(\mathbf{r})$ of the form[1]:

$$\rho(\mathbf{r}) = \sum_{l=1}^{6} \rho_l \cos(\mathbf{k}_l \cdot \mathbf{r} + \varphi_l), \tag{3.1}$$

where the six reciprocal vectors $\mathbf{k}_l$, $l = 1,\ldots,6$ connect the center of the icosahedron to its six vertices in the upper half-space ($z > 0$), and $|\mathbf{k}_l| = k_0$. More explicitly, we choose the vectors $\mathbf{k}_l$ to have the following form (see Fig. 2):

$$\begin{aligned} \mathbf{k}_1 &= k_0(0,0,1) \\ \mathbf{k}_l &= k_0\,[\sin\xi\cos(l-2)\eta, \sin\xi\sin(l-2)\eta, \cos\xi]\ l = 2,\ldots,6, \end{aligned} \tag{3.2}$$

where $\xi = \cos^{-1}(1/\sqrt{5})$ and $\eta = 2\pi/5$. The amplitudes ρ_l and the phases φ_l are real quantities. In order to preserve the icosahedral symmetry, one has to take $\rho_l = \rho_0$, $l = 1,\ldots,6$. In general, $\rho(\mathbf{r})$ should include higher harmonics of the six fundamantal reciprocal vectors $\mathbf{k}_l$. For simplicity we use an order parameter of the form (3.1), neglecting higher harmonics. Taking into account the higher harmonics would not change the results of this analysis. To study the local stability of the structure (3.1), we let it deform and calculate the free energy associated with this deformation. We consider deformations in which the phases φl become $\mathbf{r}$-dependent, keeping the amplitudes fixed. This approach is justified, since phase excitations are gapless (these are the phonon and phason modes),[2] while amplitude fluctuations involve a finite gap. The amplitude excitations may therefore be neglected in considering the local stability of the structure

[1]The density $\rho(\mathbf{r})$ has at most one point which possesses the full icosahedral symmetry. For example, if $\varphi_l = 0$, $l = 1, \ldots, 6$, this point is the origin. For a general set of phases φ_l, the density $\rho(\mathbf{r})$ does not have a center of symmetry. However, after rotating the structure around an arbitrary point, one can shift the phases φ_l such that $\rho(\mathbf{r})$ will come back to itself. In this paper, we allow such shifts in φ_l. Therefore, when we say that $\rho(\mathbf{r})$ has icosahedral symmetry, it means that it is invariant under a set of symmetry operations which are combinations of icosahedral rotations and phase translations.

[2]It has been conjectured that the phason modes are not necessarily gapless (Bak, 1985a,b; Levine *et al.*, 1985), namely that if the interaction between the six density waves is sufficiently strong, they may become locked to each other, as in one-dimensional incommensurate crystals (Peyrard and Aubry, 1983). In this analysis, it is assumed that these degrees of freedom do not exhibit a gap. Within the framework of the Landau theory, this indeed is the case.

(3.1). We proceed by inserting (3.1) into (2.1), assuming that the phases φ_l are slowly varying functions of $\mathbf{r}$. We find that the resulting free-energy density may be written as a sum of two contributions:

$$f\{\rho(\mathbf{r})\} = f_{\text{el}}\{\nabla\varphi_l\} + f_{\text{lc}}\{\varphi_l,\nabla\varphi_l,\mathbf{r}\},\ l = 1,\ldots,6, \tag{3.3}$$

where f_{el} is the elastic part of the free energy, and f_{lc} is the lock-in part. The elastic free energy f_{el} is associated with the gradient terms in (2.1). To the second order in $\nabla\varphi_l$, it takes the form:

$$f_{\text{el}} = \rho_0^2 \sum_{l=1}^{6} \left[a(\mathbf{k}_l \cdot \nabla\varphi_l)^2 + b(\nabla\varphi_l)^2 \right], \tag{3.4}$$

where a and b are constants. On the other hand, the lock-in free energy f_{lc} is associated with terms like $a_n\rho^n$ in (2.1). It consists of a sum of terms of the form:

$$- A(n)cos \left[\sum_{l=1}^{6} n_l(\mathbf{k}_l\cdot\mathbf{r} + \varphi_l) \right], \tag{3.5}$$

where n_l are integers satisfying $\Sigma\,|n_l| = n$, and to the lowest order in ρ_0, $A(n) = a_n\rho_0^n$. For phases φ_l which are not $\mathbf{r}$-dependent, these terms vanish upon integration. The reason is that the argument of the cosine term in (3.5) is nonzero for any set of integers (other than the trivial set $n_l = 0$). However, when the structure $\rho(\mathbf{r})$ is deformed and the phases φ_l become $\mathbf{r}$ dependent, such terms may yield a finite and negative contri-

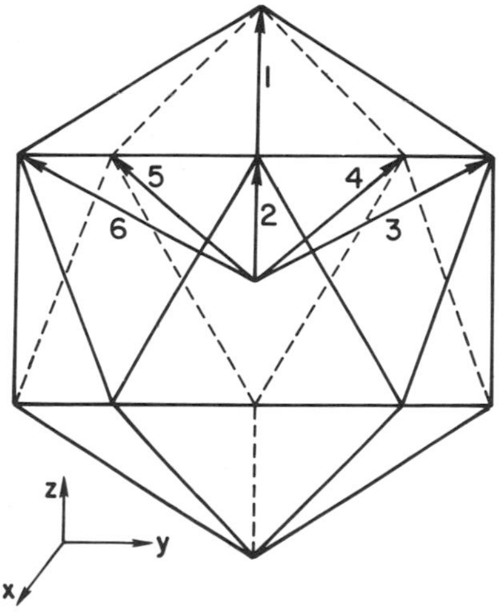

Figure 2. The six reciprocal basic vectors $\mathbf{k}_1, \ldots, \mathbf{k}_6$ associated with the icosahedral structure.

bution to the free energy. In particular, in considering distortions for which $\varphi_l = \boldsymbol{\alpha}_l \cdot \mathbf{r}$, one replaces the basic wavevectors $\mathbf{k}_l$ by the vectors

$$\mathbf{k}'_l = \mathbf{k}_l + \boldsymbol{\alpha}_l, \qquad l = 1,\ldots,6. \tag{3.6}$$

For certain choices of the vectors $\boldsymbol{\alpha}_l$, one may find nonzero integers n_l, such that $\Sigma n_l \mathbf{k}'_l = 0$. In this case, the argument of the cosine function in (3.5) becomes a constant, yielding a gain of order $-a_n\rho_0^n$ in the free energy. On the other hand, the elastic free-energy cost of this distortion is of the order of $\Sigma \mid \boldsymbol{\alpha}_l \mid^2$.

We now estimate the energy balance between the lock-in terms which favor such distortions and the elastic terms which tend to resist them. As an example, we consider a distortion of the icosahedral structure which makes it commensurate along the z direction, keeping the fivefold rotation symmetry in this direction. When such deformation takes place, the vector $\mathbf{k}_1$ which lies along the z direction becomes commensurate with the z components of the other vectors $\mathbf{k}_l$, $l = 2, \ldots 6$. The terms in the free energy associated with such deformations take the form:

$$-a_n\rho_0^n \cos\left[\left(q\mathbf{k}_1 - p\sum_{l=2}^{6}\mathbf{k}_l\right)\cdot\mathbf{r} + q\varphi_1 - p\sum_{l=2}^{6}\varphi_l\right], \tag{3.7}$$

where $n = q + 5p$. This term yields a nonzero contribution to the free energy if:

$$q\mathbf{k}_1 - p\sum_{l=2}^{6}\mathbf{k}_l = -\left(q\boldsymbol{\alpha}_1 - p\sum_{l=2}^{6}\boldsymbol{\alpha}_l\right). \tag{3.8}$$

For simplicity, we consider a distortion for which $\boldsymbol{\alpha}_1 \neq 0$ and $\boldsymbol{\alpha}_l = 0$, $l = 2,\ldots,6$. In this case:

$$\boldsymbol{\alpha}_1 = -(\sqrt{5} - 5p/q)\, k\hat{z}, \tag{3.9}$$

where $k = k_0/\sqrt{5}$, and the elastic free-energy cost associated with the distortion is of the order of:

$$|\boldsymbol{\alpha}_1|^2 = \left(\sqrt{5} - \frac{5p}{q}\right)^2 k^2. \tag{3.10}$$

In order to find the free-energy cost of this distortion, we have to estimate the difference between the irrational number $\sqrt{5}$ and the closest rational $5p/q$ for a given denominator q. Note that $\sqrt{5}$ is a quadratic irrational.[3] For any quadratic irrational τ, there is a constant C such that[4]:

[3]Quadratic irrational is an irrational number which satisfies an algebraic equation of the second degree with rational coefficients.

[4]Note that other irrational numbers do not necessarily satisfy this condition. For example, for any Liouville number ξ, and every positive integer m, there is a rational number p/q and ξ is less than $1/q^m$ (Niven, 1956).

$$|\tau - p/q| > C/q^2 \tag{3.11}$$

for all the rationals p/q (Niven, 1956). Therefore, the elastic free-energy cost associated with a commensurate deformation is of the order of:

$$|\boldsymbol{\alpha}_1|^2 > \frac{C^2k^2}{q^4} > \frac{C^2k^2}{n^4}. \tag{3.12}$$

By combining this with a gain of $-a_n\rho_0^n$, associated with the commensurability energy, we find that the free-energy difference between the icosahedral phase and the distorted phase is

$$\Delta f \sim \frac{1}{n^4} - a_n\rho_0^n. \tag{3.13}$$

For small order parameter $\rho_0 < 1$, to which this analysis applies, the gain decays exponentially with n, while the loss decays only algebraically. Therefore, Δf is expected to be positive for large n, and the icosahedral structure is stable to deformations which make it periodic along the z axis with very long periods. If the coefficients a_n are such that $\Delta f > 0$ also for small n, the icosahedral structure is stable! On the other hand, if for some small $n = n_0$, Δf becomes negative, the icosahedral phase becomes unstable, exhibiting a commensurate distortion along the z axis. Both possibilities are generic, and therefore we conclude that icosahedral structures *may* exist as stable phases.

To complete this analysis, one has to consider other possible distortions not necessarily along the fivefold axis. The elastic energy associated with such distortion takes the form:

$$\sum_{l=1}^{6} |\boldsymbol{\alpha}_l|^2 = \sum_{l=1}^{6} [|\alpha_{lx}|^2 + |\alpha_{ly}|^2 + |\alpha_{lz}|^2]. \tag{3.14}$$

Since the incommensurate ratios of the icosahedron in the x, y, and z directions are quadratic irrationals, the former analysis can be applied in each axis direction separately. Therefore, the icosahedral structure may be stable to deformations in all directions. Similar analysis for planar quasicrystals with n >6-fold rotation axis shows that these structures are also generically locally stable.

3.2 Elastic Deformations

Having demonstrated that icosahedral structures may be stable to small distortions, we study in this section the possible instabilities and deformations of these systems. We start by considering deformations which are associated with the elastic free energy $f_{\text{el}}\{(\nabla\varphi l)\}$. Since the question of commensurability does not play a role in these deformations, the analysis is similar to that of elastic transitions in ordinary crystals. The elastic free energy may be expanded in powers of $\nabla\varphi_l$, keeping those terms which are invariant under the icosahedral point group. To carry out this analysis, we first note that the phases φ_l, $l = 1, \ldots, 6$ transform as a six-dimensional representation of the icosahedral group. Following Bak (1985b), we denote this representation by Δ (see Table 1). This representation decomposes into two three-dimensional irreducible representations

$$\Delta = \Gamma_3 + \Gamma_{3'}, \tag{3.15}$$

where Γ_3 corresponds to the three phonon modes **u**, and $\Gamma_{3'}$ corresponds to the three phason modes **v**. The character table of the icosahedral group in which Γ_3 and $\Gamma_{3'}$ are defined is given in Table 1. Since the gradient operator ∇ transforms as Γ_3, the gradients $\nabla\varphi_l$, $l = 1,\ldots,6$ transforms as $\Gamma_3 \times \Delta$. The decomposition of $\Gamma_3 \times \Delta$ to irreducible representations associated with phonon and phason deformations takes the form:

$$\Gamma_3 \times \Gamma_3 = \Gamma_1 + \Gamma_3 + \Gamma_5 \tag{3.16a}$$

$$\Gamma_3 \times \Gamma_{3'} = \Gamma_4 + \Gamma_5. \tag{3.16b}$$

The representation Γ_1 is the unit representation which does not break the symmetry. It corresponds to the bulk modulus. Moreover, the representation Γ_3, which is associated with the phonon degrees of freedom, is antisymmetric and, therefore, there is no elastic energy associated with it. As a result, the only representations which correspond to elastic phase transitions are Γ_4 and Γ_5. In the following, we analyze the elastic transitions associated with these representations. We express the elastic free energy in terms of the basis functions of the corresponding representation. This analysis has previously been carried out to second order in $\nabla\varphi_l$ (Bak, 1985b; Levine *et al.*, 1985). However, in order to find the symmetries

Table 1. The character table of the icosahedral rotation group ($G = (\sqrt{5}\text{-}1)/2$ is the golden mean).

	E	$12C_5$	$12C_5^2$	$20C_3$	$15C_2$
Γ_1	1	1	1	1	1
Γ_3	3	1+G	−G	0	−1
$\Gamma_{3'}$	3	−G	1+G	0	−1
Γ_4	4	−1	−1	1	0
Γ_5	5	0	0	−1	1
Δ	6	1	1	0	−2
$[\Gamma_4^3]$	20	0	0	2	0
$[\Gamma_5^3]$	35	0	0	2	3

which may result from these transitions, one has to consider the elastic free energy to at least third order in $\nabla\varphi_l$. The number of third-order invariants associated with the representations Γ_m, $m = 4,5$, is found using standard group theoretical methods. This is done by constructing the character table of the representation $[\Gamma_m^3]$, $m = 4,5$, where the square brackets denote the symmetric part of the representation (Lyubarskii, 1960). The number of third-order invariants is given by the number of times the unit representation is contained in this representation. We find that Γ_5 has two third-order invariants, while Γ_4 has only one. In the following, we construct these invariants and analyze the elastic transitions associated with Γ_5 and Γ_4.

Transitions associated with Γ_5. The representation Γ_5 is five-dimensional, whose basic functions are the spherical harmonics Y_{2m}, $m = -2,\ldots,2$. Define the following real-basis functions:

$$\hat{\xi}_1 = \sqrt{2}Y_{20},$$

$$\hat{\xi}_2 = (Y_{22} + Y_{2-2}),$$

$$\hat{\xi}_3 = -i(Y_{22} - Y_{2-2}),$$

$$\hat{\xi}_4 = -(Y_{21} - Y_{2-1}),$$

and

$$\hat{\xi}_5 = i(Y_{21} + Y_{2-1}).$$

Here $\hat{z}$ is taken to lie along one of the fivefold axes, and $\hat{y}$ lies along one of the perpendicular twofold axes. Let $\Sigma_{l=1}^5 \xi_l \hat{\xi}_l$ be the order parameter associated with this transition. The elastic free energy to the third order in ξ_l takes the form:

$$E = \alpha \sum_{l=1}^{5} \xi_l^2 + \beta I_0 + \gamma I_1, \tag{3.17a}$$

where

$$I_0 = \xi_1 \left[\frac{1}{3}\xi_1^2 - (\xi_2^2 + \xi_3^2) + \frac{1}{2}(\xi_4^2 + \xi_5^2) \right] + \frac{\sqrt{3}}{2}\left[\xi_2(\xi_4^2 - \xi_5^2) + 2\xi_3\xi_4\xi_5 \right], \tag{3.17b}$$

$$I_1 = -\xi_1 \left[\frac{1}{3}\xi_1^2 - (\xi_4^2 + \xi_5^2) + \frac{1}{2}(\xi_2^2 + \xi_3^2) \right] + \frac{\sqrt{3}}{2}\left[\xi_4(\xi_3^2 - \xi_2^2) + 2\xi_2\xi_3\xi_5 \right]. \tag{3.17c}$$

In order to find the symmetry which results when elastic instability associated with the representation Γ_5 takes place, one has to minimize the energy (3.17) under the constraint $\Sigma\xi_l^2 = 1$. Two types of solutions are found:

(a) for $\beta\gamma < 0$ the energy is minimized by $\xi_1 \neq 0$, $\xi_l = 0$, $l = 2,\ldots,5$. In this case, the resulting symmetry is D_{5d}.

(b) for $\beta\gamma > 0$ the energy is minimized by $\xi_1 = (1 + 2G)/5$, $\xi_2 = 2\sqrt{3}(2 - G)/15$, $\xi_4 = -2\sqrt{30 + 15G}/15$, $\xi_3 = \xi_5 = 0$. In this case, the resulting symmetry is D_{3d}. Along the lines $\beta = 0$ and $\gamma = 0$ in the $\beta - \gamma$ plane, there is infinite degeneracy, and the resulting symmetry is not uniquely defined. When both β and γ are nonzero, the degeneracy is removed, and the energy is minimized by a structure whose symmetry is either D_{5d} or D_{3d}. Note that although D_{2h} is a maximal subgroup of the icosahedral group, a phase with D_{2h} symmetry is not obtained by elastic deformations.

Transitions associated with Γ_4. The representation Γ_4 is four-dimensional and has the following real-basis functions (Butler, 1981):

$$\hat{\eta}_1 = (Y_{31} - Y_{3-1}),$$

$$\hat{\eta}_2 = -i(Y_{31} + Y_{3-1}),$$

$$\hat{\eta}_3 = \left[-\sqrt{\frac{2}{5}}(Y_{32} + Y_{3-2}) + \sqrt{\frac{3}{5}}(Y_{3-3} - Y_{33})\right],$$

$$\hat{\eta}_4 = i\left[\sqrt{\frac{2}{5}}(Y_{32} - Y_{3-2}) - \sqrt{\frac{3}{5}}(Y_{33} + Y_{3-3})\right].$$

Let $\Sigma_{l=1}^4 \eta_l\hat{\eta}_l$ be the order parameter associated with the distortion. To third order in η_l, the elastic free energy takes the form

$$E = a\sum_{l=1}^{4}\eta_l^2 + b[\eta_4(\eta_1^2 - \eta_2^2) + \eta_2(\eta_4^2 - \eta_3^2) - 2\eta_1\eta_3(\eta_2 + \eta_4)]. \tag{3.18}$$

Under the constraint $\Sigma\eta_l^2 = 1$, we find that for $b > 0$ the energy is minimized by $\eta_1 = \eta_3 = 0$, $\eta_2 = -\eta_4 = -1/\sqrt{2}$, while for $b < 0$ it is minimized by $\eta_1 = \eta_3 = 0$, $\eta_2 = -\eta_4 = 1/\sqrt{2}$. In both cases, the resulting phase is cubic with the point symmetry T_h.

In summary, elastic terms in icosahedral structures may lead to the following symmetries: D_{5d}, D_{3d}, and T_h. The first two are associated with Γ_5, while the last one is associated with Γ_4. The elastic transitions from the icosahedral phase to these phases are expected to be first-order, due to the existence of third-order terms in the corresponding Landau free energy. The structures which result from elastic instabilities in the icosahedral phase are incommensurate in all directions. These structures may become locked, and thus commensurate, in one or more directions as a result of lock-in terms in the free energy. The D_{5d} structure may become commensurate in the z direction without breaking the fivefold symmetry, while the two other structures may become commensurate in all directions. The lock-in transitions are considered in the following section.

The density functional theory of Ramakrishnan and Yussouff (1979) has recently been applied by Jarić and Mohanty (1987) to calculate the elastic tensor of the icosahedral phase. In these calculations, the experimental structure factor of amorphous cobalt has been used as an input. It has been found that this tensor has one negative eigenvalue, indicating that the icosahedral structure should exhibit an elastic instability.

3.3 Lock-in Transitions

The lock-in free energy contains terms of the form (3.5) which are explicitly **r**-dependent. In the following, we consider the effect of these terms on the structure of the icosahedral phase. It is expected that small lock-in terms would not distort the icosahedral structure but rather introduce higher harmonics to the structure (3.1). However, large lock-in terms may induce distortions which make the quasicrystal periodic in one or more directions. The lock-in free energy contains infinitely many terms, each favoring a different periodic order. For simplicity, we consider one such term which tends to distort the system along a fivefold axis and study its effect on the structure of the icosahedral phase. In particular, we take a term of the form (3.5), with $n_1 = 0$ and $n_2 = \cdots = n_6 = 1$. This term tends to distort the icosahedral structure and make it periodic along the fivefold axis $\mathbf{k}_1$. Clearly, for symmetry reasons, the lock-in free energy should include six such terms, corresponding to the six fivefold axes, $\mathbf{k}_l$, which exist in the icosahedral structure. Let $\mathbf{k}_l$, $l = 1, \ldots, 6$ be the six wavevectors associated with the icosahedral structure defined by (3.2), such that $\xi = \cos^{-1}(1/\sqrt{5})$ and $n = 2\pi/5$. We define six vectors:

$$\begin{aligned}
\mathbf{p}_1 &= -\mathbf{k}_2 - \mathbf{k}_3 - \mathbf{k}_4 - \mathbf{k}_5 - \mathbf{k}_6,\\
\mathbf{p}_2 &= -\mathbf{k}_1 - \mathbf{k}_3 + \mathbf{k}_4 + \mathbf{k}_5 - \mathbf{k}_6,\\
\mathbf{p}_3 &= -\mathbf{k}_1 - \mathbf{k}_2 - \mathbf{k}_4 + \mathbf{k}_5 + \mathbf{k}_6,\\
\mathbf{p}_4 &= -\mathbf{k}_1 + \mathbf{k}_2 - \mathbf{k}_3 - \mathbf{k}_5 + \mathbf{k}_6,\\
\mathbf{p}_5 &= -\mathbf{k}_1 + \mathbf{k}_2 + \mathbf{k}_3 - \mathbf{k}_4 - \mathbf{k}_6,\\
\mathbf{p}_6 &= -\mathbf{k}_1 - \mathbf{k}_2 + \mathbf{k}_3 + \mathbf{k}_4 - \mathbf{k}_5,
\end{aligned} \tag{3.19}$$

and six phases:

$$\begin{aligned}
\psi_1 &= -\varphi_2 - \varphi_3 - \varphi_4 - \varphi_5 - \varphi_6, \\
\psi_2 &= -\varphi_1 - \varphi_3 + \varphi_4 + \varphi_5 - \varphi_6, \\
\psi_3 &= -\varphi_1 - \varphi_2 - \varphi_4 + \varphi_5 + \varphi_6, \\
\psi_4 &= -\varphi_1 + \varphi_2 - \varphi_3 - \varphi_5 + \varphi_6, \\
\psi_5 &= -\varphi_1 + \varphi_2 + \varphi_3 - \varphi_4 - \varphi_6, \\
\psi_6 &= -\varphi_1 - \varphi_2 + \varphi_3 + \varphi_4 - \varphi_5.
\end{aligned} \tag{3.20}$$

The vectors $\mathbf{p}_l$ point along the six vertices of the icosahedron in the lower half-space ($z < 0$), satisfying:

$$\mathbf{p}_l = -\sqrt{5}\cdot\mathbf{k}_l, \; l = 1,\ldots,6. \tag{3.21}$$

The phases ψ_l are linearly independent, and therefore the free energy may be expressed by the ψ_l's rather than the φ_l's. It takes the form:

$$f = \sum_{l=1}^{6}\left[\frac{1}{2}\alpha(\nabla\psi_l)^2 - V_5 \cos\,(\mathbf{p}_l\cdot\mathbf{r} + \psi_l)\right], \tag{3.22}$$

where $V_5 = a_5\rho_0^5$. Here we have neglected higher-order terms in $\nabla\psi_l$ and considered only one lock-in term, V_5. The minimization of (3.22) with respect to ψ_l, $l = 1,\ldots,6$ leads to six uncoupled sine–Gordon equations:

$$\alpha\nabla^2\psi_l - V_5 \sin(\mathbf{p}\cdot\mathbf{r} + \psi_l) = 0, \; l = 1,\ldots,6. \tag{3.23}$$

These equations have two types of solutions, depending on the ratio V_5/α. For $V_5/\alpha > v_c \equiv 5(\pi k_0/4)^2$, the phases ψ_l satisfy $\psi_l = -\mathbf{p}_l\cdot\mathbf{r}$, while for $V_5/\alpha < v_c$, the phases exhibit a soliton lattice–type behavior. In the first case, one has $\varphi_l = -\mathbf{k}_l\cdot\mathbf{r}$, and the density $p(\mathbf{r})$ becomes $\mathbf{r}$-independent, corresponding to a liquid-like phase. On the other hand, for $V_5/\alpha < v_c$, the structure is incommensurate in all directions, exhibiting six arrays of planar domain walls perpendicular to the fivefold axes of the icosahedron. Here the icosahedral symmetry is preserved. This domain wall structure has recently been discussed by Chakraborty *et al.* (1986). As the coupling parameter V_5 is increased, the distance l between the domain walls increases, satisfying $l \to \infty$ as $V_5 \to v_c$. Inside the domains, the density $p(\mathbf{r})$ approaches a constant, independent of $\mathbf{r}$. In this limit, the icosahedral phase is composed of glassy domains separated by domain walls which are arranged in an icosahedral symmetry. The lock-in term which is considered in the free energy (3.22) does not break the icosahedral symmetry. The reason is that in the free energy (3.22), the six phases ψ_l are uncoupled, and therefore there is no energy associated with domain wall crossings. However, when higher order terms in p_0 are taken into account, the phases ψ_l may become coupled. Similar considerations have previously been applied to hexagonal crystals (Bak and Mukamel, 1979; Bak *et al.*, 1979). Depending on the specific form of the higher order terms, the structure may either stay icosahedral or distort. In the first case, the phases $\psi_l(\mathbf{r})$ are related to each other by the appropriate symmetry operation of the icosahedral group, while in the second case, this symmetry is broken. For example, one may find a uniaxial structure in which $\psi_1 = \mathbf{p}_1 \cdot \mathbf{r}$, while $\psi_2, \ldots, \psi_6$ exhibit a soliton-like solution which does not break the pentagonal symmetry in the xy plane. Such a structure is periodic along the z axis, and the symmetry is D_{10h}. The icosahedral structure (3.1) has a fivefold rotation axis in the z direction. This structure is not invariant under reflections through the xy plane. When this structure is distorted and becomes locked such that $\mathbf{p}_1 \cdot \mathbf{r} + \psi_1 = 0$, keeping the fivefold axis along the z direction, the resulting vectors $\mathbf{k}_2, \ldots, \mathbf{k}_6$ form a regular pentagon in the xy plane, while $\mathbf{k}_1$ does not change. This resulting structure is invariant under reflections through the xy plane; it has a tenfold rotation axis, and the symmetry is D_{10h}. Similar results have previously been obtained (Ho, 1986; Biham *et al.*, 1986a) by using the model presented by Kalugin *et al.* (1985). This model contains two harmonics of the icosahedral structure associated with the vertices and the edges of the icosahedron. In this model, there is tenth-order term which tends to distort the structure and to make it commensurate in the z direction, such that the distorted wavevectors form a pentagonal bipyramid. This structure is equivalent to the structure obtained in the present analysis and has the symmetry D_{10h}. In fact, the present analysis is more general and is applicable to a more general class of locked structures.[5]

Therefore, it is concluded that in considering distortions of the icosahedral phase along a fivefold axis (say z), two possible structures may be found:

(a) a D_{5d} phase which may result of elastic distortions. This structure is not periodic along the z axis.

(b) a periodic structure along z whose symmetry is D_{10h}. This structure is associated with the lock-in terms in the free energy. Clearly, by con-

[5]Note that for $\cos \xi = r/s$, where r and s are integers, the structure is periodic along the z axis. By considering a structure given by the fundamental **k**-vectors as well as the harmonics, one can show that for $s \neq 5n$, where n is an integer, this structure has a D_{10h} point symmetry. However, for $s = 5n$ (where r and s have no common divisor), the symmetry is D_{5d}. In the present analysis, we consider the transition from $\xi = \pi/2$ for which the symmetry is D_{10h} to a ξ for which $\cos \xi$ is irrational with the symmetry D_{5d}.

sidering other lock-in terms, one may find other structures in which the pentagonal symmetry in the *xy* plane is broken. We do not consider these phases in the present work.

By varying the interaction parameters in the free energy of the icosahedral structure, we expect to find three phases whose symmetries are icosahedral (IC), D_{5d}, and D_{10h} (see Fig. 3). In the remainder of this section, we examine the phase transitions between these phases. As was argued in the previous section, the transition from the icosahedral to the D_{5d} phase is first-order, due to the existence of a third-order term in the Landau free energy corresponding to this transition. The transition from the icosahedral to the D_{10h} phase is also expected to be first-order. The reason is that the D_{10h} phase is induced by a fifth- (or higher) order term in the Landau free energy, which may be neglected near a second-order transition. Also note that there is no group–subgroup relation between the icosahedral group and D_{10h}. We now consider the D_{5d}–D_{10h} transition in some detail and show that this transition which is a kind of commensurate–incommensurate transition is also first-order. To analyze the D_{10h}–D_{5d} transition, we consider a density $p(\mathbf{r})$ given by Eq. (3.1) with:

$$\mathbf{k}_1^d = k_0(0,0,1)$$

$$\mathbf{k}_l^d = k_0[sin\xi cos(l - 2)\eta, sin\xi \sin(l - 2)\eta, cos\ \xi]\ l = 2,\ldots,6, \quad (3.24)$$

where $\eta = 2\pi/5$. For $\xi = \pi/2$, the vectors $\mathbf{k}_l^d$, $l = 2,\ldots,6$ form a regular pentagon in the *xy* plane and the symmetry of $\rho(\mathbf{r})$ is D_{10h}. However, for

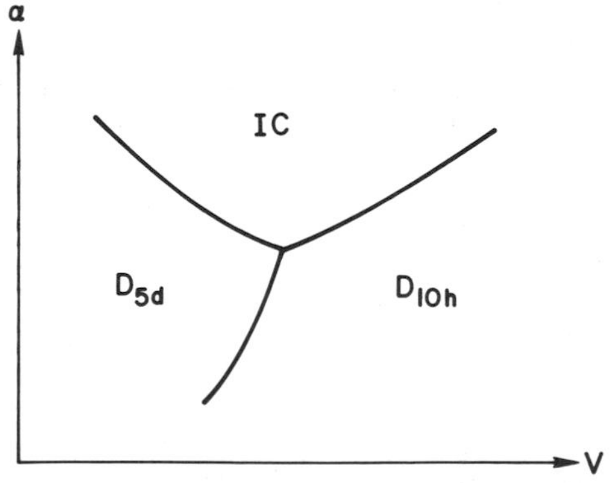

Figure 3. Schematic phase diagram: the icosahedral phase (IC) and the phases D_{5d} and D_{10h} which result from deformations of IC along the fivefold axis. The parameter V is the coefficient of the lock-in term, while α is the coefficient of the quadratic term in the elastic free energy.

$\xi \neq \pi/2$, the structure is not invariant under reflection through the xy plane, and the symmetry is D_{5d}. To study the D_{10h}–D_{5d} transition, we take $\xi = \pi/2$ and construct the Landau free energy associated with the phases $\varphi_1,\ldots,\varphi_6$. These phases form a reducible representation of the point group D_{10h}. This representation may be decomposed into six one-dimensional irreducible representations, whose basis functions are φ_1 and

$$\psi_l = \sum_{n=2}^{6} e^{iln\eta}\varphi_n, \; l = 0, \pm 1, \pm 2. \tag{3.25}$$

The elastic free energy associated with deformations along the z axis takes the form

$$f_{\text{el}} = A\left(\frac{\partial\varphi_1}{\partial z}\right)^2 + \alpha\left(\frac{\partial\psi_0}{\partial z}\right)^2 + B\left(\frac{\partial\psi_1}{\partial z}\right)\left(\frac{\partial\psi_{-1}}{\partial z}\right) + C\left(\frac{\partial\psi_2}{\partial z}\right)\left(\frac{\partial\psi_{-2}}{\partial z}\right). \tag{3.26}$$

We are interested in deformations which keep the fivefold symmetry in the xy plane. We therefore take $\psi_{\pm 1} = \psi_{\pm 2} = 0$. Moreover, $d\varphi_1/dz$ does not break the decagonal symmetry. It corresponds to dilatation along the z axis and may therefore be neglected in the present analysis. The only elastic strain which is associated with the D_{10h}–D_{5d} transition is thus $d\psi_0/dz$. The free energy corresponding to this order parameter takes the form (Luban, unpublished; Grinstein and Jayaprakash, 1982):

$$f = \frac{1}{2}\alpha\left(\frac{d\psi_0}{dz}\right)^2 + \frac{1}{4}\left(\frac{d\psi_0}{dz}\right)^4 + \frac{1}{2}\beta\left(\frac{d^2\psi_0}{dz^2}\right)^2 - V_5 cos\psi_0, \; \beta > 0, \tag{3.27}$$

where we have kept terms to fourth order in $d\psi_0/dz$, and to second order in $d^2\psi_0/dz^2$ and considered only the leading lock-in term, $V_5 = a_5\rho_0^5$. This free energy exhibits a commensurate–incommensurate transition as the parameter α is varied. Clearly, for $\alpha > 0$, the energy is minimized by the solution $\psi_0 = 0$. It may be shown that this solution minimizes the free energy even for negative but small α. However, as $-\alpha$ is increased, the system exhibits a first-order transition into an incommensurate state in which ψ_0 has a soliton lattice form. The resulting point symmetry is D_{5d}. The analysis of this transition is given in the Appendix.

The analysis presented in this section may also be carried out for lock-in terms which tend to distort the IC structure along one of the threefold axes, or along three mutually perpendicular twofold axes. It can be shown that in the first case the point symmetry may become D_{6h}, while in the second case it may become O_h. The corresponding phase diagrams are expected to exhibit IC–D_{3d}–D_{6h} and IC–T_h–O_h phases, respectively. The various transitions which take place in these phase diagrams, namely the

$IC–D_{3d}$, $IC–D_{6h}$, and $D_{6h}–D_{3d}$ in the first case, and $IC–T_h$, $IC–O_h$, and $O_h–T_h$ in the second case, are expected to be first-order.

4. Discussion

The stability of quasicrystals has been a subject of extensive theoretical studies in recent years. These studies, which were carried out mainly within the framework of the Landau and mean field theories, were reviewed in this chapter. They suggest that within the phenomenological Landau theory approach, quasicrystals may be globally and locally stable in an appropriate range of certain parameters which define the free energy. The density functional theory, however, provides a scheme for calculating the Landau free energy when the structure factor of the liquid phase is known. It has been shown that when the structure factor of amorphous cobalt is considered, the free energy is such that the icosahedral structure is locally unstable, with a negative elastic constant.

The possible symmetries which may arise when icosahedral structures become unstable were considered. It was shown that deformations associated with elastic terms in the free energy may result in quasicrystalline phases with symmetries which are subgroups of the icosahedral group. In particular, it was found that elastic distortions may lead to quasicrystals which are incommensurate in any direction with point group symmetries which are either D_{5d}, D_{3d}, or T_h. These are maximal subgroups of the icosahedral group. For example, the experimentally observed T″ phase which exhibits diffraction patterns with sixfold symmetry may be a result of such deformation, and therefore its symmetry is expected to be D_{3d}. The effect of the lock-in terms in the free energy on the structure of these quasicrystals was also considered. As a result of these terms, the structure may become locked, and thus commensurate, in one or more directions. When the structure becomes locked, the point symmetry may increase. In the case of deformations along the fivefold axis, the resulting symmetry is D_{10h}, which is consistent with the symmetry of the experimentally observed T phase. Deformations along the threefold axis may lead to a hexagonal phase with point symmetry D_{6h}. Cubic deformations may lead to a bcc structure with point symmetry O_h. The phase transitions between the various phases for each of the three types of distortions, namely the fivefold, threefold, and cubic, have been studied. It is found that all transitions are expected to be first-order.

Acknowledgments

This work was supported in part by the Israel Commission for Basic Research and by the Minerva Foundation, Munich, Germany.

Appendix

We consider the D_{10h}–D_{5d} transition in some detail and show that this transition, which is a kind of commensurate–incommensurate transition, is first-order. The free energy (3.27) associated with this transition contains four terms. The first three terms, which depend only on derivatives of ψ_0, are the elastic terms, while the last term, which depends on ψ_0 itself, is the lock-in term. For $\alpha > 0$, the free energy (3.27) is minimized by $\psi_0 = 0$. This solution corresponds to the D_{10h} structure which is commensurate along the z direction and has a tenfold rotation axis.

Below some $\alpha_0 < 0$, the D_{10h} symmetry is broken spontaneously, and the structure becomes incommensurate in the z direction. In this case, ψ_0 has the soliton lattice form, and the resulting symmetry is D_{5d}. In the general, commensurate–incommensurate transitions can be either first-order or second-order transitions. The order of the transition is determined by the long-range force between solitons (Jacobs and Walker, 1980; Schaub and Mukamel, 1985). If the solitons repel each other (as in the case of the sine–Gordon equation), their density below the transition increases continuously, and the transition is second-order. On the other hand, if the interaction is attractive, there is a discontinuity in the density of solitons at the transition, and the transition is first-order. In the following, we show that the solitons associated with (3.27) attract each other, and therefore the D_{10h}–D_{5d} transition is first-order.

The Euler equation associated with (3.27) takes the form:

$$\left[\alpha + 3\left(\frac{d\psi_0}{dz}\right)^2\right]\left(\frac{d^2\psi_0}{dz^2}\right) - V_5 sin\psi_0 = \beta\left(\frac{d^4\psi_0}{dz^4}\right). \tag{A.1}$$

This equation has three types of solutions: the trivial solution $\psi_0 = 0$, the oscillatory solution in which ψ_0 oscillates periodically, and the soliton lattice solution. In order to consider the long-range force between solitons, we linearize (A.1) around $\psi_0 = 0$ (and $d\psi_0/dz = 0$). The linear equation which describes the behavior of the tail of a soliton takes the form:

$$\alpha\left(\frac{d^2\psi_0}{dz^2}\right) - V_5\cdot\psi_0 = \beta\left(\frac{d^4\psi_0}{dz^4}\right). \tag{A.2}$$

Assuming that the solution is of the form $\psi_0 = A\cdot\exp(-kz)$, we obtain the following equation for k:

$$\beta k^4 - \alpha k^2 + V_5 = 0. \tag{A.3}$$

In the regime, where α is negative, and in which the commensurate–incommensurate transition occurs, all the possible values of k which are solutions of (A.3) have a nonzero imaginary part. As a result, the asymptotic solution ψ_0 has oscillatory behavior for all $\alpha < 0$. Jacobs and Walker (1980) considered the interaction between solitons for the case in which the tails have oscillatory form. They have shown that if the tail of the soliton is oscillatory, the energy of interaction between two solitons as a function of the distance between them is also oscillatory. Therefore, the free energy for a soliton lattice in which the distances between solitons correspond to minima of this oscillatory function is lower than for the single soliton. The interaction between solitons is thus attractive. As a result, at the transition, there is a jump in the soliton density, and the transition is first-order.

References

Alexander, S., and McTague, J. (1978). Should all crystals be bcc? Landau theory of solidification and crystal nucleation. *Phys. Rev. Lett.* **41,** 702–705.

Bak, P. (1985a). Phenomenological theory of icosahedral incommensurate ("quasiperiodic") order in Mn-Al alloys. *Phys. Rev. Lett.* **54,** 1517–1519.

Bak, P. (1985b). Symmetry, stability, and elastic properties of icosahedral incommensurate crystals. *Phys. Rev. B* **32,** 5764–5772.

Bak, P., and Mukamel, D., (1979). Phase transitions in two-dimensional modulated systems. *Phys. Rev. B* **19,** 1604–1609.

Bak, P., Mukamel, D., Villain J., and Wentowska, K. (1979). Commensurate–incommensurate transitions in rare-gas monolayers adsorbed on graphite and in layered charge-density-wave systems. *Phys. Rev. B* **19,** 1610–1613.

Bancel, P. A., Heiney, P. A., Stephens, P. W., Goldman, A. I., and Horn, P. M. (1985). Structure of rapidly quenched Al-Mn. *Phys. Rev. Lett.* **54,** 2422–2425.

Baym, G., Bethe, H. A., and Pethick, C. J. (1971). Neutron star matter. *Nucl. Phys. A* **175,** 225–271.

Bendersky, L. (1985). Quasicrystal with one-dimensional translational symmetry and a tenfold rotation axis. *Phys. Rev. Lett.* **55,** 1461–1463.

Bendersky, L., Schaefer, R. J., Biancaniello, F. S., Boettinger, W. J., Kaufman, M. J., and Shechtman, D. (1985). Icosahedral Al-Mn and related phases: resemblance in structure. *Scripta Metallurgica* **19,** 909–914.

Biham, O., Mukamel, D., and Shtrikman, S. (1986a). Symmetry and stability of icosahedral and other quasicrystalline phases. *Phys. Rev. Lett.* **56,** 2191–2194.

Biham, O., Mukamel, D., and Shtrikman, S. (1986b). Local stability of quasicrystals. *Proceedings of the Les-Houches Workshop on Aperiodic Crystals [J. Phys. (Paris) Colloq* **47, C3** 245–249].

Butler, P. H. (1981). *Point Group Symmetry Applications: Methods and Tables.* Plenum Press, New York.

Chakraborty, B., Sood A. K., and Valsakumar, M. C. (1986). Discommensurations in icosahedral phases. *Phys. Rev. B* **34,** 8202–8206.

Elser, V., and Henley, C. L. (1985). Crystal and quasicrystal structures in Al-Mn-Si alloys. *Phys. Rev. Lett.* **55,** 2883–2886.

Fung, K. K., Yang, C. N., Zhou, Y. Q., Zhau, J. G., Zhan, W. S., and Shen, B. G. (1986). Icosahedrally related decagonal quasicrystal in rapidly cooled Al-14-at.% - Fe Alloy. *Phys. Rev. Lett.* **56,** 2060–2063.

Gardner, M. (1977). Extraordinary nonperiodic tiling that enriches the theory of tiles. *Sci. Am.* **236**(1), 110–121.

Grinstein, G., and Jayaprakash, C. (1982). First-, second-, and infinite-order transitions in three-dimensional models with competing interactions. *Phys. Rev. B* **25,** 523–526.

Gronlund, L. D., and Mermin, N. D. (1987). Instability of the Kalugin–Kitaev–Levitov model of a quasicrystal. *Bull. Am. Phys. Soc.* **32**(3), 520.

Henley, C. L. (1985). Crystals and quasicrystals in the aluminum - transition metal system. *Journal of Non-Crystallographic Solids* **75,** 91–96.

Ho, T. L. (1986). Periodic quasicrystal. *Phys. Rev. Lett.* **56,** 468–471.

Horn, P. M., Malzfeldt, W., DiVincenzo, D. P., Toner J., and Gambino, R. (1986). Systematics of disorder in quasiperiodic material. *Phys. Rev. Lett.* **57,** 1444–1447.

Horn, P. M., Heiney, P. A., Bancel, P. A., Gayle, F. W., Jordan, J. L., LaPlaca, S., and Angilello J. (1987). Universal disorder in icosahedral alloys. Preprint.

Ishimasa, T., Nissen, H. U., and Fukano, Y. (1985). New ordered state between crystalline and amorphous in Ni-Cr particles. *Phys. Rev. Lett.* **55,** 511–513.

Jacobs, A. E., and Walker, M. B. (1980). Phenomenological theory of charge-density-wave states in trigonal-prismatic, transition-metal dichalcogenides. *Phys. Rev. B* **21,** 4132–4148.

Jarić, M. V. (1985). Long-range icosahedral orientational order and quasicrystals. *Phys. Rev. Lett.* **55,** 607–610.

Jarić, M. V. and Mohanty, U. (1987). "Martensitic" instability of an icosahedral quasicrystal. *Phys. Rev. Lett.* **58,** 230–233.

Jarić, M. V., Michel L., and Sharp R. T. (1984). Zeroes of covariant vector fields for the point groups: invariant formulation. *J. Phys. (Paris)* **45,** 1–27.

Kalugin, P. A., Kitaev, A. Yu. and Levitov, L. S. (1985). $Al_{0.860}Mn_{0.14}$: a six-dimensional crystal. *Zh. Eksp. Teor. Fiz. Pisma* **41,** 119 (1985) [*JETP Lett.* **41,** 145–149.].

Kramer, P., and Neri, R. (1984). On periodic and non-periodic space fillings of E^n obtained by projection. *Acta. Cryst. A* **40,** 580–587.

Levine. D., and Steinhardt, P. J. (1984). Quasicrystals: a new class of ordered structures. *Phys. Rev. Lett.* **53,** 2477–2480.

Levine. D., and Steinhardt, P. J. (1985). Quasicrystals. 1. Definition and structure. *Phys. Rev. B* **34,** 596–616.

Levine, D., Lubensky, T. C., Ostlund, S., Ramaswamy, S., Steinhardt P. J., and Toner, J. (1985). Elasticity and dislocations in pentagonal and icosahedral quasicrystals. *Phys. Rev. Lett.* **54,** 1520–1523.

Loiseau, A., and Lapasset, G. (1986). Crystalline and quasicrystalline structures in an Al-Li-Cu-Mg alloy. *Proceedings of the Les-Houches Workshop on Aperiodic Crystals [J. Phys. (Paris) Colloq* **47, C3** 331–340].

Luban, M., unpublished.

Lubensky, T. C., Socolar, J. E. S., Steinhardt, P. J., Bancel, P. A., and Heiney, P. A. (1986). Distortion and peak broadening in quasicrystal diffraction patterns. *Phys. Rev. Lett.* **57,** 1440–1443.

Lyubarskii, G. Ya. (1960). *The Application of Group Theory in Physics.* Pergamon Press, New York.

Mermin, N. D., and Troian, S. M. (1985). Mean-field theory of quasicrystalline order. *Phys. Rev. Lett.* **54,** 1524–1527.

Niven, I. (1956). *Irrational Numbers.* Mathematical Association of America, Washington D.C.

Penrose, R. (1974). The role of aesthetics in pure and applied mathematical research. *Bull. Inst. Math. Appl.* **10,** 266–271.

Peyrard, M., and Aubry, S. (1983). Critical behaviour at the transition by breaking of analyticity in the discrete Frenkel–Kontorova model. *J. Phys.C* **16,** 1593–1608.

Poon, S. J., Drehman, A. J., and Lawless, K. R. (1985). Glassy to icosahedral phase transformation in Pd-U-Si alloys. *Phys. Rev. Lett.* **55,** 2324–2327.

Ramakrishnan, T. V., and Yussouff, M. (1979). First-principles order parameter theory of freezing. *Phys. Rev. B* **19,** 2775–2794.

Sachdev, S., and Nelson, D. R. (1985). Order in metallic glasses and icosahedral crystals. *Phys. Rev. B* **32,** 4592–4606.

Schaub, B., and Mukamel, D. (1985). Phase diagrams of systems exhibiting incommensurate structures. *Phys. Rev. B* **32,** 6385–6393.

Shechtman, D., Blech, I., Gratias, D., and Cahn, J. W. (1984). Metallic phase with long range orientational order and no translational symmetry. *Phys. Rev. Lett.* **53,** 1951–1953.

Socolar, J. E. S., and Steinhardt, P. J. (1986). Quasicrystals. 2. Unit-cell configurations. *Phys. Rev. B* **34,** 617–647.

Socolar, J. E. S., Lubensky, T. C., and Steinhardt, P. J. (1986). Phonons, phasons and dislocations in quasicrystals. *Phys. Rev. B* **34,** 3345–3360.

Stephens, P. W., and Goldman, A. I. (1986). Sharp diffraction maxima from an icosahedral glass. *Phys. Rev. Lett.* **56,** 1168–1171.

Troian, S. M., (1986). Phenomenological theories of quasicrystal formation. *Proceedings of the Les-Houches Workshop on Aperiodic Crystals [J. Phys. (Paris) Colloq* **47, C3** 271–279].

Urban, K., Mayer, J., Rapp, M., Wilkens, M., Csanady, A., and Fidler, J. (1986). Studies on aperiodic crystals in Al-Mn and Al-V alloys by means of transmission electron microscopy. *Proceedings of the Les-Houches Workshop on Aperiodic Crystals [J. Phys. (Paris) Colloq* **47, C3** 465–474].

Widom, M., Strandburg, K. J., and Swendsen R. H. (1987). Quasicrystal equilibrium state. *Phys. Rev. Lett.* **58,** 706–709.

Chapter 6

Symmetry, Elasticity, and Hydrodynamics in Quasiperiodic Structures

T. C. LUBENSKY

Department of Physics
University of Pennsylvania
Philadelphia, Pennsylvania 19104

Contents

APERIODICITY AND ORDER
Introduction to
Quasicrystals

ISBN 0-12-040601-2

Some commonly used symbols:

$\rho(\mathbf{x})$	microscopic mass density at position **x**
$\rho_{\mathbf{G}}(\mathbf{x})$	order parameter field for mass density wave at wavevector **G**
$< >$	thermodynamic average
$<\rho_{\mathbf{G}}>$	scattering amplitude at wavevector **G**
$\phi_{\mathbf{G}}$	phase of the complex scattering amplitude $<\rho_{\mathbf{G}}>$
$\mathbf{u}$	displacement variable
$\mathbf{w}$	phason variable
$\mathbf{G}$	Reciprocal Lattice Vector
$\mathbf{R}$	Direct Lattice vector for periodic structures
$\mathbf{G}^{\perp}$	Complementary vector
$\tilde{\mathbf{G}}$	Reciprocal lattice vector of parent lattice
$\tilde{\mathbf{R}}$	Direct lattice vector of periodic parent lattice
D	dimension of reciprocal lattice
d	dimension of space
r	rank of lattice
$L^A(D,r)$	D-dimensional rank-r reciprocal lattice of type A ($A = P$ for periodic, I for incommensurate, or Q for quasicrystalline).
$B^A(D,r)$	basis for a D-dimensional rank-r reciprocal lattice.

1. Introduction

The discovery by Schechtman, Blech, Gratias, and Cahn[1] of an alloy of aluminum and manganese exhibiting an electron diffraction pattern of Bragg peaks with the rotational symmetry of an icosahedron challenged long-held beliefs about the nature of translational order. Periodic lattices[2] consist of regularly repeated identical unit cells. They are characterized by a shortest length equal to the smallest dimension of the unit cell. Only certain point group symmetries—the crystallographic symmetries—are compatible with the existence of a periodic lattice.[2,3] All other point symmetries—pentagonal symmetry in two dimensions and icosahedral symmetry in three dimensions, for example—are not. Periodic lattices can be created by packing together copies of a single unit cell with one of the crystallographic point group symmetries. It is, however, impossible to pack pentagonal or icosahedral cells together in a space-filling structure.[2]

It was thus thought that materials simultaneously exhibiting the rotational symmetry of an icosahedron and long-range translational order implied by the existence of Bragg peaks in a diffraction pattern could not exist. However, the possiblity of materials with icosahedral rotational order but with no long-range translational order was considered.[4]

A theoretical explanation of the existence of Bragg peaks in a system with icosahedral point symmetry was provided by Levine and Steinhardt[5,6] almost simultaneously with the experimental announcement by Schechtman *et. al.* Levine and Steinhardt showed that Penrose tilings[7] with pentagonal symmetry in two dimensions could be generalized to icosahedral symmetry in three dimensions. They further showed that the diffraction pattern of the icosahedral tiling has Bragg peaks on a dense set of points in reciprocal space with intensities that are in good agreement with those obtained from icosahedral aluminum-manganese.[1] Since the icosahedral diffraction pattern exhibits Bragg peaks at parallel wavevectors with irrational magnitude ratios, its associated translational symmetry is *quasiperiodic* rather than periodic. Because of the importance of quasiperiodicity, Levine and Steinhardt introduced the term *quasicrystals* (short for quasi-periodic crystals) for systems exhibiting diffraction patterns with noncrystallographic symmetry. They noted that even though quasicrystals have diffraction peaks that are arbitrarily close to each other in reciprocal space, atoms in their real-space structures can be separated by a minimum distance associated with the smallest dimension of a tile in the Penrose tiling. The possibility of generalizing Penrose tilings to icosahedral symmetry had been considered earlier by several groups including Amman[8], Mackay[9], Mosseri and Sadoc[10], and Kramer and Neri[11], although the key role of quasiperiodicity was not identified.

Examples of Penrose tilings are shown in Figs. 1.1 and 1.2. There are two distinct types of tiles or unit cells, rather than the single unit cell of periodic structures, that are joined together according to certain matching rules to produce a structure with pentagonal symmetry. Today there are a number of methods for generating quasi-crystalline tilings with any symmetry in two, three, or more dimensions.[11–18] The most popular are the projection [12–15] and the generalized dual methods.[11,12,16–18] Quasicrystals with icosahedral symmetry can be constructed with two or more unit cells[9,11–14,16,17], whereas the periodic lattice requires only one. The shortest length is determined by the smallest dimension of the smallest unit cell.

Materials with icosahedral and related symmetries[19–22] represent a new phase of matter with possibly unique physical properties that one would like to identify and understand. Are quasicrystals stable or only metastable? How do they respond to external forces? What is the nature of their low energy excitations? What are their transport properties? There are

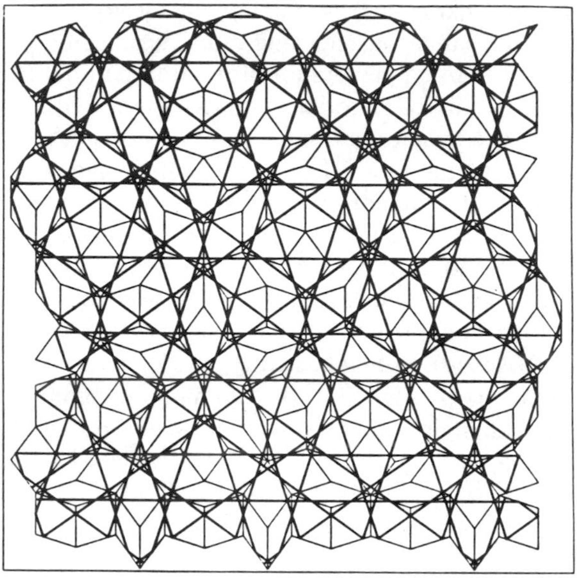

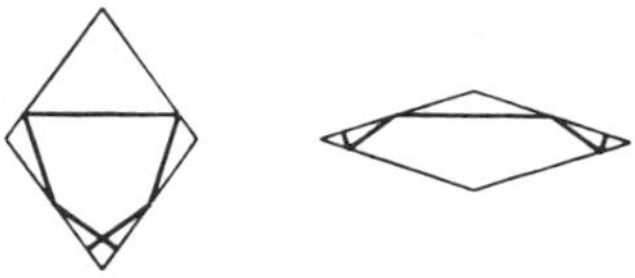

Figure 1.1. A portion of a Penrose tiling with five-fold symmetry showing the decoration of the "fat" and "skinny" rhombuses (tiles) leading to the Amman pentagrid. The Amman pentagrid consists of five sets (grids) of parallel lines with normals pointing to one of the five vertices of a pentagon. The position x_n along the grid normal of the n-th line in each grid follows a Fibonacci sequence determined by $x_n = n+\alpha+\tau^{-1}[n\tau^{-1}+\beta]$, where α and β are arbitrary phases, $[x]$ is the greatest integer less than or equal to x, and $\tau = (1+\sqrt{5})/2$ is the golden mean. There are only two possible separations, 1 and $\tau = 1+\tau^{-1}$, between adjacent lines in a grid, as is clear from the figure.

well-established methods for addressing these and related questions about periodic crystals. First, an ideal crystal structure is established by using X-ray or electron diffraction data. Then elementary excitations from the ground state (phonons) and the electronic states (band structure) are determined and used to calculate thermodynamic and transport properties. This program, unfortunately, cannot yet be applied to quasicrystals because the precise positions of atoms remains unknown.[21] Some other approach to the study of their physical properties must be sought.

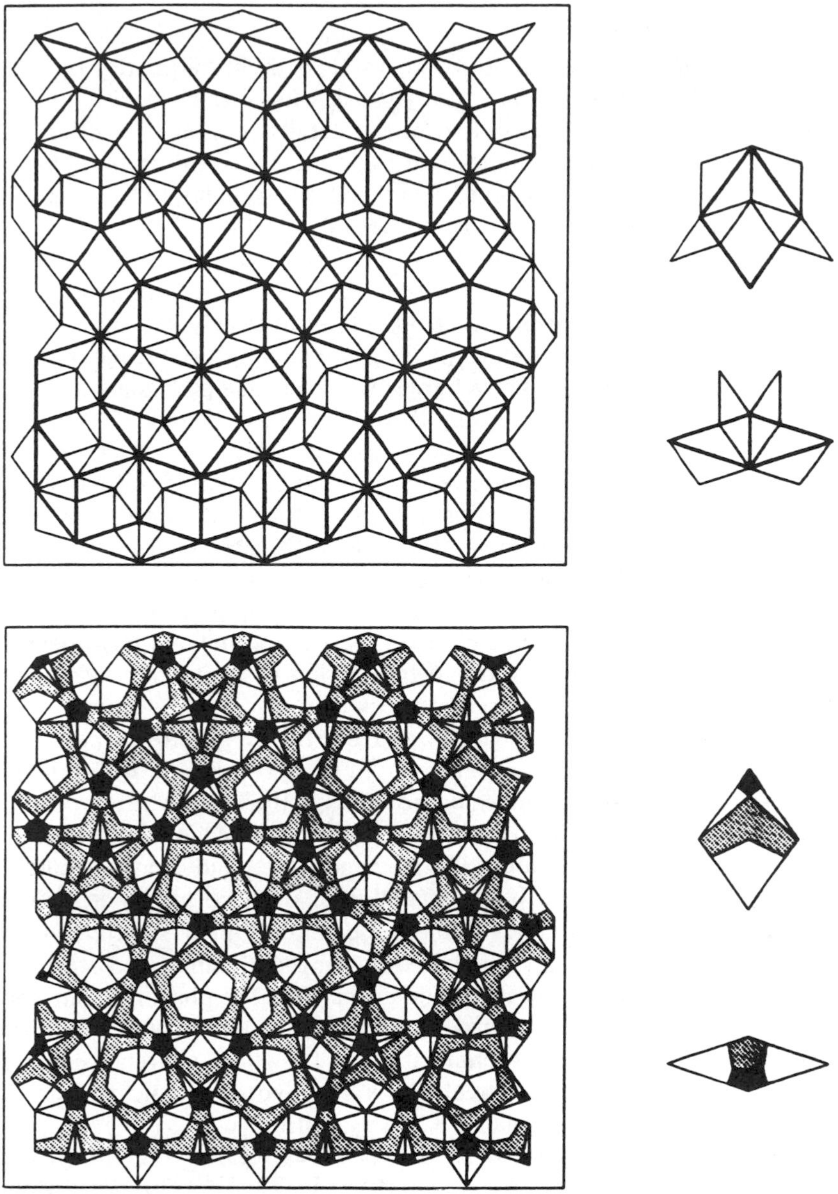

Figure 1.2. Portions of Penrose tilings showing inflation (top) and matching rules (bottom). The decoration of the tiles shown in the top figure leads to a new Penrose tiling with tiles with sides that are τ times larger than those of the original tiling. Matching rules of the Penrose local isomorphism class are enforced by the decoration shown in the bottom figure. The rules are that the dark and dotted portions of one tile must join with the same portions of all adjacent tiles.

The distinguishing feature of quasicrystals is their unique translational and rotational symmetry. It is natural to ask what properties follow from these symmetries alone and not from the details of atomic position and interatomic interactions. Phenomenological elastic and hydrodynamic theories based entirely on symmetry and conservation laws have met with enormous success in explaining properties of superfluids[28], magnets[24], liquid crystals[25], and other materials with broken continuous symmetries. These theories use only the invariances implied by the symmetries of a given phase to infer the response of that phase to slowly varying external forces (elasticity) and the nature of long-wavelength, low-frequency excitations (hydrodynamics). This chapter will present a pedagogical review of continuum elastic-hydrodynamic theory for quasicrystals[15,26–32] and related structures.

The dense set of points in the reciprocal lattice of an icosahedral quasicrystal arises because it contains sets of vectors that sum to zero with relatively irrational coefficients. Materials with reciprocal lattices with this property are called quasiperiodic or incommensurate.[33–35] Quasicrystals are a special class of quasi-periodic structures possessing a noncrystallographic point group symmetry. Quasiperiodic structures have degrees of freedom associated with relative translation of incommensurate sublattices or, equivalently, with special types of atomic rearrangements. The elementary excitations[15,36,26–32] arising from this degree of freedom are called *phasons* and are parametrized by a phason variable **w.** Variables whose spatially uniform changes do not change the free energy are called *elastic* variables. The phason variable **w,** like the displacement or phonon variable **u,** is an elastic variable. The free energy depends on the gradients of elastic variables. An analytic expansion of the free energy in powers of gradients of elastic variables leads to the continuum elastic free energy F_{el}. In periodic crystals, F_{el} depends only on the symmetrized strain u_{ij}. In quasi-periodic structures, it depends on u_{ij} and the phason strain $\nabla_i w_j$. Associated with each elastic variable appearing in an analytic expansion of the free energy is a hydrodynamical mode[37,38] whose characteristic frequency goes to zero at long wavelengths.The hydrodynamical mode associated with the phason variable is always diffusive.[39]

Though **w** is an elastic variable, it is possible for the free energy of a quasi-periodic structure to be nonanalytic in ∇w.[40,41] This analyticity breaking is unique to the broken symmetry associated with quasiperiodicity. It arises from the discrete structure of particles at the atomic scale and cannot be inferred from a continuum elastic theory. It is likely that the free energy of quasicrystals [42–44] is nonanalytic in ∇w. Conjectures about the form of the nonanalytic free energy and its physical consequences will be presented in Section 10.

The goal of this chapter is to provide a general framework for treating the elasticity and long-wavelength dynamics of quasiperiodic structures in general and icosahedral quasicrystals in particular. To prepare for this task, the characterization of periodic solids in terms of mass density-waves and the derivation of their elastic free energy is reviewed in Section 2. Then, in Section 3, elementary properties of quasi-periodic structures are reviewed with the aid of a one-dimensional example. In order to study higher dimensional systems, it is necessary first to classify different types of translationally ordered structures. A classification scheme for periodic and quasiperiodic structures is presented in Section 4. This scheme differs from others in the literature[45] but is similar to a scheme discussed in Reference 20.

Section 5 presents a brief review of the Landau theory[15,27,28,31,46–50] for ordered phases, a more detailed account of which can be found in the article by Mukamel[49] in this volume. Section 6 discusses the group theory of quasiperiodic structures and its relation to the phason variable. It also discusses how the density of a quasiperiodic structure can be obtained as a cut through a higher dimensional periodic lattice[15,51–55] and how this higher dimensional lattice can be constructed. Section 7 derives continuum elasticity[26–28,31,56,57] for quasi-periodic structures with an emphasis on pentagonal and icosahedral quasicrystals. Section 8 discusses dislocations.[26,58,59] Section 9 derives hydrodynamics for general quasiperiodic structures and the mode structure for icosahedral quasicrystals in particular.[29] It is argued that the phason variable should relax extremely slowly. Section 10 presents a brief discussion of analyticity breaking[40,41] in one-dimensional systems and quasicrystals.[42–44] It predicts an even slower relaxation of phason strains than does the hydrodynamical theory. Section 11 discusses some experimental consequences of slow phason relaxation.[60–62] Finally, the appendix presents a brute force calculation of the elastic energy of an icosahedral quasicrystal.

This chapter is mostly a pedagogical presentation of material that has already been published in the open literature. It does, however, contain some parts that are new. The classification scheme for periodic and quasiperiodic structures has not, to the author's knowledge, been presented previously, although the scheme is similar to one that will appear in an annotated collection of articles on quasicrystals edited by Steinhardt and Ostlund[20]. The identification in Section 6 of two types of incommensurate crystals is new and possibly important, since all experimental quasiperiodic crystals except quasicrystals appear to be incommensurate crystals of type I. Finally some of the speculations in Section 10 about the possibility of thermally driven depinning transitions in quasicrystals are new.

2. Periodic Crystals

This section will review the derivation of continuum elasticity for periodic crystals in a way that is easily generalized to quasiperiodic structures.

Periodic solids, unlike isotropic fluids, have a mass distribution that is periodic in space. Order parameters distinguishing periodic solids from fluids can, therefore, be defined in terms of Fourier components of the mass density. Let $\rho(\mathbf{x})$ be the mass density at position $\mathbf{x}$ in a d-dimensional space. In classical systems with a single species of atoms of mass m located at positions $\mathbf{x}^\alpha$,

$$\rho(\mathbf{x}) = \sum_\alpha m\delta(\mathbf{x} - \mathbf{x}^\alpha). \tag{2.1}$$

In quantum mechanical systems, $\mathbf{x}^\alpha$ and hence $\rho(\mathbf{x})$ is an operator. $\rho(\mathbf{x})$ is a microscopic fluctuating field. In isotropic fluids, its average $< \rho(\mathbf{x}) >$ over an equilibrium ensemble is independent of coordinate $\mathbf{x}$. In periodic solids, $< \rho(\mathbf{x}) >$ is a periodic function of $\mathbf{x}$ satisfying $< \rho(\mathbf{x} + \mathbf{R}) > = < \rho(\mathbf{x}) >$ for every vector $\mathbf{R}$ in some periodic lattice. The dimension D of the periodic lattice must be less than or equal to the spatial dimension d. In crystalline solids, $D = d$. There are, however, many interesting liquid crystal phases[25,63] for which $D < d$. Periodicity in coordinate space implies that $< \rho(\mathbf{x}) >$ can be expressed as a discrete Fourier sum

$$< \rho(\mathbf{x}) > = \sum_{\mathbf{G}} < \rho_{\mathbf{G}} (\mathbf{x}) > e^{i\mathbf{G}\cdot\mathbf{x}}, \tag{2.2}$$

where $\mathbf{G}$ is a vector in a D-dimensional periodic lattice reciprocal to the lattice of vectors $\mathbf{R}$. A two-dimensional hexagonal reciprocal lattice is shown in Fig. 2.1. The order parameters, $< \rho_{\mathbf{G}}(\mathbf{x}) > \equiv < \rho_{\mathbf{G}} >$, are mass-density-wave amplitudes that distinguish the solid from the fluid phase and are independent of $\mathbf{x}$ in an equilibrium ensemble in the absence of external stresses. In the presence of external stresses, the lattice will distort and induce a position dependence in the order parameters $< \rho_{\mathbf{G}}(\mathbf{x}) >$. Because the scattering intensity into Bragg peaks is proportional to $|< \rho_{\mathbf{G}} >|^2$, $< \rho_{\mathbf{G}} >$ is often called the scattering amplitude at reciprocal lattice vector $\mathbf{G}$.

An expression for the microscopic fluctuating field $\rho_{\mathbf{G}}(\mathbf{x})$, whose average is the scattering amplitude, can be obtained from the Fourier components,

$$\rho_{\mathbf{q}} = \int \frac{d^d x}{(2\pi)^d} e^{-i\mathbf{q}\cdot\mathbf{x}} \rho(\mathbf{x}) = m \sum_\alpha e^{i\mathbf{q}\cdot\mathbf{x}^\alpha}, \tag{2.3}$$

of the microscopic density via

$$\rho_{\mathbf{G}}(\mathbf{x}) = \int_{q\in BZ} \frac{d^d q}{(2\pi)^d} e^{i\mathbf{q}\cdot\mathbf{x}} \rho_{\mathbf{q}+\mathbf{G}}, \tag{2.4}$$

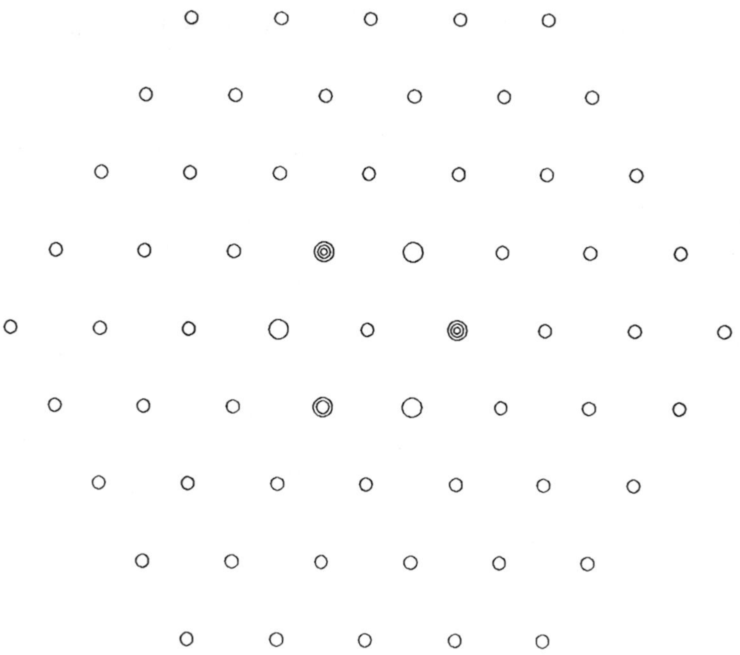

Figure 2.1. A two-dimensional hexagonal reciprocal lattice showing the symmetrized basis (large circles), the extended basis (circles with at least one concentric circle), and a basis (three concentric circles).

where *BZ* refers to the first Brillouin Zone of the reciprocal lattice. Because $\rho(\mathbf{x})$ has Fourier components restricted to the first Brillouin Zone, its spatial variation is slow on the scale of the period of the lattice. Only the $\mathbf{q} = 0$ component of $< \rho_{\mathbf{G}}(\mathbf{x}) >$ survives the equilibrium average so that $< \rho_{\mathbf{G}} >$ reduces to the familiar expression, $< \rho_{\mathbf{G}} > = m < \Sigma_{\alpha} e^{i\mathbf{G}\cdot\mathbf{x}^{\alpha}} >$, for the scattering amplitude.

$< \rho_{\mathbf{G}}(\mathbf{x}) >$ is a complex number with an amplitude and a phase,

$$< \rho_{\mathbf{G}}(\mathbf{x}) > = | < \rho_{\mathbf{G}}(\mathbf{x}) > | \, e^{i\phi_{\mathbf{G}}(\mathbf{x})}. \tag{2.5}$$

The density is a real function so that

$$\phi_{\mathbf{G}}(\mathbf{x}) = -\phi_{-\mathbf{G}}(\mathbf{x}). \tag{2.6}$$

Under a rigid translation $-\mathbf{u}$ of all particles, the average mass density changes from $< \rho(\mathbf{x}) >$ to $< \rho(\mathbf{x} + \mathbf{u}) >$. From Eq. (2.2), it follows that such a translation is equivalently described by a uniform displacement of the phases of the mass density waves according to $\phi_{\mathbf{G}} \rightarrow \phi_{\mathbf{G}} + \mathbf{G}\cdot\mathbf{u}$. This implies that the origin of the coordinate system can be chosen so that in equilibrium

$$\phi_{\mathbf{G}} = \phi_{\mathbf{G}}^{0} + \mathbf{G}\cdot\mathbf{u}. \tag{2.7}$$

A uniform rotation through an angle $\mathbf{\Omega}$ with components Ω_i takes x_i to $R_{ij}x_j$, where R_{ij} is the rotation matrix associated with $\mathbf{\Omega}$. (Here, and in what follows, the Einstein convention on repeated indices is understood.) For small rotation angles $\delta\mathbf{\Omega}$, $R_{ij} = \delta_{ij} + \delta\Omega_{ij}$, where $\delta\Omega_{ij}$ is the antisymmetric matrix $\varepsilon_{ijk}\Omega_k$. Thus an infinitesimal rigid rotation takes $< \rho(x_i) >$ to $< \rho(x_i + \delta\Omega_{ij}x_j) >$ and is equivalently described by the transformation of phases: $\phi_{\mathbf{G}} \rightarrow \phi_{\mathbf{G}} + G_i\delta\Omega_{ij}x_j$. It follows from Eq. (2.7) that a rotation is described by a $u_i = \delta\Omega_{ij}x_j$.

In the absence of external stresses, the free energy F of any system is invariant with respect to uniform translations and rigid rotations. This means that F for a periodic crystal cannot depend on the spatially uniform displacement vector $\mathbf{u}$ or on its antisymmetric derivative $\nabla_i u_j - \nabla_j u_i$. Spatially nonuniform distortions of the displacement vector defined via Eq. (2.7) do, however, increase the free energy. Thus the free energy arising from slow spatial variations of $\mathbf{u}(\mathbf{x})$ must tend to zero with the symmetrized strain[64]

$$u_{ij}(\mathbf{x}) = \frac{1}{2}(\nabla_i u_j + \nabla_j u_i). \tag{2.8}$$

(In order to ensure invariance of the free energy with respect to arbitrary as opposed to infinitesimal rotations, a nonlinear term $\nabla_i u_k \nabla_j u_k$ must be added to this equation.[64] Nonlinear effects will not be considered in this chapter, and Eq. (2.8) will be used as a definition of the strain.) An analytic expansion of the free energy in powers of u_{ij} yields the elastic free energy F_{el}. Such an analytic expansion is possible except in exceptional cases.[65] Since the system is assumed to be in equilibrium in the absence of external stress, there is no linear term in the expansion, and the lowest order contribution is quadratic in u_{ij}:

$$F_{\mathrm{el}} = \frac{1}{2}\int d^d x\, K_{ijkl}u_{ij}u_{kl}. \tag{2.9}$$

This is the harmonic elastic free energy for a periodic solid. K_{ijkl} is the elastic constant tensor that, depending on the dimension and symmetry of the crystal, has at least two, and usually more, independent components. In general F_{el} should include anharmonic terms in u_{ij}, but they will not be considered here. The response of the strain to an external stress σ_{ij} is

$$\frac{\delta F_{\mathrm{el}}}{\delta u_{ij}} = \sigma_{ij}. \tag{2.10}$$

This equation shows clearly how a uniform stress leads to a uniform strain.

F_{el} can be expressed in terms of the Fourier transform,

$$\mathbf{u}(\mathbf{q}) = \int d^d x e^{-i\mathbf{q}\cdot\mathbf{x}} u(\mathbf{x}),$$

of $\mathbf{u}(\mathbf{x})$ rather than in terms of the strain:

$$F_{\text{el}} = \frac{1}{2}\int \frac{d^d q}{(2\pi)^d} K_{ij}(\mathbf{q}) u_i(\mathbf{q}) u_j(\mathbf{q}), \tag{2.11}$$

where

$$K_{ij}(\mathbf{q}) = K_{ikjl} q_k q_l. \tag{2.12}$$

This form is particularly useful, as will be discussed below, for calculating fluctuations in $u(\mathbf{x})$.

Closely associated with the elastic free energy is an elastic Hamiltonian H_{el}, describing the energy increase arising from spatial variations of the phase $\psi_{\mathbf{G}}(\mathbf{x})$ of the microscopic fields $\rho_{\mathbf{G}}(\mathbf{x})$. Under a uniform translation, $\psi_{\mathbf{G}}$ transforms in the same way as does the phase $\phi_{\mathbf{G}}$ of the averaged field. Thus a displacement field $\mathbf{u}'(\mathbf{x})$ can be introduced via $\psi_{\mathbf{G}} = \psi_{\mathbf{G}}^0 + \mathbf{G} \cdot \mathbf{u}'$, where $\psi_{\mathbf{G}}^0$ is the value of $\psi_{\mathbf{G}}$ in the coordinate system in which $\mathbf{u}'$ is zero. In the harmonic approximation, the elastic Hamiltonian is identical to the elastic free energy F_{el}. It is common practice not to distinguish between the fields $\mathbf{u}$ and $\mathbf{u}'$ nor between H_{el} and F_{el}. This convention will be used throughout this chapter. Thus fluctuations in any function $Q(\mathbf{u})$ of the displacement can be calculated via

$$< Q(\mathbf{u}) > = Z_{\text{el}}^{-1} \int D\mathbf{u}\, e^{-F_{\text{el}}(\mathbf{u})/k_B T} Q(\mathbf{u}), \tag{2.13}$$

where

$$Z_{\text{el}} = \int D\mathbf{u} e^{-F_{\text{el}}(\mathbf{u})/k_B T} \tag{2.14}$$

is the partition function, k_B the Boltzman constant, and T the temperature.

The mean square fluctuation in $\mathbf{u}(\mathbf{q})$ follows directly from Eqs. (2.11) and (2.13):

$$< u_i(\mathbf{q}) u_j(-\mathbf{q}) > = k_B\, T\, V K_{ij}^{-1}(\mathbf{q}), \tag{2.15}$$

where V is the volume and the inverse sign signifies a matrix inverse of $K_{ij}(\mathbf{q})$. Similarly, the Debye–Waller factor $\exp(-2W_{\mathbf{G}})$ measuring the reduction in the scattering amplitudes at $\mathbf{G}$ arising from fluctuations in $\mathbf{u}$ is

$$e^{-2W_{\mathbf{G}}} = | < e^{i\mathbf{G}\cdot\mathbf{u}(\mathbf{x})} > |^2 = \exp[- G_i G_j < u_i(\mathbf{x}) u_j(\mathbf{x}) >]. \tag{2.16}$$

The local average fluctuations in $\mathbf{u}(\mathbf{x})$ follow from Eq. (2.16):

$$< u_i(\mathbf{x})u_j(\mathbf{x}) > \; = \; k_B \, T \int \frac{d^d q}{(2\pi)^d} K_{ij}^{-1}(\mathbf{q}). \tag{2.17}$$

When $D = d$, $K_{ij}(\mathbf{q}) \sim q^2$ so that Eq. (2.17) is finite for all $D = d > 2$. Thus the Debye–Waller factor is nonzero for all periodic solids in spatial dimension greater than two, provided the dimension of the reciprocal lattice is equal to the dimension of space.

Electron and X-ray diffraction probe the Fourier transform of the density–density correlation function. The elastic scattering intensity from a sample of volume V is

$$I(\mathbf{q}) = \int d^d x \int d^d x' \; e^{-i\mathbf{q}\cdot(\mathbf{x}-\mathbf{x}')} < \rho(\mathbf{x})\rho(\mathbf{x}')> = VS(\mathbf{q}) + I^{\text{coh}}(\mathbf{q}), \tag{2.18}$$

where

$$I^{\text{coh}} \; = \; V^2 \sum_{\mathbf{G}} \mid < \rho_{\mathbf{G}} > \mid^2 \delta_{\mathbf{q},\,\mathbf{G}} \tag{2.19}$$

is the coherent scattering intensity, and

$$S(\mathbf{q}) \; = \; V^{-1}[< \rho_{\mathbf{q}}\rho_{-\mathbf{q}} > \; - \; < \rho_{\mathbf{q}} > \; < \rho_{-\mathbf{q}} >] \tag{2.20}$$

is the structure function. The scattering intensity at Bragg peaks at reciprocal lattice vector **G** arises from the coherent scattering amplitude and is proportional to $\mid < \rho_{\mathbf{G}} > \mid^2$. The fluctuation term proportional to $S(\mathbf{q})$ gives rise to an incoherent background in the scattering amplitude. Strain fluctuations reduce the weight into the Bragg peaks by the Debye–Waller factor of Eq. (2.16). There is an associated increased in the incoherent background scattering.

The purpose of this section was to review the derivation and implications of the elasticity of periodic solids in a form that will generalize to the quasiperiodic systems that are the major subject of this chapter. The procedure for obtaining a generalized theory of elasticity for any system is as follows: (1) identify those variables whose spatially uniform changes leave the energy invariant. These variables are the *elastic* variables; (2) construct the elastic free energy as the most general quadratic form in the gradients of the elastic variables. The elastic energy can then be used to calculate responses to external forces or statistical fluctuations leading, for example, to generalized Debye–Waller factors.

3. One-Dimensional Quasiperiodic Structures

This section will introduce many of the properties distinguishing periodic crystals from quasiperiodic structures through the study of a one-dimensional example. Quasiperiodic structures are by definition structures with relatively irrational vectors in their reciprocal lattices. They are often called incommensurate systems in the literature[33–35] because of the existence of relatively irrational lengths.

To be concrete, consider a system [Fig. 3.1] composed of two parallel linear sublattices, a and b. Sublattice $a(b)$ consists of balls of mass $m_a(m_b)$ connected by identical springs and constrained to have an average separation $l_a(l_b)$. The ratio l_a/l_b of average separations in the two sublattices is constrained to be an irrational number. The balls in one sublattice interact via short-range forces with those of the other so that, in equilibrium, the positions of the balls in each sublattice are modulated at a spatial frequency determined by the average spacing of the balls in the other sublattice. More precisely, the equilibrium position of the n-th ball in the two sublattices is

$$x_{an} = nl_a + u_a + f_a\,(nl_a + u_a - u_b),$$
$$x_{bn} = nl_b + u_b + f_b\,(nl_b + u_b - u_a), \tag{3.1}$$

where f_a and f_b are periodic functions with respective periods l_b and l_a. The parameters u_a and u_b define the origin of the two chains. A uniform increment of u_a describes a uniform displacement of chain α. The mass density of this system is simply

$$\rho(x) = \sum_{n,m} [m_a\delta(x - x_{an}) + m_b\delta(x - x_{bm})]. \tag{3.2}$$

$< \rho(x) >$ can be expressed as a double discrete Fourier transform:

$$< \rho(x) > = \sum_{p,q} < \rho_{pq} > e^{i(pk_a x + qk_b x)}, \tag{3.3}$$

where $k_a = 2\pi/l_a$ and $k_b = 2\pi/l_b$, and where p and q are integers. In other words, the reciprocal lattice consists of one-dimensional vectors,

$$G = pk_a + qk_b \equiv G_a + G_b, \tag{3.4}$$

generated by the two basis vectors k_a and k_b. Each vector in the reciprocal lattice is uniquely expressible as a sum of two vectors G_a and G_b, respectively, in the periodic reciprocal lattices generated by the bases k_a

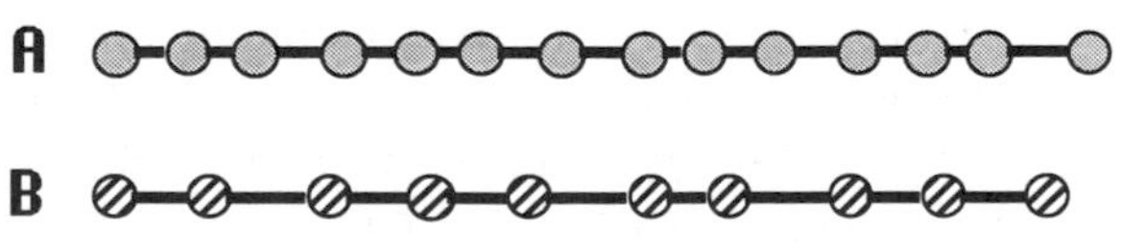

Figure 3.1. A model quasi-periodic structure formed by two sublattices of balls and springs. The average separation of balls in sublattice A is l_a and that in sublattice B is l_b. The ratio l_a/l_b is fixed and irrational. The equilibrium positions of the balls follow from Eq. (3.1).

and k_b. Two important properties distinguishing a reciprocal lattice, L^Q, of a quasiperiodic structure from the reciprocal lattice, L^P, of a periodic crystal emerge from Eq.(3.4). First, there are linear combinations of vectors in L^Q with relatively irrational coefficients that equal zero. In particular, $k_a - (l_b/l_a)k_b = 0$. Second, there is no minimum length vector in L^Q because, as a consequence of the irrationality of k_a/k_b, it is always possible to find integers p and q such that $|pk_a - qk_b|$ is less than any preassigned number. These properties will be used to generalize quasiperiodic reciprocal lattices to higher dimensions in the next section.

As in the periodic case, $<\rho_{pq}> \equiv <\rho_G>$ is a complex number with an amplitude and a phase that can depend on the coordinate x when the positions of the balls are allowed to fluctuate away from their ideal positions described by Eq. (3.1):

$$< \rho_G(x) > = | < \rho_G(x) > | \; e^{i\phi_G(x)}. \tag{3.5}$$

Uniform displacements of the two chains through respective distances u_a and u_b lead to a change in the phase ϕ_G by an amount $G_a u_a + G_b u_b$ so that in analogy with the periodic case,

$$\phi_G(x) = G_a u_a(x) + G_b u_b(x) + \phi_G^0(x), \tag{3.6}$$

where the possibility of spatially nonuniform increments $u_a(x)$ and $u_b(x)$ is allowed. This is the exact analog of Eq. (2.5) for the periodic crystal. There are, however, $2D = 2$ rather than $D = 1$ displacement fields.

u_a and u_b describe separate displacements of the two sublattices. If both are displaced by the same amount, u_a and u_b are equal. To treat this situation, it is useful to introduce alternate coordinates via

$$u_a = u - \alpha s w, \qquad u_b = u + \alpha(1 - s)w. \tag{3.7}$$

$u = (1 - s)u_a + s u_b$ describes displacement of the two chains together, while $w = \alpha^{-1}(u_b - u_a)$ describes relative displacement of the two sublattices. The displacement u is sometimes called the *phonon variable* because it is associated with phonon excitations. w is called the *phason variable*[36,33–36] because it is associated with the phason mode, to be discussed in Section 7 that arises from the relative displacement of two sublattices in quasiperiodic systems. Note that the parameter s is totally arbitrary so that the relation between u and u_a and u_b is not unique. Similarly, the scale factor α is arbitrary. It is introduced to facilitate comparison with other descriptions of the phason variable.

The phase equation can be expressed in terms of u and w,

$$\phi_G(x) = \phi_G^0 + Gu(x) + \alpha G^{\perp} w(x), \tag{3.8}$$

where

$$G^{\perp} = sG_a + (1 - s)G_b \tag{3.9}$$

is a *complementary* vector[11–14] indexed by the same integers p and q as G. Thus, as in the periodic case, there is a part of the phase of the mass-density wave describing translation. There is, in addition, a part describing relative motion or local *rearrangements* of atoms in two sublattices. $u(x)$ and $w(x)$ are independent variables describing translations and rearrangements of the relative positions of atoms. It is natural to treat them as components of a two-component vector

$$\tilde{\mathbf{u}} = u \oplus w. \tag{3.10}$$

Then the introduction of the two-component reciprocal lattice vector

$$\tilde{\mathbf{G}} = G \oplus \alpha\, G^{\perp} \tag{3.11}$$

allows Eqs. (3.6) and (3.8) to be rewritten as

$$\phi_G = \phi_G^0 + \tilde{\mathbf{G}} \cdot \tilde{\mathbf{u}}. \tag{3.12}$$

This form as well as that of Eq. (3.8) will always generalize to higher dimension.

Because k_a/k_b is irrational, the energy of the model ball and spring system cannot change if either of the sublattices is translated uniformly by an arbitrary amount. Thus, the free energy F can only depend on the spatial gradients of u_a and u_b. If F is analytic in du_a/dx and du_b/dx, then to lowest order, the elastic free energy F_{el} has the harmonic form

$$F_{\text{el}} = \frac{1}{2}\int dx \left[K_{aa}\left(\frac{du_a}{dx}\right)^2 + K_{bb}\left(\frac{du_b}{dx}\right)^2 + 2K_{ab}\frac{du_a}{dx}\frac{du_b}{dx}\right]. \tag{3.13}$$

F_{el} can alternatively be expressed in terms of the strain du/dx and the phason strain dw/dx:

$$F_{\text{el}} = \frac{1}{2}\int dx \left[K_{uu}\left(\frac{du}{dx}\right)^2 + K_{ww}\left(\frac{dw}{dx}\right)^2 + 2K_{uw}\frac{du}{dx}\frac{dw}{dx}\right], \tag{3.14}$$

where K_{uu}, K_{ww}, and K_{uw} can be expressed in terms of K_{aa}, K_{bb}, and K_{ab} using Eq. (3.7). Equation (3.14), but not always Eq. (3.13), generalizes to higher dimensional quasiperiodic systems. The elastic free energies of Eqs. (3.13) and (3.14) result from the assumption of analyticity in the phason strain. This assumption can break down in strongly coupled quasiperiodic systems, as will be discussed in Section 10.

Fluctuations in u_a and u_b are easily calculated from Eq. (3.14):

$$< u_\alpha(q) u_\beta(-q) > = k_B T\, V K^{-1}_{\alpha\beta} q^{-2}, \qquad \alpha,\beta = a,b. \tag{3.15}$$

This equation implies that $<u_\alpha^2(x)>$ is infinite for all $T>0$ as expected in one-dimensional systems. Thus the Debye–Waller factor is zero, and the scattering amplitudes of one-dimensional quasiperiodic structures are zero for all $T>0$. Long-range quasiperiodicity, like long-range periodicity, cannot exist at any nonzero temperature in one dimension.

As discussed in the article by P. Bak in this volume[53], the density of any quasiperiodic structure can be obtained as a cut[15,31,51–55] through a higher dimensional periodic structure. The periodic structure is embedded in a hyperspace. The hyperspace from which the quasiperiodic structure is generated will be called a *parent* space to emphasize the special connection between the two. It will be useful for future discussion to review some aspects of this procedure as applied to the one-dimensional case. The vector $\tilde{\mathbf{G}}$ introduced in Eq. (3.11) is a member of a two-dimensional periodic reciprocal lattice and, like the vector G, is uniquely specified by the two integers p and q. There is, therefore, a one-to-one correspondence between $\tilde{\mathbf{G}}$ and G. By construction, $\tilde{\mathbf{G}}$ has orthogonal components G and $\alpha G^\perp$ and can be written as $G\,\tilde{\mathrm{e}}_\parallel + G\,\tilde{\mathrm{e}}_\perp$ where $\tilde{e}_\parallel$ and $\tilde{e}_\perp$ are orthogonal unit vectors in the parent space. Now introduce a two-dimensional vector $\tilde{x} = (x_\parallel, x_\perp)$. The set of vectors $\{\tilde{\mathbf{G}}\}$ then define a two-dimensional periodic crystal with density

$$< \tilde{\rho}(\tilde{\mathbf{x}}) > = \sum_{p,q} | < \rho_G > | \, e^{i\phi^0_G} e^{i\tilde{\mathbf{G}}\cdot\tilde{\mathbf{x}}}. \tag{3.16}$$

The periodic crystal in the parent space will be called the *parent lattice*. The density of the one-dimensional quasiperiodic structure is exactly reproduced by the density of the two-dimensional quasiperiodic structure in the parent space along the cut [Fig. 3.2] defined by $x_\parallel = x + u$ and $x_\perp = w$: $<\rho(\mathrm{x})> = <\rho_P(\mathrm{x}+\mathrm{u},\mathrm{w})>$. Thus the space spanned by $\tilde{e}_\parallel$ is the *physical* space. The space spanned by $\tilde{e}$ is orthogonal to the physical space and will be called the *complementary* space. Translations in the physical space correspond to displacements u, whereas the translations in the orthogonal complementary space correspond to displacements of the phason variable w.

An alternate representation of the parent space with orthogonal axes along directions associated directly with the a and b sublattices is instructive. Consider a transformation of the parent lattice to a new coordinate system in which $\tilde{\mathbf{G}} = (c_a G_a, c_b G_b)$ and choose c_a and c_b so that the transformation corresponds to a simple rotation through an angle $-\theta$. In this case,

$$G = G_a + G_b = \cos\theta\; c_a\, G_a + \sin\theta\; c_b\, G_b, \tag{3.17a}$$

$$\alpha G^\perp = -\alpha s G_a + \alpha(1-s) G_b = -\sin\theta\; c_a\, G_a + \cos\theta\; c_b\, G_b. \tag{3.17b}$$

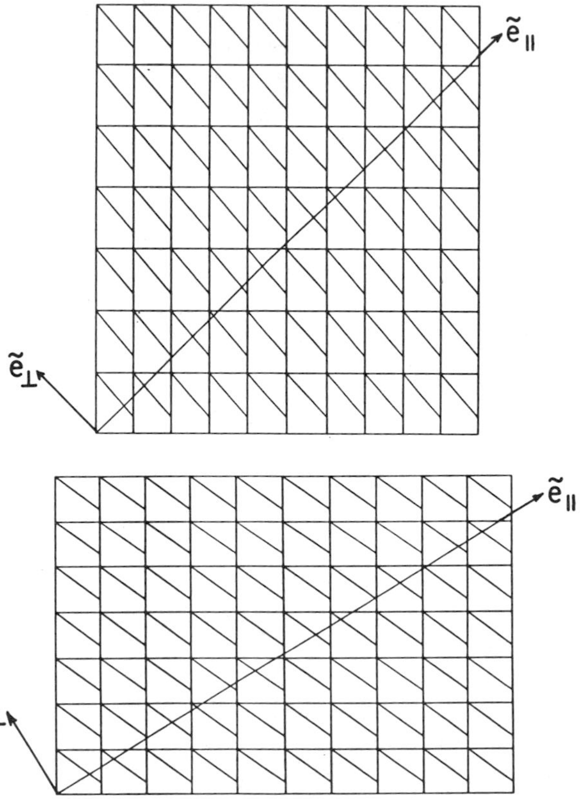

Figure 3.2 A linear quasi-periodic structure obtained from cuts through two-dimensional periodic parent structures. The physical space is the line parallel to $\tilde{e}_{\parallel}$. The complementary space is along the direction perpendicular to the physical space indicated by $\tilde{e}_{\perp}$. Each cell in the two-dimensional structures contains a region of high mass density represented by a slanting line. The intersection of the slanted lines with the physical space represents atoms in the physical space. The density along the physical line is the same in both figures even though the periodic parent lattices are different.

These equations determine c_a, c_b, and θ in terms of α and s for arbitrary G_a and G_b:

$$\begin{aligned} c_a^{-1} &= (1-s)^{1/2} = \cos\theta, \\ c_b^{-1} &= s^{1/2} = \sin\theta, \\ \alpha &= (\sin\theta\cos\theta)^{-1}. \end{aligned} \tag{3.18}$$

In this coordinate system, $\tilde{\mathbf{u}} = (c_a{}^{-1}u_a, c_b{}^{-1}u_b) \equiv c_a{}^{-1}u_a \oplus c_b{}^{-1}u_b$, so that

$\tilde{\mathbf{G}} \cdot \tilde{\mathbf{u}} = G_a u_a + G_b u_b$, as required by Eq. (3.6). The above construction shows that the parent lattice is a rectangular lattice with a reciprocal lattice generated by the vectors $\tilde{\mathbf{k}}_a = (c_a k_a, 0)$ and $\tilde{\mathbf{k}}_b = (0, c_b k_b)$ and with associated real space primitive translation vectors $\tilde{\mathbf{R}}_a = (l_a/c_a, 0)$ and $\tilde{\mathbf{R}}_b = (0, l_b/c_b)$. The parent lattice, as shown in Fig. 3.2, is not unique. The same irrational ratio of lengths along the physical direction $\tilde{\mathbf{e}}_{\parallel}$ can be obtained in an infinite number of ways by varying the lattice constants and the projection angle θ. For each value of θ, however, there is a unique scale factor α such that the phason variable is exactly a translation perpendicular to $\tilde{\mathbf{e}}_{\perp}$. The nonuniqueness of the parent lattice is a direct result of the nonuniqueness of the definition of u in terms of u_a and u_b.

4. Classification of Translationally Ordered Structures

4.1 Definitions

The last two sections discussed how to describe periodic and one-dimensional quasiperiodic crystals in terms of their mass density waves and how invariances associated with the translations of the whole lattice or relative translations of sublattices lead to an elastic free energy. In order to generalize these results to higher dimensional quasiperiodic systems and to quasicrystals, it is useful to provide a more precise characterization of general quasiperiodic systems. This section, which relies heavily on Reference 66, will develop a classification scheme for periodic and quasiperiodic systems.

First, it is necessary to introduce some definitions. A *lattice* is any countable set of vectors closed under addition and subtraction, i.e., if $\mathbf{G}_1$ and $\mathbf{G}_2$ are two vectors in the lattice, then $-\mathbf{G}_1$, $-\mathbf{G}_2$ and $\pm\mathbf{G}_1 \pm\mathbf{G}_2$ are also in the lattice. A D- dimensional *reciprocal lattice* L is a lattice of D-dimensional wavevectors. Each reciprocal lattice is invariant under some point group $\mathcal{G}_R$ consisting of rotations, reflections, and inversions. Thus if $g \in \mathcal{G}_R$ and $\mathbf{G} \in L$, then the image $g\mathbf{G}$ of $\mathbf{G}$ under g is also a member of the reciprocal lattice. Since the negative of any vector in a reciprocal lattice is also in the lattice, $\mathcal{G}_R$ must contain the inversion operation. A *periodic* lattice[2] is one with a shortest-length vector. It necessarily has one of the *crystallographic point group* symmetries[2,3] that includes the inversion operation. Associated with each periodic reciprocal lattice L^P is a *direct lattice* consisting of vectors $\mathbf{R}$ satisfying

$$e^{i\mathbf{G}\cdot\mathbf{R}} = 1 \tag{4.1}$$

for every $\mathbf{G} \in L^P$. The direct lattice is periodic.

A *quasi-periodic* reciprocal lattice is a reciprocal lattice without a shortest-length vector. This condition is satisfied by any lattice with two or more colinear vectors, the ratio of whose magnitudes is irrational. More generally, it is satisfied by any lattice containing a set of not necessarily colinear vectors that sum to zero with relatively irrational coefficients. Finally a *quasicrystalline* reciprocal lattice is one with a noncrystallographic point group symmetry. It cannot have a shortest-length vector, since if it did, it would have a crystallographic symmetry. A quasicrystalline reciprocal lattice is, thus, quasiperiodic.

A *structure* will refer to any material or system with a mass density $\rho(\mathbf{x})$ at position $\mathbf{x}$ in a d-dimensional space. The Fourier transform $\rho_{\mathbf{q}}$ of the mass density is a function of the wavevector $\mathbf{q}$ in a d-dimensional wavevector space. In general, $\rho(\mathbf{x})$ and $\rho_{\mathbf{q}}$ undergo thermal fluctuations about some mean value, and it is their averages, $<\rho(\mathbf{x})>$ and $<\rho_{\mathbf{q}}>$, over some ensemble that characterize the translational order of a structure. A *translationally ordered structure* (TOS) is any structure for which $<\rho_{\mathbf{q}}>$ is nonzero only if $\mathbf{q}$ is a vector in some reciprocal lattice L. This means that the average mass density of a TOS can be written as

$$< \rho(\mathbf{x}) > \; = \sum_{\mathbf{G} \in L} < \rho_{\mathbf{G}}(\mathbf{x}) > e^{i\mathbf{G}\cdot\mathbf{x}}, \tag{4.2}$$

where in equilibrium $<\rho_{\mathbf{G}}(\mathbf{x})>$ is independent of $\mathbf{x}$ and gives the scattering amplitude at wavevector $\mathbf{G}$ in L. Of course, $<\rho_{\mathbf{G}}>$ may be zero for some subset of vectors in L. The coherent scattering amplitudes give rise to Bragg peaks with intensities proportional to $| <\rho_{\mathbf{G}}> |^2$. The pattern of scattering intensities formed by decorating each point in L with a spot of brightness proportional to $| <\rho_{\mathbf{G}}> |^2$ is invariant under the Laue group $\mathcal{G}_L$, which must be a subgroup of $\mathcal{G}_R$.

$<\rho(\mathbf{x})>$ is invariant under some space group consisting of translations and rotations about certain symmetry axes passing through special symmetry points of the structure.[67] In symmorphic systems, the space group is the direct product[3] of a translation group and a point group $\mathcal{G}_P$ consisting only of rotations, inversions, and reflections about specific symmetry axes in the system. In nonsymmorphic systems, there are additional symmetry elements associated with glide planes and screw axes. This chapter will consider only symmorphic groups. The invariance of $<\rho(\mathbf{x})>$ under $\mathcal{G}_P$ implies as a result of Eq. (4.2) that the scattering amplitudes are also invariant under $\mathcal{G}_P$. Since the scattering amplitudes can be viewed as decorations of the reciprocal lattice, it is clear that the point group $\mathcal{G}_P$ of the real-space structure is a subgroup of both the point group $\mathcal{G}_R$ of the reciprocal lattice and the Laue group $\mathcal{G}_L$. $\mathcal{G}_P$ differs from $\mathcal{G}_L$ only if it does not contain the inversion operation. Point group operations performed about points other than the special symmetry points leave the

$|\langle\rho_G\rangle|^2$ unchanged but will induce changes in the phases ϕ_G. This point will be discussed further in the following sections.

TOSs are divided into *periodic crystals* and *quasiperiodic structures,* depending on whether their reciprocal lattice is periodic or quasiperiodic. Quasiperiodic structures are further divided into *incommensurate crystals* and *quasicrystals* according to whether their point group $\mathcal{G}_P$ is crystallographic or noncrystallographic. All quasicrystals must have a quasicrystalline reciprocal lattice. The converse, however, is not necessarily true. It is possible to imagine a structure with a quasicrystalline reciprocal lattice L^Q invariant under $\mathcal{G}_R$ but with scattering amplitudes invariant only under some crystallographic subgroup of $\mathcal{G}_R$. This case is extremely unlikely, however, since the lower symmetry of the amplitudes will lead to a lower symmetry reciprocal lattice via distortions of the reciprocal lattice vectors induced by coupling to the strain.

Any vector in a periodic lattice can be expressed as an integral linear combination of vectors in a basis. To discuss how this property generalizes to quasiperiodic reciprocal lattices, a few more definitions are useful. A set of vectors is *integrally independent* if any linear combination with integral coefficients vanishes only when all of the coefficients vanish. A lattice is said to be *generated* by a set of vectors S, called a *generating set,* if every vector in the lattice can be expressed as an integral linear combination of the vectors in S. The *rank* r of the lattice is the smallest number of integrally independent vectors in any generating set. Clearly r must be greater than or equal to the dimension D of the lattice. Finally, a *basis* $\{\mathbf{k}_1, \ldots, \mathbf{k}_r\}$ is any generating set consisting of exactly r vectors. Any vector $\mathbf{G}$ is uniquely specified by the r-tuple of integers $(a_1, \ldots, a_r)$ appearing in its decomposition,

$$\mathbf{G} = \sum_{i=1}^{r} a_i \mathbf{k}_i, \tag{4.3}$$

in terms of the basis vectors.

In a square lattice in two dimensions, the two vectors $\hat{e}_x = (1,0)$ and $\hat{e}_y = (0,1)$ are integrally independent and form a basis. The rank of the basis is equal to the dimension of the lattice. Any set containing $\hat{e}_x$ and $\hat{e}_y$, for example the set consisting of $\pm e_x$ and $\pm\hat{e}_y$, constitutes a generating set. In the one-dimensional quasiperiodic structure discussed in the last section, k_1 and k_2 are integrally independent vectors because their ratio is irrational. They also form a basis, so that the rank $r = 2$ of the lattice they generate is greater than the dimension $D = 1$ of the lattice. These two examples illustrate that vectors can be integrally independent, either because they have components in orthogonal directions in a D-dimensional space or because they sum to zero with relatively irrational coefficients.

A reciprocal lattice is periodic if and only if its rank is equal to its dimension. It is easy to see that the lattice is quasiperiodic if $r>D$. The set of the r- basis vectors must contain a subset of D vectors that span the D-dimensional space with real (not necessarily integer) coefficients. Thus, the other $r-D$ members can be expressed as linear combinations with real coefficients of the D that span the space. Since the basis vectors are integrally independent, some of the real coefficients must be irrational, and the lattice is quasiperiodic. It is intuitively obvious but somewhat more tedious to show that if $r = D$, there is no shortest-length vector and the lattice is periodic.

In what follows, $L^A(D,r)$ will be used to denote a D-dimensional rank-r, reciprocal lattice of type A, where $A = P$, $A = I$, and $A = Q$ signify, respectively, periodic, incommensurate, and quasicrystalline. Similarly, a basis for $L^A(D,r)$ will be denoted by $\mathrm{B}_a{}^A(\mathrm{D,r})$, where the subscript a will be used to designate other properties of the basis, such as the lengths of its constituent vectors.

It is convenient, as will be seen below, to give names to several important sets of vectors. The image of any reciprocal lattice vector $\mathbf{G}$ under all point group operations in G_R is called the *star* of $\mathbf{G}$. The *symmetrized basis* B_{srm} is generated by applying all operations in $\mathcal{G}_{\mathrm{R}}$ to the basis vectors. Since $\mathcal{G}_{\mathrm{R}}$ contains the inversion operation, B_{sym}will always contain an even number $2b$ of elements with $b \geqslant r$. Thus B_{sym} can be represented as the union of a set $\{\mathbf{G}_n\}$ of b vectors $\mathbf{G}_n$, $n = 0,1,\ldots, b-1$, with its image $\{-\mathbf{G}_n\}$ under inversion. The set of positive vectors $\{\mathbf{G}_n\}$ is the *extended basis*. The simplest nontrivial example of a lattice for which the basis and extended basis differ is the triangular lattice in two dimensions. The extended basis consists of the three vectors,

$$\mathbf{G}_n = G\left[\cos\frac{2\pi n}{3}, \sin\frac{2\pi n}{3}\right], \qquad n = 0,1,2, \tag{4.4}$$

whereas a basis is any two of the three vectors in the extended basis, since $\Sigma_n \mathbf{G}_n = 0$. A basis, the symmetrized basis, and an extended basis for the triangular lattice are indicated in Fig. 2.1.

4.2 Incommensurate Crystals

Incommensurate crystals are quasiperiodic structures with crystallographic point group symmetry. The simplest example of an incommensurate crystal is one with a one-dimensional rank-two reciprocal lattice $L^I(1,2)$ considered in Section 3. Its basis $B^I(1,2)$ is the union, $B^P(1,1) \cup B^P(1,1)$ of the bases for two rank-one one-dimensional periodic lattices. This result generalizes to higher dimensional and higher rank incommensurate crystals. The basis $B^I(D,r)$ of a D-dimensional rank-r reciprocal lattice may be expressed as the union of the bases $B^A(D_i,r_i)$ of lower rank reciprocal lattices:

$$B^I(D,r) = \bigcup_{\Sigma_i r_i = r} B^A(D_i,r_i). \tag{4.5}$$

It is always possible to choose this decomposition such that all of the constituent bases are periodic. In this case, $A = P$ for every term in the above. It is often convenient, however, to express the basis of an incommensurate lattice in terms of incommensurate bases of lower rank and dimension.

It is instructive to consider in detail the example of incommensurate crystals with two-dimensional rank-four reciprocal lattices. As discussed above, the basis for these lattices can be expressed as the union of two crystallographic bases oriented at arbitrary angles. Let $B_a^P(2,2)$ and $B_b^P(2,2)$ be bases consisting, respectively, of the vectors

$$\begin{aligned} \mathbf{k}_{an} &= k_a[\cos(n\theta_0 - \psi), \sin(n\theta_0 - \psi)], \\ \mathbf{k}_{bn} &= k_b[\cos(n\theta_0 + \psi), \sin(n\theta_0 + \psi)], \qquad n = 0,1. \end{aligned} \tag{4.6}$$

The incommensurate basis $B^I(2,4) = B_a^P(2,2) \cup B_b^P(2,2)$ can have a variety of symmetries depending on the values of k_a, k_b, θ_0, and ψ. Representative incommensurate reciprocal lattices with crystallographic symmetry are depicted in Fig. 4.1. Lattices with four- and six-fold rotation axes but no other symmetries [C_4 and C_6 point groups in the Schoenflies notation] are generated by $k_a \neq k_b$, $\psi \neq 0$, and, respectively, by $\theta_0 = \pi/2$ and $\theta_0 = \pi/3$. Lattices with C_2 symmetry are generated by $k_a \neq k_b$, $\psi \neq 0$, and $\theta_0 \neq \pi/4$ or $\pi/3$. When $\psi = 0$ and k_a is an irrational multiple of k_b, $B^I(2,4)$ can alternatively be represented as the union of two one-dimensional rank-two incommensurate bases oriented along the (1,0) and ($\cos\theta_0$, $\sin\theta_0$) axes. Lattices with reflection lines producing C_{2v}, C_{4v}, and C_{6v} symmetry can be generated with $k_a = k_b$ and the appropriate choices of θ_0 and ψ. As discussed earlier, there are various real-space symmetries compatible with a given symmetry of scattering intensities. Thus, for example, it is possible for an incommensurate crystal to have a reciprocal lattice invariant under C_{6v} but a density invariant only under C_{3v}.

Three-dimensional incommensurate crystal lattices can be constructed in a similar way. An interesting set of lattices can be constructed from the six vectors

$$\begin{aligned} \mathbf{k}_{x\sigma} &= k(\cos\sigma\psi, 0, \sin\sigma\psi), \\ \mathbf{k}_{y\sigma} &= k(0, \cos\sigma\psi, \sin\sigma\psi), \\ \mathbf{k}_{z\sigma} &= k(\sin\sigma\psi, 0, \cos\sigma\psi), \end{aligned} \tag{4.7}$$

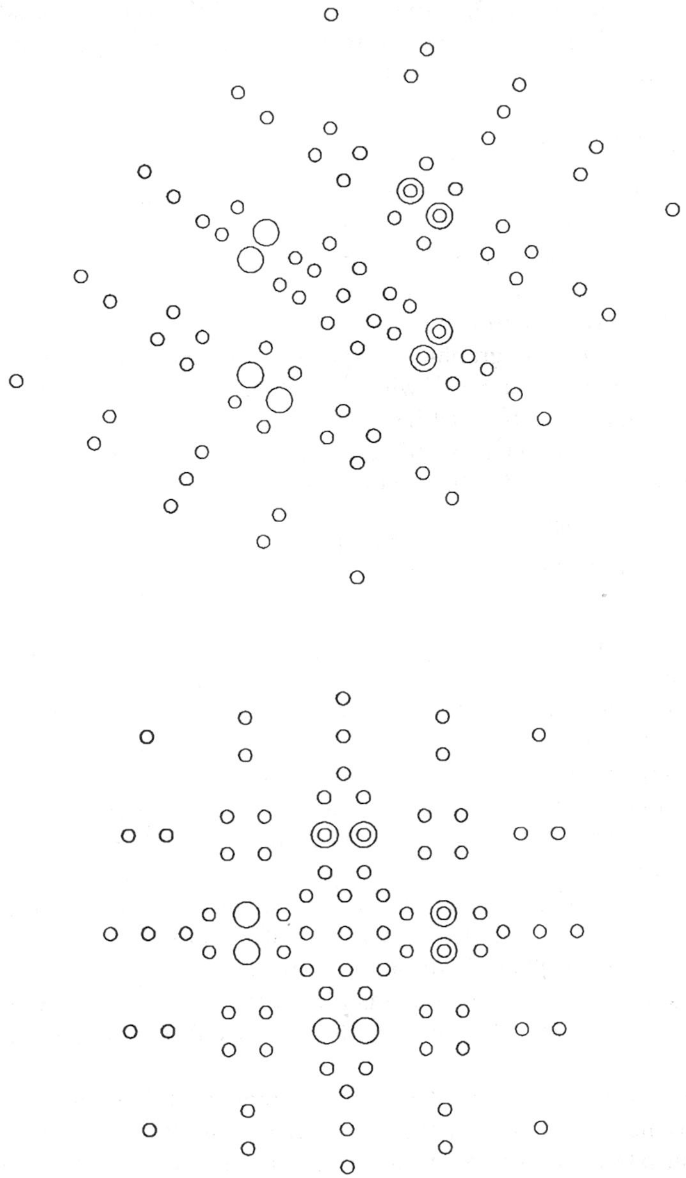

(a)

(b)

Figure 4.1. (a) Points in an incommensurate reciprocal lattice with C_2 symmetry generated by Eq. (4.6) with $\theta = 0.05\times 2\pi$, $\psi = 0.11\times 2\pi$, and $k_b/k_a = 1.23$. (b) Points in an incommensurate reciprocal lattice with C_{4v} symmetry generated with $\theta = \pi/2$, $\psi = 0.03\times 2\pi$, and $k_a = k_b$. The symmetrized bases are indicated by large circles and the bases, which are equivalent to the extended bases, by concentric circles.

with $\sigma = \pm 1$. For $\psi \neq 31.72°$, the point group of this reciprocal lattice is the tetrahedral group plus inversions T_h. The real space point group $\mathcal{G}_P$ can be either T_h or the tetrahedral group T.

4.3 *Quasicrystals*

Quasicrystals are quasiperiodic structures with a noncrystallographic point group symmetry. Since G_P is a subgroup of G_R, it immediately follows that G_R must be a noncrystallographic group. In two dimensions, the only crystallographic groups are those which have two-, four- or sixfold symmetry.[3] All other symmetries are incompatible with a regular periodic lattice and are noncrystallographic. The prototypical two-dimensional quasicrystal is the Penrose tiling[6–12] with fivefold symmetry. Since reciprocal lattices in two dimensions must have an evenfold symmetry, the reciprocal lattice for pentagonal Penrose tilings has C_{10} symmetry [Fig. 4.2] and is generated by the extended basis consisting of the vectors

$$\mathbf{G}_n = \left[\cos\frac{2\pi n}{5}, \sin\frac{2\pi n}{5}\right], \qquad n = 0, \ldots, 4. \tag{4.8}$$

It is clear that these vectors generate a lattice with tenfold symmetry. Furthermore, the two vectors $\mathbf{G}_0$ and $\mathbf{G}_1 + \mathbf{G}_4$ are colinear and have a magnitude ratio equal the to golden mean $\tau = (1+\sqrt{5})/2 = 1.6180\ldots$, so that the associated structure is quasiperiodic. This extended basis contains only four integrally independent vectors since $\Sigma_n \mathbf{G}_n = 0$. Thus the pentagonal reciprocal lattice is a special case of the more general quasiperiodic lattices considered in the last subsection. It is generated by the basis of Eq. (4.6), with $k_a = k_b$, $\theta_O = \pi/5$, and $\psi = \pi/10$. Quasicrystals with a reciprocal lattice with C_{10v} symmetry can have C_5, C_{5v}, C_{10}, or C_{10v} symmetry. There are Penrose tilings with all of these symmetries. Lattices with noncrystallographic C_8 and C_{12} symmetry are also special cases of the general $D = 2$, $r = 4$ incommensurate lattices generated respectively with $k_a = k_b$, $\theta_O = \pi/4$, $\psi = \pi/8$ and $\theta_O = \pi/6$ and $\psi = \pi/12$ in the basis of Eq. (4.6).

The most important noncrystallographic point group in three dimensions is the icosahedral group. A generating set for an icosahedral reciprocal lattice consists of the twelve vectors pointing to the vertices of an icosahedron. Of these, six are integrally independent and constitute a basis for the icosahedral lattice. The particular set of six chosen for the basis is somewhat arbitrary, and at least two choices appear in the literature.[26,28,68] The set of six vectors pointing to a vertex of an icosahedron and its five neighboring vertices is called the "umbrella set" and will be

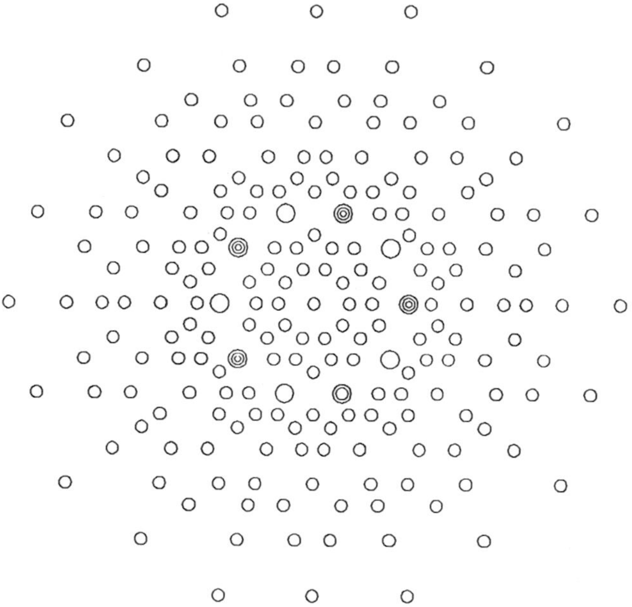

Figure 4.2. Points in a reciprocal lattice with C_{10v} symmetry. The symmetrized basis is indicated by large circles, the extended basis by circles with at least one concentric circle, and a basis by three concentric circles.

used here. With the z-axis chosen along the central vertex and the projection of one of the other vertices along the z-axis, the fundamental set can be parametrized as

$$\mathbf{G}_5 = G(0,0,1); \mathbf{G}_n = G[\sin\beta\cos n\theta, \sin\beta\sin n\theta, \cos\beta], n = 0,\ldots,4, \quad (4.9)$$

where $\theta = 2\pi/5$, $\cos\beta = 1/\sqrt{5}$, and $G = |\mathbf{G}_n|$. Alternatively, a coordinate system with the x-, y-, and z-axes normal to orthogonal edges of the icosahedron can be chosen, as shown in Fig 4.3. This coordinate system is obtained from the first via a rotation through an angle $\alpha = \sin^{-1}[(1+\tau^2)^{-1/2}] = 31.72°$ about the y-axis (τ is the golden mean). In this coordinate system, the extended basis is

$$\begin{aligned} \mathbf{G}_0 &= \bar{G}[1,0,\tau], & \mathbf{G}_1 &= \bar{G}[0,\tau,1], \\ \mathbf{G}_2 &= \bar{G}[-\tau,1,0], & \mathbf{G}_3 &= \bar{G}[-\tau,-1,0], \\ \mathbf{G}_4 &= \bar{G}[0,-\tau,1], & \mathbf{G}_5 &= \bar{G}[-1,0,\tau], \end{aligned} \quad (4.10)$$

where $\bar{G} = (1+\tau^2)^{-1/2}G$ and $G = |\mathbf{G}_n|$. It is clear from this form that the icosahedral reciprocal lattice is a special case of the rank-six lattices generated by the vectors in Eq. (4.7) with $\psi = \alpha$.

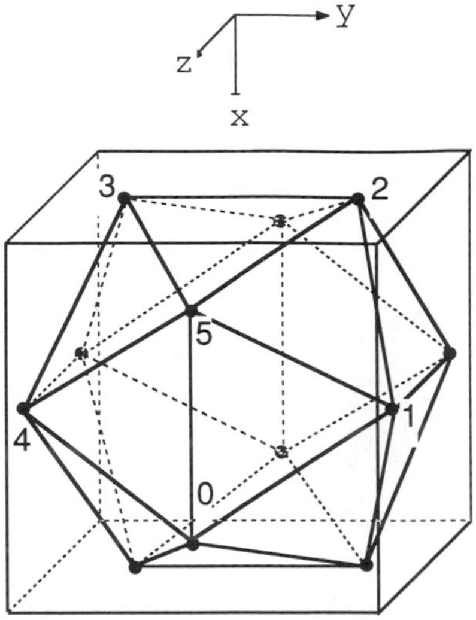

Figure 4.3. An icosahedron oriented as described in Eq. (4.10). The labeled vertices consititute a basis (and an extended basis) for the icosahedral reciprocal lattice.

The vectors defined in Eqs. (4.9) and (4.10) generate a primitive icosahedral reciprocal lattice that is the analog of the simple cubic lattice. The icosahedral analogs of BCC and FCC reciprocal lattices also exist.[66]

The bases of the pentagonal and icosahedral lattices just discussed contain the minimum number four or six of vectors needed to generate a lattice with the given noncrystallographic symmetry. They are, thus, *simple* quasicrystalline reciprocal lattices. Quasicrystalling lattices with bases containing more than the minimum number of vectors needed to generate a given noncrystallographic symmetry are *modulated* quasicrystalline lattice. An example of a rank-twelve modulated icosahedral lattice is the one generated by two sets of icosahedral vertex vectors sharing common axes with a magnitude ratio of, say, $\sqrt{2}$.

5. Landau Theory

A great deal can be learned about the thermodynamics and low energy excitations of condensed phases from phenomenological models constructed to incorporate all relevant symmetries of the real system. Such models, originally introduced by Landau, have been applied with enormous success to the study of phase transitions[69] and properties of ordered phases such as superconductors.[70] These models are particularly useful for studying complicated materials such as quasicrystals[15,26,46–49] for which there is an experimentally determined or theoretically presumed symmetry that is not easily predicted by a microscopic Hamiltonian. This section will show how to develop a Landau theory for translationally ordered structures and will discuss some of its consequences.

The order parameters distinguishing a TOS from a homogeneous fluid are the average mass density wave amplitudes $<\rho_{\mathbf{G}}(\mathbf{x})>$. In equilibrium, the values of these order parameters are those which minimize the free energy $F(\{<\rho_{\mathbf{G}}(\mathbf{x})>\})$:

$$\frac{\delta F}{\delta<\rho_{\mathbf{G}}(\mathbf{x})>} = 0. \tag{5.1}$$

In general, F is a complicated, possibly nonanalytic function of the order parameters. Landau mean field theory results when the free energy $F(\{<\rho_{\mathbf{G}}(\mathbf{x})>\})$ is approximated by a power series in the order parameters. The symmetry of the TOS places restrictions on the nature of this power series. In particular, F must be invariant under uniform translations of the system. Section 2 showed that translations lead to phase shifts $\mathbf{G}\cdot\mathbf{u}$ in the complex amplitudes $<\rho_{\mathbf{G}}>$. Though this result was derived for periodic structures, it follows from translational invariance and is valid for all translationally ordered structures. Since F cannot depend on $\mathbf{u}$, any p^{th} order term in the expansion of F in powers of $<\rho_{\mathbf{G}}(\mathbf{x})>$ must have the form

$$\int d^d x \Delta\Big(\sum_i^p \mathbf{G}_i\Big) \prod_{i=1}^{p} <\rho_{\mathbf{G}_i}(\mathbf{x})>$$
$$= \int d^d x \Delta\Big(\sum_i^p \mathbf{G}_i\Big) \cos\Big[\sum_i^p \phi_{\mathbf{G}_i}(\mathbf{x})\Big] \prod_i^p | < \rho_{\mathbf{G}_i}(\mathbf{x})> | , \tag{5.2}$$

where $\Delta(\mathbf{a}) \equiv \delta_{\mathbf{a},0}$. Since $-\mathbf{G}$ is in the reciprocal lattice if $\mathbf{G}$ is, the Landau expansion always includes a quadratic term for every $\mathbf{G}$:

$$F_2 = \frac{1}{2}\int d^d x \sum_{\mathbf{G}} A_{\mathbf{G}} \, | <\rho_{\mathbf{G}}(\mathbf{x})> |^2, \tag{5.3}$$

where $A_{\mathbf{G}}$ is a temperature-dependent phenomenological parameter. Third-order terms arise from closed triangles of vectors in the reciprocal lattice and have the form

$$F_3 = \frac{1}{3!}\int d^d x \sum_{\mathbf{G}_1\cdot\mathbf{G}_2\cdot\mathbf{G}_3}{}' A_{\mathbf{G}_1\cdot\mathbf{G}_2\cdot\mathbf{G}_2} |<\rho_{\mathbf{G}_1}>| \, |<\rho_{\mathbf{G}_2}>| \, |<\rho_{\mathbf{G}_3}>| \cos(\phi_{\mathbf{G}_1}+\phi_{\mathbf{G}_2}+\phi_{\mathbf{G}_3}), \tag{5.4}$$

where the prime on the sum signifies that $\mathbf{G}_1+\mathbf{G}_2+\mathbf{G}_2$ must be zero. Similar expressions for higher order terms in the expansion of F follow directly from Eq. (5.2).

Terms such as those in Eqs. (5.3) and (5.4) in the expansion of F depend on the order parameters only at a single point in space. They constitute the local part of F_L of the free energy. In addition, F will have a nonlocal part, F_{NL}, coupling $<\rho_{\mathbf{G}}(\mathbf{x})>$ at different points in space. In the limit of slow spatial variations, F_{NL} is a function only of the gradient of $<\rho_{\mathbf{G}}(\mathbf{x})>$ at a single point in space. For the present purposes, the most important term in F_{NL} has the form

$$F_{NL} = \int d^d x \sum_{\mathbf{G}} G_i G_j \nabla_i <\rho_{\mathbf{G}}(\mathbf{x})> \nabla_j <\rho_{-\mathbf{G}}(\mathbf{x})>$$
$$= \int d^d x \sum_{\mathbf{G}} G_i G_j \Big[\nabla_i \mid <\rho_{\mathbf{G}}> \mid \nabla_j \mid <\rho_{\mathbf{G}}> \mid + \mid <\rho_{\mathbf{G}}> \mid^2 \nabla_i \phi_{\mathbf{G}} \nabla_j \phi_{\mathbf{G}} \Big]. \tag{5.5}$$

It is this term and others, which are second-order in the gradient operator but of higher order in $\mid <\rho_{\mathbf{G}}> \mid$, that determine the elastic free energy.

Any vector $\mathbf{G}$ in the reciprocal lattice can be expressed as a unique integral linear combination of the basis vectors according to Eq. (4.3). This implies that for every $\mathbf{G}$ not in the basis there will be terms in F_L, depending on $\cos(\phi_{\mathbf{G}} - \Sigma_{i=1}^{r} a_i \phi_{\mathbf{k}_i})$, where $\phi_{\mathbf{k}_i}$ is the phase associated with the vector $\mathbf{k}_i$ in the basis. Thus, if

$$\phi_{\mathbf{G}} = \phi_{\mathbf{G}}^0 + \sum_{i=1}^{r} a_i \phi_{\mathbf{k}_i}, \tag{5.6}$$

the argument of the cosine in Eq. (5.2) will depend on $\phi_{\mathbf{G}}^0$ and not on the phases $\phi_{\mathbf{k}_i}$. In addition, $\Sigma\phi_{\mathbf{G}_a} = \Sigma\phi_{\mathbf{G}_a}^0$ for every set of vectors $\{\mathbf{G}_a\}$ satisfying $\Sigma\mathbf{G}_a = 0$. Thus, the local free energy depends only on the phases $\phi_{\mathbf{G}}^0$ for $\mathbf{G}$ not in the basis; it is independent of the r phases $\phi_{\mathbf{k}_i}$ associated with vectors in the basis. This means that there are exactly r phases that remain completely undetermined by the equilibrium equations of state in the absence of external forces. These undetermined phases will be called free phases in what follows. In this case, the order parameters $<\rho_{\mathbf{G}}>$ are independent of $\mathbf{x}$, and the equations of state are

$$\frac{1}{V} \frac{\partial F_L}{\partial \mid <\rho_{\mathbf{G}}> \mid} = 0, \qquad \frac{1}{V}\frac{\partial F_L}{\partial \phi_{\mathbf{G}}} \equiv \frac{1}{V}\frac{\partial F_L}{\partial \phi_{\mathbf{G}}^0} = 0, \qquad \mathbf{G} \in L, \tag{5.7}$$

where V is the volume of the structure.

Once all $<\rho_{\mathbf{G}}>$ have been determined by Eq. (5.7), they can be used in Eq. (5.5) for F_{NL} to determine how the free energy increases if the phases $\phi_{\mathbf{G}}(\mathbf{x})$ vary slowly in space. F_{NL}, unlike F_{L}, depends on the phases

$\phi_{\mathbf{k}_i}$. Thus, spatial variations in the $\phi_{\mathbf{k}_i}$ lead to an increase in the free energy whereas spatially uniform changes in $\phi_{\mathbf{k}_i}$ do not. The phases $\phi_{\mathbf{k}_i}$ are elastic variables whereas the phases ϕ_G^0 reach some preferred value in equilibrium and are not.

Though the r phases $\phi_{\mathbf{k}_i}$ associated with vectors in a basis could be chosen as independent elastic variables, it is preferable to parameterize these phases in terms of displacement and phason variables such as those introduced in the discussion of one-dimensional quasiperiodic structures. This parameterization will be discussed in the next section.

6. Group Theory and the Phason Variable

6.1 *Summary of Results*

It was just argued that, in equilibrium, there are exactly r free phases in a rank-r quasi-periodic structure. In order to obtain a parameterization of the r free phases that reflects the point group symmetry of the structure, it is useful to consider not the set of phases $\phi_{\mathbf{k}_i}$ associated with basis vectors but the possibly larger set of b phases ϕ_n in the extended basis. Then $b - r$ linear combinations of these b phases will be determined by the equilibrium conditions [Eq. (5.7)], and the remaining r linear combinations corresponding to elastic variables can be parameterized by the D-component displacement $\mathbf{u}$ and by a $(r - D)$-component phason variable $\mathbf{w}$.

This section will use group theory to derive parameterizations for ϕ_n in terms of $\mathbf{u}$ and $\mathbf{w}$. The general result is that the phases in the extended basis can be written in the form [Eq. (3.8)] suggested by the one-dimensional example:

$$\phi_n = \mathbf{G}_n \cdot \mathbf{u} + \alpha \mathbf{G}_n^{\perp} \cdot \mathbf{w} + \phi_n^0. \tag{6.1}$$

$\mathbf{G}^{\perp}$ depends on the symmetry of the quasiperiodic structure and will be calculated with the aid of group theory for a few specific examples in the section. When $b = r$ (i.e., when the extended basis is identical to the basis), the phase ϕ_n^0 is zero. When $b > r$, the phases ϕ_n^0 are determined by the equilibrium equations [Eq. (5.7)]. In systems in which $|\mathbf{G}_n|$ is independent of n, ϕ_n^0 must be independent of n in order for the structure to be invariant under the group $\mathcal{G}_P$. Equation (6.1) for the phases in the extended basis determines the functional relationship between $\mathbf{u}$ and $\mathbf{w}$ and the phase ϕ_G for arbitrary $\mathbf{G}$ not in the extended basis. Any vector $\mathbf{G}$ in the reciprocal lattice can be expressed as an integral linear combination (not necessarily unique), $\Sigma_n c_n \mathbf{G}_n$, of the vectors in the extended basis. This suggests, following the discussion of the previous section, that $\phi_{\mathbf{G}}$ be written

$$\phi_{\mathbf{G}} = \mathbf{G} \cdot \mathbf{u} + \alpha \mathbf{G}^{\perp} \cdot \mathbf{w} + \phi_{\mathbf{G}}^{0}, \tag{6.2}$$

where the $(r - D)$-dimensional complementary vector

$$\mathbf{G}^{\perp} = \sum_{n} c_n \, \mathbf{G}_n^{\perp} \tag{6.3}$$

is uniquely determined by $\mathbf{G}$. In analogy with the one-dimensional case, r-dimensional reciprocal lattice vectors

$$\tilde{\mathbf{G}} = \mathbf{G} \oplus \alpha \mathbf{G}^{\perp} \tag{6.4}$$

and an r-dimensional displacement variable

$$\tilde{\mathbf{u}} = \mathbf{u} \oplus \mathbf{w} \tag{6.5}$$

can be introduced, allowing Eq. (6.3) to be reexpressed as

$$\phi_{\mathbf{G}} = \tilde{\mathbf{G}} \cdot \tilde{\mathbf{u}} + \phi_{\mathbf{G}}^{0}. \tag{6.6}$$

With this parameterization of $\phi_{\mathbf{G}}$, all sums $\Sigma \phi_{\mathbf{G}_\alpha}$ are independent of $\mathbf{u}$ and $\mathbf{w}$, and F_L is independent of $\mathbf{u}$ and $\mathbf{w}$. Thus, $\mathbf{u}$ and $\mathbf{w}$ are elastic variables. The phases ϕ_n^0 and $\phi_{\mathbf{G}}^0$ are fixed by the equilibrium conditions and are not elastic variables. Equations (6.1) to (6.6) are the most important results of this section.

Once the complementary vectors have been determined, an r-dimensional periodic parent lattice can be constructed by a direct generalization of the discussion in Section 3. The vectors $\tilde{\mathbf{G}}$ form an r-dimensional reciprocal lattice under the definitions of Section 3. This lattice must be periodic because the generating set consisting of the vectors $\tilde{\mathbf{G}}_n = \mathbf{G}_n \oplus \alpha \mathbf{G}_n^{\perp}$ contains exactly r integrally independent vectors. The density

$$< \tilde{\rho}(\tilde{\mathbf{x}}) > = \sum_{\tilde{\mathbf{G}}} < \rho_{\tilde{\mathbf{G}}} > e^{i\tilde{\mathbf{G}} \cdot \tilde{\mathbf{x}}} \equiv < \tilde{\rho}(\mathbf{x}_{\parallel}, \mathbf{x}_{\perp}) > \tag{6.7}$$

is then a periodic function of the r-dimensional vector $\tilde{\mathbf{x}} \equiv \mathbf{x}_{\parallel} \oplus \mathbf{x}_{\perp}$, where $\mathbf{x}_{\parallel}$ is a D-dimensional and $\mathbf{x}_{\perp}$ an $(r - D)$-dimensional vector. (Strictly speaking, the construction of $\tilde{\mathbf{x}}$ just described applies only to the case when the dimension d of space is equal to the dimension D of the reciprocal lattice. The generalization to the case $d > D$ is obvious.) With $< \rho_{\tilde{\mathbf{G}}} > = | < \rho_{\mathbf{G}} > | \, e^{i\phi_{\mathbf{G}}^{0}}$, the density of the periodic parent crystal along the cut $\mathbf{x}_{\parallel} = \mathbf{x} + \mathbf{u}, \mathbf{x}_{\perp} = \mathbf{w}$ is equal to the density of the r-dimensional structure:

$$< \rho(\mathbf{x}) > = <\tilde{\rho}(\mathbf{x} + \mathbf{u}, \mathbf{w}) > \equiv \sum_{\tilde{\mathbf{G}}} < \rho_{\tilde{\mathbf{G}}} > e^{i\mathbf{G} \cdot \mathbf{x}} \, e^{i\tilde{\mathbf{G}} \cdot \tilde{\mathbf{u}}}. \tag{6.8}$$

The set of vectors $\tilde{\mathbf{R}}$ in the direct lattice of the parent crystal is defined via Eq. (4.1), which in the present case becomes

$$e^{i\tilde{\mathbf{G}}\cdot\tilde{\mathbf{R}}} = 1. \tag{6.9}$$

Whenever $\tilde{\mathbf{u}} = \tilde{\mathbf{R}}$, the lattice point $\tilde{\mathbf{R}}$ lies in the physical space, and the density $<\rho(\mathbf{x})>$ is identical to what it is when $\tilde{\mathbf{u}} = 0$. Operations in $\mathcal{G}_P$ about symmetry axes passing through the point $\tilde{\mathbf{R}}$ in the physical space do not change $<\rho(\mathbf{x})>$. Thus the symmetry points in a quasiperiodic structure are lattice sites in an r-dimensional parent space. These symmetry points cannot be accessed by simple physical space translations of the structure as they can be in periodic crystals. Their access requires translations in the complementary space as well, i.e., changes in the phason variable, which involve rearrangements of atoms. Since the energy of a quasiperiodic structure is independent of $\mathbf{w}$, there is no reason for $\mathbf{w}$ to have a value such that the physical space passes through a symmetry point. As a result, the real space density of a quasiperiodic structure will not, in general, be invariant under G_p. The vectors $\tilde{\mathbf{R}}$ in the direct parent lattice of some quasiperiodic structures will be derived in Section 8 where dislocations are discussed.

6.2 Representations and Characters

As discussed in Section 4, a structure is invariant with respect to any operations in the point group $\mathcal{G}_P$ performed about special symmetry points. In periodic crystals, these special symmetry points are precisely the lattice sites in the direct lattice. In quasiperiodic structures, the concept of a symmetry point must be generalized as discussed above. This implies that the phases associated with all vectors in the reciprocal lattice that can be transformed into each other under the operations of $\mathcal{G}_P$ must be equal when the origin is a symmetry point. If point group operations are performed about a point other than a symmetry point, the phases in a given star in reciprocal space will transform among themselves under some representation of $\mathcal{G}_P$. In particular, the phases in the symmetrized basis will transform under a $2b$-dimensional representation of $\mathcal{G}_P$. Because of the constraint $\phi_{\mathbf{G}} = -\phi_{-\mathbf{G}}$, the phases in the extended basis transform under a b-dimensional representation of $\mathcal{G}_P$. It is the decomposition of this b-dimensional representation into lower dimensional (possibly reducible) representations that will determine the form of $\mathbf{G}^{\perp}$ and $\phi^0_{\mathbf{G}}$.

Any k-dimensional representation[3] $\Gamma^{(k)}$ of a group of finite order h can be decomposed into the irreducible representations $\Gamma_0^{(j)}$,

$$\Gamma^{(k)} = \sum_j a_j \Gamma_0^{(j)}. \tag{6.10}$$

The coefficients a_j can be determined from the characters $\chi^{(k)}(C)$ of the k-dimensional representation and the characters $\chi_0^{(j)}(C)$ of all irreducible representations of the group

$$a_j = \frac{1}{h} \sum_C n_C \chi^{(k)}(C) \chi^{(j)}(C), \tag{6.11}$$

where C is a class, and n_C is the number of elements in class C. It should be noted that the sum of squares of the dimensionalities of the irreducible representations must equal the order h of the group. This fact and a few other simple properties of groups usually suffice to determine the dimensionalities of all representations of a group.

6.3 *Periodic Crystals*

In many crystallographic lattices, for example, the square and simple cubic lattice $b = r = D$, and the extended basis is equivalent to the basis. In this case, the r independent phases can be parameterized directly by the D - dimensional vector $\mathbf{u}$ via $\phi_i = \mathbf{k}_i \cdot \mathbf{u}$ for $i = 1, \ldots, D$, where the vectors $\mathbf{k}_i$ form the basis. The considerations of the last section then immediately lead to Eq. (2.7) for the phase associated with any vector $\mathbf{G}$.

The extended basis of some periodic reciprocal lattices, such as the hexagonal lattice in two dimensions and the FCC lattice in three dimensions, contain more than D elements. To see how group theory can be used to treat this case, consider the hexagonal reciprocal lattice in two dimensions. The three vectors $\mathbf{G}_n$ defined in Eq. (4.4) form an extended basis for this lattice. The three phases $\phi_n = \phi_{\mathbf{G}n}$ then transform under a three-dimensional representation $\Gamma^{(3)}$ of the group G_p. As discussed earlier, this group can be C_3, C_{3v}, C_6, or C_{6v} if the amplitudes $| \langle p_{\mathbf{G}n} \rangle |$ are equal for all n. For the purposes of discussion, consider the case of C_{3v} symmetry. C_{3v} has six elements: the identity, rotations by $\pm 2\pi/3$, and three reflections through mirror planes, and it has three classes: the identity class E with one element, a class $C_{2\pi/3}$ with the two rotation elements, and a class σ_d with the three mirror operations. Finally, there are three irreducible representations of respective dimensionality 1, 1, and 2. The character table[3] for C_{3v} is shown in Table 6.1.

Table 6.1 C_{3v} Character Table

	E	$2C_{2\pi/3}$	$3\sigma_d$
$\chi_r^{(1)}$	1	1	1
$\chi_r^{(1)}$	1	1	-1
$\chi_r^{(2)}$	2	-1	0

The characters of the three classes for $\Gamma^{(3)}$ are easily calculated to be 3, 0, and 1 for E, $C_{2\pi/3}$, and σ_d. Thus, Eq. (6.5) implies

$$\Gamma^{(3)} = \Gamma_0^{(2)} + \Gamma_0^{(1)}. \tag{6.12}$$

The vector or two-dimensional representation appears once, and the identity representation appears once. The vectors $\mathbf{G}_n$ form a two-dimensional representation of C_{3v}, so that

$$\phi_n = \mathbf{G}_n + \mathbf{u} + \gamma/3, \tag{6.13}$$

where

$$\gamma = \phi_0 + \phi_1 + \phi_2.$$

The fact that ϕ_n has a part proportional to $\mathbf{G} \cdot \mathbf{u}$ ensures that all ϕ_G will have the form of Eq. (2.7) as expected. The identity representation arises from the fact that $\mathbf{G}_0 + \mathbf{G}_1 + \mathbf{G}_2 = 0$, so that F can depend on $\cos(\Sigma\phi_n) = \cos\gamma$. There is a preferred equilibrium value for γ determined by the potentials in F or, alternatively, by the initial microscopic Hamiltonian and the temperature. The nature of symmetry points about which operations in $\mathcal{G}_p$ leave the density invariant is clear from Eq. (6.8). In a coordinate system with $\mathbf{u} = 0$, all ϕ_n are equal to $\gamma/3$. Then the phases, and thus the density, are invariant under all operations in the group C_{3v}. If γ is zero or π, all of the phases ϕ_n and $-\phi_n$ in the symmetrized basis consisting of the six vectors $\mathbf{G}_n$ and $-\mathbf{G}_n$ are equal mod 2π. In this case, the phases $\{\phi_n\} \cup \{\phi_{-n}\}$ in the symmetrized basis are invariant under C_{6v} rather than C_{3v}. The density of the whole structure, however, will be invariant under C_6 only if the phases in all stars of $\mathbf{G}$, in particular those containing twelve (the number of elements in C_{6v}) vectors, are also invariant under C_{6v}.

6.4 *Incommensurate Crystals*

Now turn to incommensurate crystals, and consider, for simplicity, only crystals whose basis can be represented as a union of only two periodic bases, $B_a^P(2,2)$ and $B_b^P(2,2)$ [See Eq. (4.6)]. In this case, the symmetrized basis can be decomposed into two sets of vectors: the sets $S_a = \{\mathbf{G}_{an}\}$ and $S_b = \{\mathbf{G}_{bn}\}$ generated by applying the point group operations respectively to $B_a^P(2,2)$ and $B_b^P(2,2)$. Two classes of incommensurate crystals can immediately be distinguished. In *type I* incommensurate crystals, the sets S_a and S_b are disjoint: $S_a \cap S_b = 0$. An example of a type I incommensurate crystal is the two-dimensional rank-four system with C_4 symmetry discussed in Section 4.3. Here S_a consists of the vectors $\pm\mathbf{k}_{an}$ and S_b of the vectors $\pm\mathbf{k}_{bn}$. The only symmetry operations are rotations by $2\pi p/4$ for $p = 0,1,2,3$. Since a vector in $B_a^P(2,2)$ is never tranformed into a vector in $B_b^P(2,2)$ under these operations, $S_a \cap S_b = 0$.

In *type II* incommensurate crystals, the sets S_a and S_b are not disjoint: $S_a \cap S_b \neq 0$. This means that the point group operations transform some of the vectors in $B_a^P(2,2)$ into vectors in $B_b^P(2,2)$. An example of a type II incommensurate crystal is the system with C_{4v} symmetry discussed in Section 4.3. As in the case of C_4 symmetry, S_a consists of the vectors $\pm\mathbf{k}_{an}$ and S_b of the vectors $\pm\mathbf{k}_{bn}$. Reflection about the x-axis takes $\mathbf{k}_{a0}$ into $\mathbf{k}_{b0}$; reflection about the y-axis takes $\mathbf{k}_{a1}$ into $\mathbf{k}_{b1}$; reflection through the line $x = y$ takes $\mathbf{k}_{a0}$ into $\mathbf{k}_{b1}$ into $\mathbf{k}_{a1}$; and so on. Thus, in this case, S_a and S_b are clearly not disjoint.

It appears that all experimentally observed incommensurate phases in dielectrics and others materials[33,35] are type I crystals.

6.4.1 Type I Incommensurate Crystals. In type I incommensurate crystals, the sets S_a and S_b are totally independent and can be parameterized as though they arose from two independent periodic crystals. Thus the phases ϕ_{an} and ϕ_{bn} associated with the vectors in S_a and S_b do not mix under $\mathcal{G}_P$ and have independent representations:

$$\phi_{an} = \mathbf{G}_{an} \cdot \mathbf{u}_a + \phi_{an}^0, \qquad \phi_{bn} = \mathbf{G}_{bn} \cdot \mathbf{u}_b + \phi_{bn}^0, \tag{6.14}$$

where ϕ_{an}^0 and ϕ_{bn}^0 arise if there are more than 2 D elements in S_a or S_b, as there would be for example if the structure had C_3 or C_6 symmetry. Equation (6.14) and the equilibrium conditions [Eq. (5.7)] then imply that the phase associated with an arbitrary $\mathbf{G} = \mathbf{G}_a + \mathbf{G}_b$ can be written as

$$\phi_{\mathbf{G}} = \mathbf{G}_a \cdot \mathbf{u}_a + \mathbf{G}_b \cdot \mathbf{u}_b + \phi_{\mathbf{G}}^0. \tag{6.15}$$

This equation is simply a D-dimensional generalization of Eq. (3.6). As in the one-dimensional case, displacement and phason variables $\mathbf{u}$ and $\mathbf{w}$ can be introduced,

$$\mathbf{u}_a = \mathbf{u} - \alpha s\mathbf{w}, \qquad \mathbf{u}_b = \mathbf{u} + \alpha(1-s)\mathbf{w}. \tag{6.16}$$

$\mathbf{u}$ describes displacement of the whole structure, whereas $\mathbf{w}$ describes relative translations of the two sublattices associated with the periodic bases $B_a^P(2,2)$ and $B_b^P(2,2)$. Agains α and s are arbitrary, and the relation between $\mathbf{u}$ and $\mathbf{u}_a$ and $\mathbf{u}_b$ is not unique. In terms of $\mathbf{u}$ and $\mathbf{w}$, Eq. (6.15) becomes

$$\phi_{\mathbf{G}} = \mathbf{G} \cdot \mathbf{u} + \alpha\mathbf{G}^{\perp}\mathbf{w} + \phi_{\mathbf{G}}^0, \tag{6.17}$$

where

$$\mathbf{G}^{\perp} = -s\mathbf{G}_a + (1-s)\mathbf{G}_b \tag{6.18}$$

is the complementary space vector associated with $\mathbf{G}$. As in the one-dimensional case, $\tilde{\mathbf{G}}$ and $\tilde{\mathbf{u}}$ can be expressed in a rotated coordinate system where they respectively become $c_a\mathbf{G}_a \oplus C_b\mathbf{G}_b$ and $c_a^{-1}\mathbf{u}_a \oplus c_b^{-1}\mathbf{u}_b$, where c_a and c_b satisfy Eq. (3.17).

6.4.2 Type II Incommensurate Crystals. Type II incommensurate crystals are more complicated. It is necessary to find linear combinations of the phases ϕ_n in the extended basis that transform among themselves under irreducible representations of the point group. The rank-four structures with C_{4v} symmetry discussed in Section 4.2 provide a simple example of a type II incommensurate crystal. C_{4v} is an eight element group consisting of the identity, rotations through $2p\pi/4$, $p = 1,2,3$ and four reflection planes. It has five classes, four one-dimensional representations and one two-dimensional representation. Its character table[3] is displayed in Table 6.2.

Table 6.2 C_{4v} Character Table

	E	$C_{2\pi/2}$	$2C_{2\pi/4}$	$2\sigma_v$	$2\sigma_d$
$\chi_r^{(1)}$	1	1	1	1	1
$\chi_r^{(\bar{1})}$	1	1	1	-1	-1
$\chi_r^{(1')}$	1	1	-1	1	-1
$\chi_r^{(\bar{1}')}$	1	1	-1	-1	1
$\chi_r^{(2)}$	2	-2	0	0	0

The phases ϕ_{an} and ϕ_{bn}, $n = 0,1$ transform under a four-dimensional representation $\Gamma^{(4)}$ of C_{4v}. The character of any representation of a group for a given class is simply the number of elements of the representation that are not changed by group operations in the class. The characters of $\Gamma^{(4)}$ for the five classes C_{4v} are easily calculated. For example, the operation of reflection through the axis $x = y$ takes ϕ_{a0} into ϕ_{b1} and ϕ_{a1} into ϕ_{b0}. This is an element in the class σ_d so that $X^{(4)}(\sigma_d) = 0$. Similarly, reflection through the x-axis interchanges ϕ_{b0} and ϕ_{a0} and takes ϕ_{a1} and ϕ_{b1}, respectively, into $-\phi_{b1}$ and $-\phi_{a2}$, so that $\chi^{(4)}(\sigma_v) = 0$. Proceeding in a similar way, one can show that $\chi^{(4)}(E) = 4$, $\chi^{(4)}(C_{2\pi/2}) = -4$, and $\chi^{(4)}(C_{2\pi/4}) = 0$. Then, from the character table [Table 6.2] and Eq. (6.5),

$$\Gamma^{(4)} = 2\Gamma_0^{(2)}. \tag{6.19}$$

The vectors $\mathbf{G}_{\alpha n} \equiv k_{\alpha n}$, $\alpha = a,b$, with $k_a = k_b \equiv G$ in the extended basis defined by Eq. (4.6), transform under a two-dimensional representation of C_{4v}. A second set of four vectors $\mathbf{G}_{\alpha n}^{\perp}$ (with $\alpha = a,b$),

$$\mathbf{G}_{a0}^{\perp} = \mathbf{G}_{a0}, \quad \mathbf{G}_{b0}^{\perp} = -\mathbf{G}_{b0}, \quad \mathbf{G}_{a1}^{\perp} = -\mathbf{G}_{b1}, \quad \mathbf{G}_{b1}^{\perp} = \mathbf{G}_{a1}, \tag{6.20}$$

also transforming under the two-dimensional representation of C_{4v}, can be constructed such that $\mathbf{G}_{\alpha n}$ and $\mathbf{G}^{\perp}_{\alpha n}$ satisfy the orthogonality relations

$$\sum_{\alpha n} G_{\alpha n,i}\, G_{\alpha n,j} = \sum_{\alpha n} G^{\perp}_{\alpha n,i}\, G^{\perp}_{\alpha n,j} = \delta_{ij},$$

$$\sum_{\alpha n} G_{\alpha n,i}\, G^{\perp}_{\alpha n,j} = 0. \tag{6.21}$$

The phases in the extended basis can now be written as

$$\phi_{\alpha n} = \mathbf{G}_{\alpha n} \cdot \mathbf{u} + \alpha \mathbf{G}^{\perp}_{\alpha n} \cdot \mathbf{w}. \tag{6.22}$$

As before, $\mathbf{u}$ is the displacement variable, and $\mathbf{w}$ is the phason variable. Unlike the previous example, however, $\mathbf{w}$ does not have a simple interpretation in terms of the relative displacement of two constituent sublattices. $\mathbf{w}$ and $\mathbf{u}$ are (apart from a trivial arbitrariness of scale) a unique linear combination of the phases $\phi_{\alpha n}$:

$$w_i = \alpha^{-1} G^{-2} \sum_{\alpha n} G^{\perp}_{\alpha n,i} \phi_{\alpha n}, \qquad u_i = G^{-2} \sum_{\alpha n} G_{\alpha n,i} \phi_{\alpha n}. \tag{6.23}$$

The free energy is invariant with respect to uniform increments in $\mathbf{u}$ and $\mathbf{w}$. In the present case, *both* $\mathbf{u}$ and $\mathbf{w}$ transform under the same representation of the point group.

An analysis almost identical to that above applies to the three-dimensional reciprocal lattice introduced in Eq. (4.7) in the last section. Here, the point group is some version of the tetrahedral group.

6.5 *Quasicrystals*

Quasicrystals are similar to type II incommensurate crystals in that the group structure leads to essentially unique definitions of $\mathbf{u}$ and $\mathbf{w}$. However, in quasicrystals, unlike in type II incommensurate crystals, $\mathbf{u}$ and $\mathbf{w}$ transform under different representations of the point group.

To see how this comes about, consider first the example of pentagonal quasicrystals in two dimensions. As discussed in Section II, the reciprocal lattice has C_{10v} symmetry, but the point group can be of lower symmetry—C_{5v}, for example. The group C_{5v} has ten elements: the identity, four rotations through multiples of $2\pi/5$, and five reflections. It has four classes: the identity (E) with one element, rotations through $2\pi/5$ ($C_{2\pi/5}$) with two elements, rotations through $4\pi/5$ ($C^2_{2\pi/5}$) with two elements, and reflections (σ_v) with five elements. The character table[3] for C_{5v} is reproduced in Table 6.3.

Table 6.3 C_{5v} Character Table

	E	$2C_{2\pi/5}$	$2C^2_{2\pi/5}$	$2\sigma_v$
$\chi_r^{(1)}$	1	1	1	1
$\chi_r^{(\bar{1})}$	1	1	1	-1
$\chi_r^{(2)}$	2	τ^{-1}	$-\tau$	0
$\chi_r^{(\bar{2})}$	2	$-\tau$	τ^{-1}	0

Note that there are two-dimensional representations of the group: one, $\Gamma_0^{(2)}$, is the vector representation, the other, $\Gamma_0^{(\bar{2})}$, is not. The orthogonality relations for the characters can be verified using $\tau^2 = \tau + 1$, where $\tau = (1+\sqrt{5})/2$ is the golden mean.

The five phases $\phi_n \equiv \phi_{\mathbf{G}n}$ associated with the vectors $\mathbf{G}_n$ [Eq. (4.8)] in the extended basis transform under a five-dimensional representation $\Gamma^{(5)}$ of C_{5v}. The characters of this representation are $\chi^{(5)}(E) = 5$, $\chi^{(5)}(C_{2\pi/5}) = \chi^{(5)}(C^2_{2\pi/5}) = 0$, and $\chi^{(5)}(\sigma_v) = 1$, implying

$$\Gamma^{(5)} = \Gamma_0^{(2)} + \Gamma_0^{(\bar{2})} + \Gamma_0^{(1)}. \tag{6.24}$$

The vectors $\mathbf{G}_n$ transform under $\Gamma_0^{(2)}$, and the complementary vectors $\mathbf{G}_n^{\perp} = \mathbf{G}_{<3n>_5}$, where the $< 3n >_5 \equiv 3n$ mod 5 transform under $\Gamma_0^{(\bar{2})}$. (An alternate and equally acceptable choice is $\mathbf{G}_n^{\perp} = \mathbf{G}_{<2n>_5}$.) Therefore,

$$\phi_n = \mathbf{G}_n \cdot \mathbf{u} + \mathbf{G}_n^{\perp} \cdot \mathbf{w} + \gamma/5, \tag{6.25}$$

where $\gamma = \Sigma_n \phi_n$ transforms under the identity representation, and as before, $\mathbf{u}$ and $\mathbf{w}$ are the displacement and phason variables. $\mathbf{G}_n$ and $\mathbf{G}_n^{\perp}$ satisfy orthogonality relations,

$$\sum_n G_{ni}\, G_{nj} = \sum_n G_{ni}^{\perp}\, G_{nj}^{\perp} = G^2 \delta_{ij},$$

$$\sum_n G_{ni}\, G_{nj}^{\perp} = 0, \tag{6.26}$$

like those of Eq. (6.21). Equation (6.25) has exactly the same form as Eq. (6.1) except that $\mathbf{G}_n^{\perp}$ does not transform under a vector representation of $\mathcal{G}_p$. It again follows that $\phi_{\mathbf{G}}$ for general $\mathbf{G}$ satisfies Eq. (6.2) with $\mathbf{G}^{\perp}$ determined by Eq. (6.3). The phase γ is determined by the details of the interaction potentials as in the case of hexagonal crystals. If $\gamma = 0$, the system can have C_{10v} rather than C_{5v} symmetry if $\mathbf{w} = 0$. Other values of γ preclude C_{10v} symmetry.

Each of the five Amman grids generating the Penrose tiling of Fig. 1.1 are determined by two phases. Of the total of ten phases determining a given Penrose pattern, four linear combinations determine $\mathbf{u}$ and $\mathbf{w}$, and one determines γ. The other five independent phases affect the phases of scattering amplitudes in stars other than the star containing the symmetrized basis. When all phases are zero, the Amman grid has tenfold symmetry; when all phases but γ are zero, the grid has only fivefold symmetry. Penrose tilings (Fig. 1.1) are dual lattices of the Amman grid. If the restriction of only two types of tiles are lifted (e.g., if decagonal tiles are permitted), it is possible to construct a tiling that has tenfold symmetry[71] when $\gamma = 0$. If only "fat" and "skinny" tiles that obey matching rules are allowed, then the Penrose tiling associated with the $\gamma = 0$ grid does not have perfect tenfold symmetry.[71] Each value of γ determines a *local isomorphism* class.[6,17] Local isomorphism classes to not depend on $\mathbf{u}$ or $\mathbf{w}$ but will in general depend on all of the nonelastic phases $\phi_{\mathbf{G}}^{0}$. Different local isomorphism classes have different arrangements of tiles. A given structure is expected to have a particular local isomorphism class determined by its free energy minimum.[42]

Density wave images provide an easily constructed visual representation of the effect of changes in $\mathbf{w}$ and γ. To construct these images, consider the simple fivefold symmetric function,

$$\rho(\mathbf{x}) = \sum_{n} \cos(\mathbf{G}_n \cdot \mathbf{x} + \mathbf{G}^{\perp} \cdot \mathbf{w} + \gamma/5), \qquad (6.27)$$

obtained by superposing density waves along the five directions in the extended basis. For each point on a grid, place a black dot if the function is greater than some cutoff and a white dot if it is less than that cutoff. A series of these images is presented in Figs. 6.1 and 6.2. Note that the pattern for $\mathbf{w} = 0$ and $\gamma = 0$ has C_{10v} symmetry. The pattern with nonzero γ has only fivefold symmetry. Though there is no global center of tenfold symmetry for $\gamma = 0$ when $\mathbf{w}$ is nonzero, there are regions of arbitrary but finite size with this symmetry. The local patterns are essentially the same for fixed γ as $\mathbf{w}$ varies. This is the reflection of the fact that $\mathbf{w}$ does not change the local isomorphism class, whereas γ does. Different local isomorphism classes allow distinct local arrangements of tiles in pentagonal tilings. These different local arrangements are reflected in the different local patterns in the density wave images for different values of γ. Note the light bands in the density wave pictures forming grids identical to the Amman grids in the Penrose tilings of Fig. 1.1.

As discussed in Section 4.5, the symmetrized basis of an icosahedral reciprocal lattice consists of the twelve vectors pointing to the vertices of an icosahedron. The symmetry operations of the reciprocal lattice include all of the operations in the group Y of rotations of an icosahedron plus the group i of inversions through the origin. Its symmetry group is, therefore, the full icosahedral group $Y_F = Y \oplus i$. The point group G_P can be either Y_F or Y if all the amplitudes $| \langle \rho_{\mathbf{G}} \rangle |$ on a given star are equal. The case where $\mathcal{G}_P$ is the group Y of icosahedral rotations will be

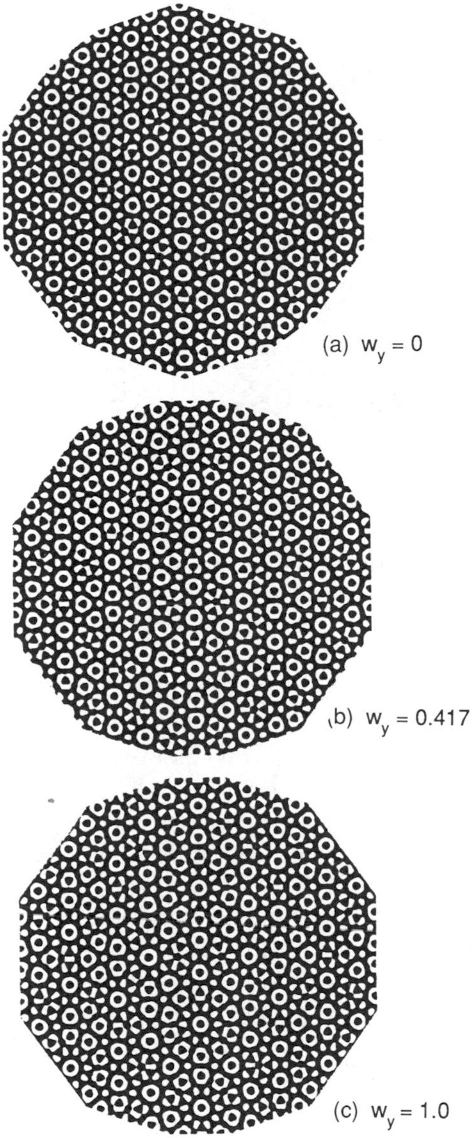

Figure 6.1. A sequence of figures (a)–(c) with $\gamma = 0$ [generated according to the instructions following Eq. (6.1)] illustrates the effect of adding a phase determined by $\tilde{u} = (0,0,0,w_y)$. $w_y = 0$, $0.417R$, and R, respectively, in (a), (b), and (c). When w_y is zero, the pattern has perfect tenfold symmetry. For small values of w_y, tenfold symmetry of the origin is destroyed, and no new center of symmetry is produced. When $\tilde{u}$ equals a vector $\tilde{R}$ of the periodic parent lattice, tenfold symmetry is regained about a different point in real space, as illustrated in (c) with $w_y = R$, where R is the length of a lattice vector in the parent lattice [see Eq. (8.5)]. Note the sequences of dark lines visible at grazing angle along the five symmetry directions. These lines are essentially identical to the lines in the Amman pentagrid of Fig. 1.1.

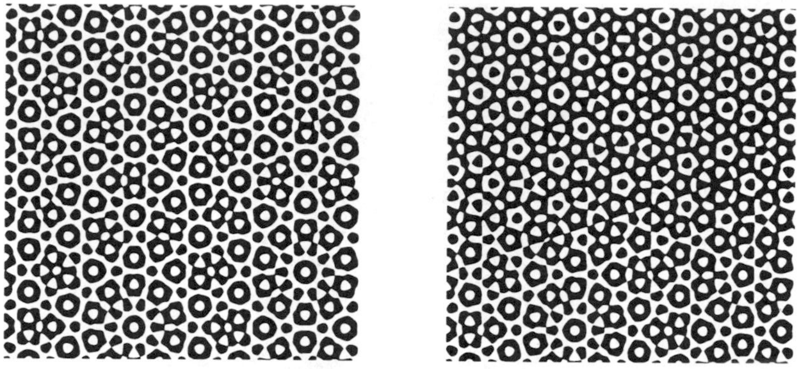

(a) (b)

Figure 6.2. (a) A density wave image with $\gamma \neq 0$ and $\mathbf{w} = 0$. This pattern has five rather that ten-fold symmetry about the origin. (b) The effect of a spatially varying γ, which is zero at the top of the figure and increases toward the bottom. The change from local patterns similar to those of Fig. 6.1 to those of Fig. 6.2 (b) is apparent.

presented here. All results discussed here, however, also apply to the full group.

The icosahedral group Y is a 60-element group consisting of five conjugacy classes. The classes are 1) the class E of the identity element, 2) the class C_π of 15 rotations about axes through the edges of the icosahedron, 3) the class $C_{2\pi/3}$ of 20 rotations of $\pm 2\pi/3$ about the threefold axes through the faces of the icosahedron, 4) the class $C_{2\pi/5}$ of 12 rotations of $\pm 2\pi/5$ about the fivefold axes through the vertices of the icosahedron, and 5) the class $C_{4\pi/5}$ of 12 rotations of $\pm 4\pi/5$ about axes again through the vertices of the icosahedron. Y has a one-dimensional irreducible representation $\Gamma_0^{(1)}$, two distinct three-dimensional irreducible representations $\Gamma_0^{(3)}$ and $\Gamma_0^{(\bar{3})}$, one four-dimensional representation $\Gamma_0^{(4)}$, and a five-dimensional representation $\Gamma_0^{(5)}$ $(60 = 1^2 + 3^2 + 3^2 + 4^2 + 5^2)$. From this, the character table for the icosahedral group, shown in Table 6.4, can be constructed.

Table 6.4 Icosahedral Character Table

	E	$15C_\pi$	$20C_{2\pi/3}$	$12C_{2\pi/5}$	$12C_{4\pi/5}$
$\chi_r^{(1)}$	1	1	1	1	1
$\chi_r^{(3)}$	3	-1	0	τ	$-\tau^{-1}$
$\chi_r^{(\bar{3})}$	3	-1	0	$-\tau^{-1}$	τ
$\chi_r^{(4)}$	4	0	1	-1	-1
$\chi_r^{(5)}$	5	1	-1	0	0

The six phases ϕ_n associated with the six vectors in the icosahedral basis [Eq. (4.10)] transform under a six-dimensional representation $\Gamma^{(6)}$ of Y. The characters of this representation are easily calculated: $\chi^{(6)}(E) = 6$, $\chi^{(6)}(C_\pi) = -2$, $\chi^{(6)}(C_{2\pi/3}) = 0$, $\chi^{(6)}(C_{2\pi/5}) = -1$, and $\chi^{(6)}(C_{4\pi/5}) = -1$. Thus, from the icosahedral character table and Eq. (6.11),

$$\Gamma^{(6)} = \Gamma^{(5)} \oplus \Gamma^{(\bar{3})}. \tag{6.28}$$

Again, the vectors $\mathbf{G}_n$ transform under the three-dimensional vector representation $\Gamma_0^{(3)}$. The set of complementary vectors,

$$\mathbf{G}_5^\perp = -\mathbf{G}_5, \qquad \mathbf{G}_n^\perp = \mathbf{G}_{<3n>_5}, \qquad n = 0,\ldots,4, \tag{6.29}$$

transform under $\Gamma_0^{(\bar{3})}$ and can be constructed using the icosahedral character table and orthogonality relations among representations. Thus again, $\phi_n = \mathbf{G}_n \cdot \mathbf{u} + \mathbf{G}_n^\perp \cdot \mathbf{w}$, and phases for general $\mathbf{G}$ in the reciprocal lattice satisfy Eq. (6.2). As in previous examples, $\mathbf{G}_n$ and $\mathbf{G}_n^\perp$ satisfy the orthogonality relations, Eq. (6.26).

7. Elasticity

Section 6 showed that exactly r phases of a TOS with a rank-r reciprocal lattice are not fixed by the equations determining thermal equilibrium. The r phases can be parameterized by displacement and phason variables $\mathbf{u}$ and $\mathbf{w}$. Uniform increments of $\mathbf{u}$ and $\mathbf{w}$ do not change the free energy of the quasiperiodic structure; spatially nonuniform $\mathbf{u}$ and $\mathbf{w}$ do, however. Nonuniform $\mathbf{u}$ describes modulations of positions of all mass points, whereas $\mathbf{w}$ describes relative rearrangements of atoms. The nature of rearrangements produced by nonuniform $\mathbf{w}$ is not in all cases intuitive. Figures 7.1 and 7.2 show density wave pictures of the type discussed in the last section for pentagonal quasicrystals with spatial variation in the displacement and phason variables. Note that a sinusoidal modulation of $\mathbf{u}$ leads, as expected, to a waviness of the whole structure, whereas a uniform phason strain produces mismatches in the Amman grids. These mismatches will be discussed in more detail in Section 10.

If an analytic expansion of the free energy is possible, its first term at long wavelength will be quadratic in the spatial gradients of $\mathbf{u}$ and $\mathbf{w}$ or, equivalently, in the r-dimensional displacement $\tilde{\mathbf{u}}$ with components $\tilde{\mathbf{u}}_\alpha$, $\alpha = 1, \ldots, r$. As in periodic crystals, the antisymmetric derivative of $\mathbf{u}$ describes rigid rotations that do not change the energy. The elastic Hamiltonian will, therefore, depend only on the symmetrized strain u_{ij}. In type I incommensurate crystals, $\mathbf{w} = \mathbf{u}_a - \mathbf{u}_b$. The antisymmetric derivatives of $\mathbf{u}_a$ and $\mathbf{u}_b$ describe separate uniform rotations of the a and b sublattices so that the antisymmetric derivative of $\mathbf{w}$ describes relative rotation of the two sublattices. Since there is an energy cost associated with such a

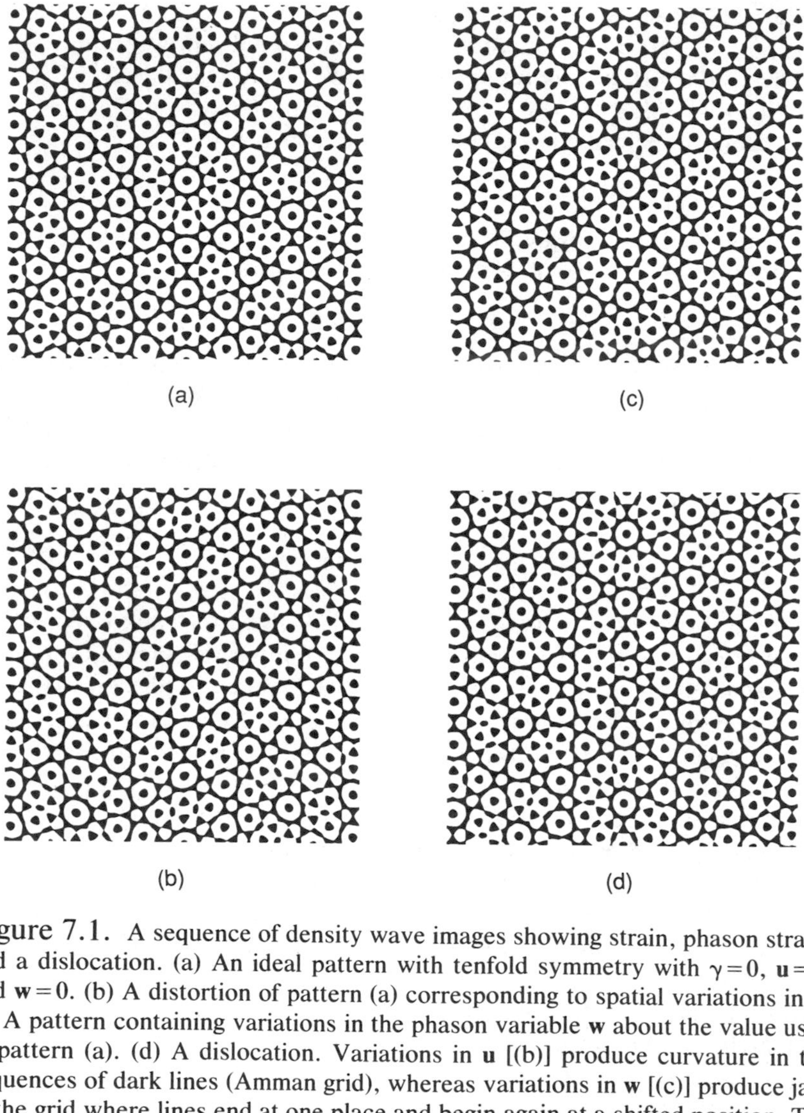

Figure 7.1. A sequence of density wave images showing strain, phason strain, and a dislocation. (a) An ideal pattern with tenfold symmetry with $\gamma=0$, $\mathbf{u}=0$, and $\mathbf{w}=0$. (b) A distortion of pattern (a) corresponding to spatial variations in $\mathbf{u}$. (c) A pattern containing variations in the phason variable $\mathbf{w}$ about the value used in pattern (a). (d) A dislocation. Variations in $\mathbf{u}$ [(b)] produce curvature in the sequences of dark lines (Amman grid), whereas variations in $\mathbf{w}$ [(c)] produce jags in the grid where lines end at one place and begin again at a shifted position. The dislocation shows both curvature and jags indicating the presence of both strain and phason strain.

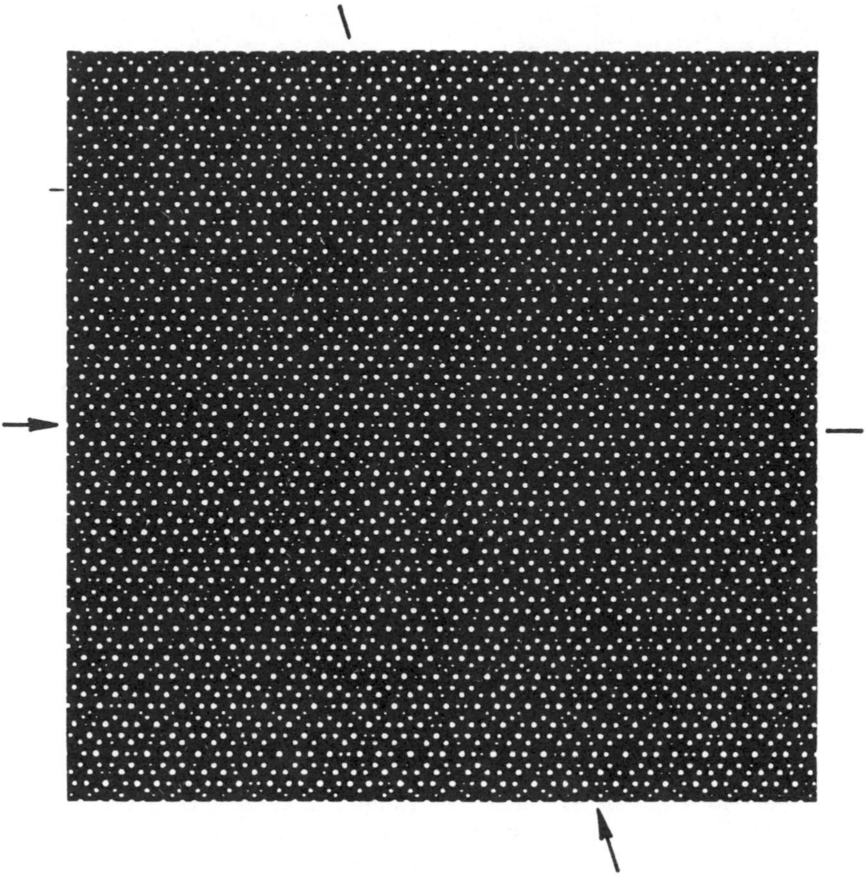

Figure 7.2. A density wave image showing a linear strain in **w** over a large area. The arrow on the left-hand side points along a straight line of white regions that form an Amman grid without jags. The arrow on the lower right-hand side points to a line of white regions that form part of an Amman grid with jags. This figure should be compared with the high-resolution micrographs of Reference 69.

rotation, the elastic energy will depend on all components of the gradient of **w**. This statement applies to all quasiperiodic structures.

The elastic free energy will, therefore, have the form

$$F_{\rm el} = \frac{1}{2}\int d^d x\, [K^{uu}_{ijkl} u_{ij} u_{kl} + K^{ww}_{ijkl}\, \nabla_i w_j\, \nabla_k w_l + 2K^{uw}_{ijkl} u_{ij}\, \nabla_k w_l],$$

$$\equiv \int d^d x\, f_{\rm el}(\nabla_i \tilde{\mathbf{u}}_\alpha) = F_{uu} + F_{uw} + F_{ww}, \qquad (7.1)$$

where K^{uu}_{ijkl}, K^{ww}_{ijkl}, and K^{uw}_{ijkl} are generalized elastic constant tensors and f_{el} is the elastic free energy density. In Fourier space, the elastic free energy is

$$F = \frac{1}{2}\int \frac{d^d q}{(2\pi)^d}[K^{uu}_{ij}(\mathbf{q})u_i(\mathbf{q})u_j(-\mathbf{q})\text{-} + K^{ww}_{ij}(\mathbf{q})w_i(\mathbf{q})w_j(-\mathbf{q}) \qquad (7.2)$$

$$+ 2K^{uw}_{ij}(\mathbf{q})u_i(\mathbf{q})w_j(-\mathbf{q})]$$

$$= \frac{1}{2}\int \frac{d^d q}{(2\pi)^d}\tilde{\mathbf{K}}_{\alpha\beta}(\mathbf{q})\tilde{\mathbf{u}}_\alpha(\mathbf{q})\tilde{\mathbf{u}}_\beta(-\mathbf{q}), \qquad (7.3)$$

where α and β run over the r components of $\tilde{\mathbf{u}}$.

The number of independent components in the elastic constant tensors can be calculated reasonably easily with the aid of group theory. F_{el} must be invariant under all of the point group operations of the structure. This implies that each individual term in F_{el} must transform under the identity representation of $\mathcal{G}_P$. The products $u_{ij}\, u_{kl}$, $\nabla_i w_j \nabla_k w_l$, and $u_{ij}\nabla_l w_k$ transform under multidimensional representations, which tell how the components of the products transform. There are certain combinations of these components which are invariant under the group operations. Each of the combinations can enter F_{el} with an independent elastic constant. Thus, the number of independent elastic constants is equal to the number of such invariant combinations, which is just the number of times the identity representation occurs in each of the products of u_{ij} and $\nabla_i w_j$. $\nabla_i u_j$ and $\nabla_i w_j$ transform under direct product representations of $\mathcal{G}_P$. Since the gradient operator (or the vector $\mathbf{q}$) transforms under a D-dimensional vector representation of the group, therefore $\nabla_i u_j$ transforms under D^2-dimensional and $\nabla_i w_j$ under a $[D(r-D)]$-dimensional representation of the group. u_{ij} is symmetric under interchange of i and j. It, therefore, transforms under a $[D(D+1)/2]$-dimensional representation of $\mathcal{G}_P$.

Direct product representations can be decomposed into irreducible representations using the rule[3]

$$\Gamma^{(i)}_0 \otimes \Gamma^{(j)}_0 = c_{ijk}\Gamma^{(k)}_0, \qquad (7.4)$$

where

$$c_{ijk} = \frac{1}{h}\sum_C n_C\, \chi^{(i)}_0(C)\chi^{(j)}_0(C)\chi^{(k)}_0(C). \qquad (7.5)$$

Thus, the number of times the identity representation appears in the direct product representations, $u_{ij}u_{kl}$, $u_{ij}\nabla_k w_l$, and $\nabla_i w_j\ \nabla_k w_l$, can be calculated by first decomposing u_{ij} and $\nabla_i w_j$ into irreducible representations using Eq. (7.4) and then decomposing the products, again using Eq. (7.4).

To illustrate the above procedures, consider first the example of pentagonal quasicrystals with C_{5v} symmetry.[26,28] In this case, ∇_i and u_i transform under the two-dimensional $\Gamma_0^{(2)}$ representation of C_{5v}. $\nabla_i u_j$ transforms under the representation $\Gamma_0^{(2)} \otimes \Gamma_0^{(2)}$ which, from Eq. (7.5) and Table 6.3, decomposes into $\Gamma_0^{(\bar{2})} + \Gamma_0^{(\bar{1})} + \Gamma_0^{(1)}$. [Note the appearance of $\Gamma_0^{(\bar{2})}$ rather than the vector representation $\Gamma_0^{(2)}$.] The antisymmetric part of $\nabla_i u_j$ transforms under the $\Gamma_0^{(\bar{1})}$ representations so that u_{ij} transforms under $\Gamma_0^{(\bar{2})} \oplus \Gamma_0^{(1)}$. w_i transforms under the $\Gamma_0^{(\bar{2})}$ representation of C_{5v}. Thus, again from Eq. (7.5) and the C_{5v} character table, $\nabla_i w_j$ transforms under $\Gamma_0^{(2)} \otimes \Gamma_0^{(\bar{2})} = \Gamma_0^{(2)} \oplus \Gamma_0^{(\bar{2})}$. The number of constants in the strain elastic constant K_{ijkl}^{uu} is the number of times the identity representation appears in $(\Gamma_0^{(\bar{2})} \oplus \Gamma_0^{(1)}) \otimes (\Gamma_0^{(\bar{2})} \oplus \Gamma_0^{(1)})$. The identity does not appear at all in the cross terms mixing the one- and two-dimensional representations. It appears once in each of $\Gamma_0^{(1)} \otimes \Gamma_0^{(1)}$ and $\Gamma_0^{(2)} \otimes \Gamma_0^{(2)}$. There are, therefore, two elastic constants appearing in K_{ijkl}^{uu}. This is exactly the number of constants of an isotropic or hexagonal two-dimensional solid.

A similar analysis shows that there are two independent elastic constants in K_{ijkl}^{ww}, since there is no identity representation in $\Gamma_0^{(2)} \otimes \Gamma_0^{(\bar{2})}$. Finally, only one elastic constant, arising from the $\Gamma_0^{(\bar{2})} \otimes \Gamma_0^{(\bar{2})}$ cross terms in $u_{ij} \nabla_k w_l$, appears in K_{ijkl}^{uw}. There are thus a total of five[28] elastic constants for pentagonal quasicrystals with C_{5v} symmetry. As will be shown shortly, one of the constants in K_{ijkl}^{ww} multiplies a perfect derivative that can be integrated to the boundaries in the absence of dislocations.[26] In this case, there are effectively only four elastic constants. The precise form of the elastic free energy will be discussed shortly. The number of elastic constants depends on the point group symmetry $\mathcal{G}_P$ and not on the point group symmetry $\mathcal{G}_R$ of the reciprocal lattice. For pentagonal quasicrystals with C_5 rather than C_{5v} symmetry, there are two rather than one independent elastic constants in K_{ijkl}^{uw} for a total of six rather than five constants.[26]

The determination of the number of elastic constants for icosahedral quasicrystals[26,28,31,57,72] proceeds exactly as for the pentagonal example. Consider first the case where G_P is the group Y (no reflections). u_{ij} transforms under the six-dimensional representation $\Gamma_0^{(1)} \oplus \Gamma_0^{(5)}$ of Y, and $\nabla_i w_j$ under the nine-dimensional representation $\Gamma_0^{(3)} \oplus \Gamma_0^{(\bar{3})} = \Gamma_0^{(4)} \oplus \Gamma_0^{(5)}$. From this it is straightforward to verify that there are two elastic constants in K_{ijkl}^{uu}, two in K_{ijkl}^{ww}, and one in K_{ijkl}^{uw}, for a total of five constants. The same results is true[20] for structures with full icosahedral symmetry Y_F.

The determination of explicit forms for the elastic free energy is much more complicated than counting the number of independent elastic constants because it requires a knowledge of the precise components of u_{ij} and $\nabla_i w_j$ that transform under the different constituent representations. These components can be calculated for pentagonal quasicrystals with C_{5v}

symmetry using the Abelian representations of C_5. Since $\mathbf{u}$ transforms under the vector representation, its x and y components can be expressed as $u_x = u_+ + u_-$ and $u_y = i^{-1}(u_+ - u_-)$, where $u_\pm \rightarrow e^{\pm i\theta} u_\pm$ under rotation through $\theta = 2\pi/5$. A similar decomposition applies to the wavevector $\mathbf{q}$. The phason field, on the other hand, transforms under the $\Gamma_0^{(2)}$ representation and can be expressed as $w_x = w_+ + w_-$ and $w_y = i^{-1}(w_+ - w_-)$, where $w_\pm \rightarrow e^{\pm 3i\theta} w_\pm$ under rotation through θ. Using this, it is straightforward to determine that the x and y components of the $\Gamma_0^{(2)}$ representation of u_{ij} are respectively $u_{xx} - u_{yy}$ and $2u_{xy}$. Similarly, the x and y components of the $\Gamma_0^{(2)}$ representation of $\nabla_i w_j$ are respectively $(\nabla_x w_x - \nabla_y w_y)$ and $(\nabla_x w_y + \nabla_y w_x)$, and those of the $\Gamma_0^{(2)}$ representation $(\nabla_x w_x + \nabla_s w_y)$ and $(\nabla_s w_y - \nabla_y w_x)$. The elastic free energy[73] density for the pentagonal quasicrystal with C_{5v} symmetry is, therefore,

$$\begin{aligned} f_{\text{el}} = \frac{1}{2}\lambda(\boldsymbol{\nabla}\cdot\mathbf{u})^2 + \mu u_{ij}u_{ij} + \frac{1}{2}K_1^w \nabla_i w_j \nabla_i w_j \\ + K_2^w[\nabla_x w_x \nabla_y w_y - \nabla_y w_x \nabla_x w_y] \\ + K^{uw}[(u_{xx} - u_{yy})(\nabla_x w_x + \nabla_y w_y) + 2u_{xy}(\nabla_x w_y - \nabla_y w_x)]. \end{aligned} \tag{7.6}$$

In the absence of dislocations, the terms following K_2^w reduce to zero after integration by parts. In this case, thermodynamic stability requires $\mu > 0$, $\lambda + \mu > 0$, and $\mu K_1^w > 2(K^{uw})^2$. In crystals with C_5 symmetry, there is an additional term[26] proportional to $2u_{xy}(\nabla_x w_x + \nabla_y w_y) - (u_{xx} - u_{yy})(\nabla_x w_y - \nabla_y w_x)$ in f_{el}.

For icosahedral quasicrystals, the unit representation appearing in u_{ij} is simply $\boldsymbol{\nabla}\cdot\mathbf{u}$, and the five-dimensional representation, as in isotropic solids, is the symmetric traceless tensor $u_{ij} - \frac{1}{3}\delta_{ij}$. The determination of the four- and five-dimensional parts of $\nabla_i w_j$ is more complicated. A "brute force" method for obtaining this decomposition is outlined in the Appendix. A more elegant derivation is given in Reference 72. The resulting elastic free energy is most compactly expressed in wavevector space. In the coordinate system of Fig. 4.3.,

$$F_u = \int \frac{d^3q}{(2\pi)^3}\left[\frac{1}{2}(\lambda + 2\mu)q_iq_j + \frac{1}{2}\mu(q^2\delta_{ij} - q_iq_j)\right]u_i(\mathbf{q})u_j(-\mathbf{q}), \tag{7.7}$$

$$F_w = \int \frac{d^3q}{(2\pi)^3}[K_1q^2\delta_{ij} + K_{ij}(\mathbf{q})]w_i(\mathbf{q})w_j(\mathbf{q}), \tag{7.8}$$

$$F_{uw} = \int \frac{d^3q}{(2\pi)^3}K_{ij}^{uw}(\mathbf{q})u_i(\mathbf{q})w_j(\mathbf{q}), \tag{7.9}$$

where

$$K_{ij}(\mathbf{q}) = K_2\tau^{-2}\begin{bmatrix} \tau^4 q_x{}^2 + q_y{}^2 + \tau^2 q_z{}^2 & -2\tau^2 q_y q_z & -2\tau^2 q_x q_z \\ -2\tau^2 q_y q_z & q_x{}^2 + \tau^2 q_y{}^2 + \tau^4 q_z{}^2 & 2\tau^2 q_x q_y \\ -2\tau^2 q_x q_z & 2\tau^2 q_x q_y & \tau^2 q_x{}^2 + \tau^4 q_y{}^2 + q_z{}^2 \end{bmatrix}, \tag{7.10}$$

$$K_{ij}^{uw}(\mathbf{q}) = K_3\tau^{-2} \begin{bmatrix} 2\tau q_x q_z & 2\tau^3 q_x q_y & -\tau^2 q_x{}^2 - \tau q_y{}^2 + \tau^3 q_z{}^2 \\ -2\tau^3 q_y q_z & \tau^3 q_x{}^2 - \tau^2 q_y{}^2 - \tau q_z{}^2 & -2\tau q_x q_y \\ \tau q_x{}^2 - \tau^3 q_y{}^2 + \tau^2 q_z{}^2 & -2\tau q_y q_z & 2\tau^3 q_x q_z \end{bmatrix} \tag{7.11}$$

Stability requires $\mu > 0$, $\lambda > 0$, $K_1 + K_2 > 0$, and $2\mu(K_1 + K_2) - K_3^2 > 0$.

The average mass density $< \rho >$, like the scattering amplitudes $| < \rho_G > |$, attains some optimum value ρ_0 in equilibrium. Deviations $\delta\rho = < \rho > - \rho_0$ from the equilibrium value occur in response to external forces. $\delta\rho$ is a scalar and transforms under the identity representation of any point group. It will therefore couple to the parts of the strain and phason strain that transform like a scalar. Since **u** and ∇ always transform under a vector representation, $\nabla \cdot \mathbf{u}$ is always a scalar. In incommensurate crystals, **w** also transforms like a vector, so $\nabla \cdot \mathbf{w}$ is also a scalar. In quasicrystals, **w** does not transform like a vector, and $\nabla \cdot \mathbf{w}$ is not a scalar. In fact, in none of the quasicrystals considered so far has there been a part of the phason strain $\nabla_i w_j$ transforming under the identity representation. This appears to be a general property of quasicrystals. For small deviations from equilibrium, the total free energy of a one-component system is thus

$$F(\delta\rho, u_{ij}, \nabla_i w_j, T) = F_0 + F_\rho + F_{el}, \tag{7.12}$$

where F_0 is the free energy when $\delta\rho$, u_{ij} and $\nabla_i w_j$ are zero, and where

$$F_\rho = \int d^dx \left[\frac{1}{2} A \left(\frac{\delta\rho}{\rho_0}\right)^2 + B\frac{\delta\rho}{\rho_0}\nabla \cdot \mathbf{u} + B_w \frac{\delta\rho}{\rho_0}\nabla \cdot \mathbf{w}\right] \tag{7.13}$$

describes the free energy cost of deviations of the density from ρ_0. As discussed above, the coefficient B_w is nonzero in incommensurate crystals but zero in quasicrystals. In multicomponent systems, there would be additional terms involving the densities of each of the atomic species.

The elastic free energy of Eqs. (7.7)–(7.11) is quadratic in the wave-vector and the displacement and strain variables. This implies that $< u_i(\mathbf{x}) u_j(\mathbf{x}) >$, $< w_i(\mathbf{x}) w_j(\mathbf{x}) >$, and $< u_i(\mathbf{x}) w_j(\mathbf{x}) >$ are all finite and that the

Debye–Waller factor is nonzero. Thus, phason fluctuations do not destroy long-range icosahedral order. As in periodic crystals, there is a diffuse scattering background arising from both phason and phonon fluctuations.[56]

An interesting consequence[29,74] of the general quasiperiodic elastic free energy of Eq. (7.1) is that the response of the strain to external stress is determined by an effective elastic constant tensor that is nonanalytic in the gradient operator. In an experiment at constant density, the equations determining equilibrium in the presence of an external stress σ_{ij} are

$$\frac{\delta F_{\text{el}}}{\delta u_i} = \nabla_j \sigma_{ij} \tag{7.14a}$$

and

$$\frac{\delta F_{\text{el}}}{\delta w_i} = 0. \tag{7.14b}$$

Equation (7.14b) determines $\mathbf{w}$ if there is any strain present. From Eq. (7.2), $\mathbf{w}(\mathbf{q})$ will equal $-(\mathbf{K}^{ww})^{-1}\mathbf{K}^{uw}\,\mathbf{u}(\mathbf{q})$, where $\mathbf{K}$ is the matrix with components K_{ij} and matrix multiplication is understood. The effective elastic energy in terms of the displacement alone, after $\mathbf{w}$ has relaxed to its equilibrium form in the presence of $\mathbf{u}$, is thus

$$F_{\text{el}}^{\text{eff}} = \frac{1}{2}\int \frac{d^d q}{(2\pi)^d}\mathbf{u}(\mathbf{q})[\mathbf{K}^{uu}(\mathbf{q}) - \mathbf{K}^{uw}(\mathbf{q})(\mathbf{K}^{ww}(\mathbf{q}))^{-1}\,\mathbf{K}^{uw}(\mathbf{q})]\mathbf{u}(-\mathbf{q}), \tag{7.15}$$

and the equation determining $\mathbf{u}(\mathbf{x})$ in response to external stresses becomes

$$\frac{\delta F_{\text{el}}^{\text{eff}}}{\delta u_i(\mathbf{x})} = \nabla_j \sigma_{ij}. \tag{7.16}$$

The combination $\mathbf{K}^{uw}(\mathbf{K}^{ww})^{-1}\mathbf{K}^{uw}$ tends to zero quadratically as $\mathbf{q}\to 0$, but its limiting value depends on the angle that $\mathbf{q}$ makes with symmetry axes of the structure. Thus, the response of quasiperiodic structures to external stresses can be very unusual. After an external stress is applied, the induced strain will be determined by the elastic constant tensor $\mathbf{K}^{uu}$ for times short compared to the relaxation time τ_{phason} for phason strains and long compared to the very short relaxation time for strains. At times longer than τ_{phason}, the strain will be determined by the angle-dependent effective elastic constant appearing in Eq. (7.15). It is likely that τ_{phason} for icosahedral quasicrystals will be much longer than any laboratory time, as will be discussed in the next two sections. The amplitude[33,35] of the incommensurate modulations in incommensurate crystals is very small so that the angle-dependent elastic constant may be difficult to see even in situations where τ_{phason} is fairly small.

Jarić[57] has calculated the full elastic constant tensor for icosahedral cobalt using a Ramakrishnan–Yussouff theory[75] with the structure factor for amorphous Co rather than that of an equilibrium liquid. He finds that the elastic constant tensor has a negative eigenvalue indicating a martensitic instability for the icosahedral phase of Co.

8. Dislocations

The mass density amplitudes $<\rho_{\mathbf{G}}>$ in any TOS must be single-valued functions of the coordinate **x**. This implies that $<\rho_{\mathbf{G}}(\mathbf{x})>$ must return to its initial value in a complete circuit around any closed loop in space. Changes in the phase $\phi_{\mathbf{G}}$ by integral multiples of 2π leave $<\rho_{\mathbf{G}}>$ unchanged. Thus, even though $<\rho_{\mathbf{G}}>$ must be single-valued, $\phi_{\mathbf{G}}$ may not be. Dislocations in periodic solids are defects[76] associated with variations in the elastic variable **u** leading to changes in some or all of the phases $\phi_{\mathbf{G}}$ by integral multiples of 2π around some closed loop. In two dimensions, the loop encloses a point core where the values of the order parameters $<\rho_{\mathbf{G}}>$ go to zero. In three dimensions, the core becomes a line that either forms a closed loop or terminates at the boundary of the solid. There is only a discrete set of changes in $\mathbf{u} = \mathbf{R}$ compatible with $2\pi k$ changes in $\phi_{\mathbf{G}}$ determined by Eq. (4.1). Thus the set of *Burgers* vectors indexing the strength of dislocations in periodic crystals is identical to the direct lattice of the crystal. This fact makes possible the Volterra construction[76] and the traditional interpretation of a dislocation as the insertion or removal of one or more layers of atoms from the ideal structure.

The generalization of dislocations to quasiperiodic structures is straightforward. Changes in the r-dimensional elastic variable $\tilde{\mathbf{u}} = \tilde{\mathbf{R}}$ that satisfy Eq. (6.8) leave the mass density amplitudes $<\rho_{\mathbf{G}}>$ unchanged. Thus the Burgers vector lattice of a quasiperiodic structure is identical to the direct lattice of the parent lattice of the structure.

Vectors $\tilde{\mathbf{G}}$ in the parent space reciprocal lattice of type I incommensurate crystals can be expressed as a direct sum $c_a\mathbf{G}_a \oplus c_b\mathbf{G}_b$ of vectors $c_a\,\mathbf{G}_a$ and $c_b\,\mathbf{G}_b$ in the scaled periodic reciprocal lattice of the $a-$ and $b-$sublattices [see Section 6.4.1]. Thus, the direct lattice primitive vectors $\tilde{\mathbf{R}}$ of the parent lattice can be expressed as a direct sum $c_a^{-1}\mathbf{R}_a \oplus c_b^{-1}\mathbf{R}_b$ of vectors $c_a^{-1}\mathbf{R}_a$ and $c_b^{-1}\mathbf{R}_b$ in periodic direct lattices associated with sublattices a and b. This implies that there are dislocations corresponding to the removal or insertion of an extra layer of atoms in the a sublattice only without any change in the b sublattice. Similarly, there are dislocations corresponding to the insertion of extra layers in the b sublattice but not in the a sublattice. In type II incommensurate crystals and in quasicrystals, the constraints on the form of $\mathbf{G}^{\perp}$ imposed by symmetry precludes this simple interpretation. The vectors in the Burgers vector lattice always have projections in the spaces spanned by any of the component bases.

There are well-established procedures for determining the lattice of vectors defined by Eq. (4.1) if the vectors $\tilde{\mathbf{G}}$ are specified. It is instructive, however, to determine the Burgers vector lattice for a few specific examples. The icosahedral quasicrystal is one of the simple examples.[26] The vectors $\tilde{\mathbf{G}}_n$ of Eq. (6.3), with $\alpha = 1$ and $\mathbf{G}_n$ and $\mathbf{G}_n^{\perp}$ given by Eqs. (4.10) and (6.29), form a basis for the parent space reciprocal lattice. It is straightforward to verify, using Eqs. (6.3), (4.10), and (6.29), that

$$\tilde{\mathbf{G}}_n \cdot \tilde{\mathbf{G}}_m = G^2 \delta_{nm}, \qquad n,m = 0, \ldots, 5, \tag{8.1}$$

so that the parent lattice is a six-dimensional simple cubic lattice. The direct lattice is simple cubic and is generated by the basis vectors

$$\tilde{\mathbf{R}}_n = \pi \tilde{\mathbf{G}}_n / \mathbf{G}^2, \qquad \tilde{\mathbf{R}}_n \cdot \tilde{\mathbf{R}}_m = \frac{2\pi^2}{G^2} \delta_{nm} \tag{8.2}$$

The Burgers vector lattice associated with the alternate reciprocal lattices discussed in Section 6 are in the six-dimensional generalization of the FCC and BCC lattices, respectively.[66]

The vectors $\tilde{\mathbf{G}}_n$ for pentagonal quasicrystals are not orthogonal [in the sense of Eq. (8.1)], and the parent lattice is not simple cubic. Before finding the vectors $\tilde{\mathbf{R}}$ in the pentagonal parent lattice, it is useful to recall how the direct lattice for hexagonal crystals can be derived from the nonorthogonal vectors $\mathbf{G}_n$ [Eq. (4.4)] of the extended basis of the hexagonal reciprocal lattice. The vectors

$$\mathbf{R}_n = R\left[-\sin\frac{2\pi n}{3}, \cos\frac{2\pi n}{3}\right] \tag{8.3}$$

are orthogonal to the vectors $\mathbf{G}_n$ of Eq. (4.4). The magnitude R is set by the condition [Eq. (4.1)] that $\mathbf{G}_n \cdot \mathbf{R}_m = 2\pi p$, where p is an integer. The choice $p = 1$ for $m = \langle n+1 \rangle_3$ yields $R = 2\pi/[G \sin(2\pi/3)]$ and $p = -1$ for $m = \langle n-1 \rangle_3$. The vectors $\mathbf{R}_n$ form an extended basis for the direct lattice.

It is now clear how to find the pentagonal Burgers vector lattice. First construct the vectors

$$\mathbf{R}_n = R\left[-\sin\frac{3\pi n}{5}, \cos\frac{3\pi n}{5}\right] \tag{8.4}$$

orthogonal to the extended basis vectors $\mathbf{G}_n$ in Eq. (4.8). Next construct four-dimensional vectors $\tilde{\mathbf{R}}_n = \tilde{\mathbf{R}}_n \oplus \beta \mathbf{R}_n^{\perp}$ from $\mathbf{R}_n$, and the vectors $\mathbf{R}_n^{\perp} = \mathbf{R}_{\langle 3n \rangle_5}$ orthogonal to the complementary vectors $\mathbf{G}_n^{\perp}$. Then

$$\phi_{n,m} \equiv \tilde{\mathbf{G}}_n \cdot \tilde{\mathbf{R}}_m = GR\,\{\sin[\frac{2\pi}{5}(n-m)] + \alpha\beta\sin[\frac{6\pi}{5}(n-m)]\}. \quad 8.5)$$

R, α, β are determined by requiring that $\phi_{n,(n+2)} = -\phi_{n,(n+3)} = 2\pi$ and $\phi_{n,(n+1)} = -\phi_{n,(n+4)} = 0$. This yields

$$GR = \frac{8\pi}{5}\sin\frac{4\pi}{5}, \qquad \alpha\beta = \frac{\sin(2\pi/5)}{\sin(4\pi/5)} = \tau. \quad (8.6)$$

Thus, the factors α and β are arbitrary, but their product is fixed. A density wave image of a dislocation is shown in Fig. 7.1(d). Note the curvature and jaggs in the dark Amman lines visible at grazing angles. This indicates that a dislocation engenders both strain and phason strain.

The construction of the icosahedral and pentagonal Burgers vector lattice generalizes directly to type II incommensurate crystals. After the vector $\mathbf{R}_n^{\perp}$ perpendicular to the complementary vectors $\mathbf{G}_n$ has been constructed, the extended basis of the Burgers vector lattice is obtained by forming the direct sum of vectors $\mathbf{R}_n$ and $\mathbf{R}_n^{\perp}$.

The energy of a dislocation can be calculated using the elastic free energy of the last section. Since the free energy is quadratic in both gradients and elastic variables **u** and **w**, it is clear that the general form of dislocation energies will be the same as it is in periodic crystals.[76] Thus, in three dimensions, the energy per unit length of an isolated dislocation is proportional to ln L, where L is a typical dimension of the sample. Similarly, the interaction potential between two dislocation lines is logarithmic in their separation. Explicit calculations of dislocation energies for pentagonal quasicrystals in two dimensions have been carried out by De and Pelcovitz.[73] If phasons are pinned, the energy of dislocations is expected to be much higher[42], as discussed in Section 10.

The discussion of dislocations presented here is perfectly general. It does not, however, treat changes in Penrose tilings arising from dislocations. Some aspects of this problem will be addressed in Section 10. For more details, see References 42, 58, and 59. Reference 59 also discusses disclinations.

9. Hydrodynamics

9.1 Introduction

This section will derive the equations governing hydrodynamics in quasiperiodic structures. These equations are quite general and apply to all nonmagnetic quasiperiodic structures, including incommensurate crystals.

The modes predicted by these equations will be analyzed in detail only for icosahedral quasicrystals in three dimensions.[29]

Thermodynamic equilibrium in any system is established and maintained by collisions among particles or elementary excitations. Let τ_c be the average time and λ be the average distance (mean free path) between collisions. Now consider excitations from the ideal equilibrium state with frequency ω and wavenumber q. If $\omega\tau << 1$, many collisions will take place in a single oscillation of the excitation, and the system always remains close to thermodynamic equilibrium. Most excitations have characteristic frequencies that are of order τ_c^{-1}. There are, however, classes of variables, called *hydrodynamic* variables, whose excitation frequency goes rigorously to zero with wavenumber. In this case, for $q\lambda << 1$, $\omega\tau_e$ must also be much less than one. Hydrodynamic variables[37,38] arise either from conservation laws or from broken symmetries such as the broken translational symmetry of a periodic solid. It is easy to see how conservation laws lead to the condition $\omega \to 0$ as $q \to 0$. Let A be a conserved variable and $a(\mathbf{x},t)$ its associated density. Since A is conserved, its time derivative is zero, and there must be a current $\mathbf{j}_a$ such that $\partial a/\partial t + \nabla \cdot \mathbf{j}_a = 0$. Fourier transformation of this equation in space and time yields $\omega a = \mathbf{q} \cdot \mathbf{j}_a$, so that ω must go to zero with q. This is merely a restatement of the fact that the $q \to 0$ limit of $a(\mathbf{q},t)$ is the variable A which is constant in time. Broken continuous symmetries give rise to elastic variables whose spatially uniform ($\mathbf{q} = 0$) changes leave the free energy unchanged and, therefore, maintain the state of thermodynamic equilibrium. Since equilibrium is by definition a steady state, the frequency associated with a $\mathbf{q} = 0$ displacement of an elastic variable is zero. Thus, the frequency associated with nonuniform $\mathbf{q} \neq 0$ distortions of elastic variables should tend to zero with q unless there is some mechanism causing a breakdown in the continuity of ω versus q.

To determine the hydrodynamic equations for any system, all of the conserved and broken symmetry hydrodynamic variables must be identified and their long-wavelength dynamical equations established. In the case of conserved variables, the dynamical equations are merely the conservation equations. The equations for elastic variables follow almost by definition and will be discussed below. Once the hydrodynamic variables have been identified, the fundamental thermodynamic relation governing the change in entropy resulting from changes in the hydrodynamic variables from their equilibrium values must be derived. This thermodynamic relation is important because all hydrodynamic processes occur near local thermodynamic equilibrium because $\omega\tau_e << 1$. There are by now several established formalisms[37,38,78] for actually implementing the above procedure. That most closely related to the spirit of the description just given will be applied to quasiperiodic structures below.

Hydrodynamic equations predict a set of low frequency modes that are generally either *propagating* (soundlike) with a dispersion relation $\omega \sim cq$, where c is a velocity, or *diffusive* with $\omega \sim iDq^2$, where D is a diffusion constant. In general, there is one mode[37] associated with each hydrodynamic variable. In this counting, a propagating mode is counted as two modes since it always results from the coupling of two hydrodynamic variables.

Energy, mass, and momentum are always conserved variables. Thus, the energy density ε, mass density ρ [the distinction between ρ and $\langle\rho\rangle$ will not be made here], and momentum density **g** are hydrodynamic variables in quasiperiodic structures as they are in liquids and periodic solids. If there are N rather than one species of atoms, there will be N mass conservation laws and hydrodynamic mass densities. This gives a total of $N + d + 1$ conserved hydrodynamic variables. Quasiperiodic structures have in addition D elastic strain variables and $r - D$ phason variables. If all of the phason variables are unpinned, then there are a total of $N + d + r + 1$ hydrodynamic variables in a quasiperiodic structure composed of N species of atoms. The phason modes are *always* diffusive in the hydrodynamic limit.[39] The $N + r + 4$ hydrodynamic modes of a rank-r three-dimensional quasiperiodic structure are, therefore, longitudinal sound (2 modes), transverse sound (two branches for a total of 4 modes), heat diffusion (1 mode), mass diffusion (N modes), and phason diffusion ($r-3$ modes). Thus there are the same sound modes in a quasiperiodic structure as there are in periodic solids. One of the mass diffusion modes is a vacancy diffusion mode involving the migration of missing atoms in some ideal structure. For one-component systems, this is the only mass diffusion mode. In multicomponent systems, the other mass diffusion modes involve relative motion of two or more species of atoms.

9.2 *Modes of Icosahedral Quasicrystals*[29]

The sound wave frequencies for icosahedral quasicrystals follow directly from the mode equations to be developed below. The results are that the sound velocities are *isotropic* but their damping reflects the icosahedral symmetry of the structure. This is not surprising since sound velocities are sensitive to the fourth-rank elasticity tensor which has the same form in an icosahedral quasicrystal as it does in an isotropic solid. The complex longitudinal sound wave frequency for a one-component icosahedral quasicrystal is

$$w = c_L q - \frac{i}{2}[\eta_L q^2 + \frac{\Gamma^w K_3^2}{\rho_0 c_L^{\,2}}(10q^{-4}I_6 - 9q^2)], \qquad (9.1)$$

where c_L is the sound velocity, η_L is the longitudinal viscosity, Γ^w is the kinetic coefficient for phasons, K_3 is the phason-strain elastic constant of Eq. (7.11), and

$$I_6 = q_x^6 + q_y^6 + q_z^6 + \frac{15\tau^4}{(1+\tau^6)}(q_x^4 q_y^2 + q_y^4 q_z^2 + q_z^4 q_z^2) + \frac{15\tau^2}{(1+\tau^6)}(q_x^2 q_y^4 + q_y^2 q_z^4 + q_z^2 q_z^4), \tag{9.2}$$

is the sixth-order icosahedral invariant expressed in the coordinate system to Eq. (4.7). The transverse sound frequencies are

$$w_\alpha(\mathbf{q}) = c_T q - \frac{1}{2} i\gamma_\alpha(\hat{q}) q^2, \tag{9.3}$$

where α refers to the polarization of the transverse wave and $c_T = (\mu/\rho_0)^{1/2}$ is the shear sound velocity. The damping coefficients $\gamma_\alpha(\hat{q})$ are functions of the unit vector $\hat{q} = \mathbf{q}/q$ and involve icosahedral invariants of sixth, eighth, and higher order satisfying

$$\frac{1}{2}[\gamma_1(\hat{q}) + \gamma_2(\hat{q})] = \Gamma_u \rho_0 c_T^2 + \eta_\tau + \frac{\Gamma^w K_3^2}{2\rho_0 c_T^2}(13 - 9q^{-6} I_6), \tag{9.4}$$

where η_T is the transverse viscosity and Γ^w is the kinetic coefficient for vacancy diffusion. The ratio of the anisotropic to isotropic pieces of the sound wave damping is of order $\Gamma^w K_3^2/\rho_0 \eta_L c_L^2$. By dimensional analysis, K_3 should be of order or less than the shear elastic constants μ, i.e., of order $\rho_0 c_T^2 < \rho_0 c_L^2$. Indeed, the calculations of Reference 57 predict that all of the eigenvalues of general icosahedral elasticity matrix differ in magnitude by less than a factor of 8. Γ^w is of order Γ^u, so that $\Gamma^w/\rho_0 c_L^2 \approx \Gamma^u/K_3$. Thus, the ratio of the anisotropic to isotropic parts of the damping is of order D/η, where $D = \Gamma^u \mu$ is the vacancy diffusion constant, and η is a typical viscosity. This ratio is likely to be exceedingly small (of order 10^{-10}) since D is of order 10^{-10} cm^2/sec and η of order 1 cm^2/sec in typical metals.[79]

Vacancies and phasons are coupled—indeed they involve fundamentally the same process, so that it is not really meaningful to speak of them separately. Their motion is governed by an anisotropic 4×4 diffusivity tensor. Provided the phasons are not pinned, typical diffusion constants for vacancy and phason diffusion in icosahedral alloys should be comparable to those for vacancy diffusion in ordinary phases composed of the same materials. As discussed above, these diffusion constants are typically very small so that phason diffusion is expected to be very slow even if the hydrodynamic description presented in this section (as opposed to the pinned dynamics discussed in the next section) holds. The relaxation time for phasons is of order $\tau_{\text{phason}} \sim DL_{\text{min}}^2$, where L_{min} is the smallest linear dimension of the sample. For typical diffusion constants, $D \approx 10^{-8}$ to 10^{-13} cm^2/sec and a sample with $L_{\text{min}} \approx 1$ cm, τ_{phason} is of order 3 to 300,000 years.

The slowness of the phason mode has important consequences for dislocation motion.[30] Since any dislocation gives rise to both strain and phason strain, a moving dislocation must carry both fields with it. The phason field adds tremendously to the drag on dislocations, making it a factor of order 10^{10} or more than in clean periodic crystals. This suggest that a perfect quasicrystal should be impervious to plastic deformations. Dislocations should behave as though pinned in a work-hardened material.

9.3 Hydrodynamic Equations

The conserved variables in a one-component TOS, as in fluids, are the mass density ρ, the energy density ε, and the momentum density **g**. They satisfy the local conservation laws,

$$\partial_t \rho + \nabla \cdot \mathbf{g} = 0, \tag{9.5}$$
$$\partial_t \varepsilon + \nabla \cdot \mathbf{j}^{\varepsilon} = 0, \tag{9.6}$$
$$\partial_t g_i + \nabla_j \sigma_{ij} = 0, \tag{9.7}$$

where $\mathbf{j}^{\varepsilon}$ is the energy current and σ_{ij} is the stress tensor. Note that the current **g** for the mass density is itself a conserved quantity. The elastic hydrodynamic variables in a quasiperiodic structure are the displacement vector **u** and the phason field **w**. If the whole structure is translated with velocity **v**, the time derivative of the displacement **u** in a fixed laboratory coordinate system will be equal to **v**. The phason field, since it describes relative dispacements of atoms, is independent of **v**. The equations of motion for **u** and **w** can, therefore, be written as

$$\partial_t \mathbf{u} = \mathbf{v} - \mathbf{X}^u, \tag{9.8}$$
$$\partial_t \mathbf{w} = -\mathbf{X}^w, \tag{9.9}$$

where $\mathbf{X}^u$ and $\mathbf{X}^w$ are "currents" associated with the variables **u** and **w**. They have the same footing in the hydrodynamical theory to be derived as the currents of the conserved variables.

The next essential ingredient of hydrodynamics is a description of small deviations from thermodynamic equilibrium. The conserved variables are fixed in equilibrium. Changes from their equilibrium values will result in a decrease in the entropy. Similarly, increases in $\nabla_i u_j$ and $\nabla_i w_j$ from their equilibrium values of zero will increase the entropy. The fundamental thermodynamic relation describing these increases is

$$Tds = d\varepsilon - \mu_\rho d\rho - \mathbf{v}\cdot d\mathbf{g} - h_{ij}^u d(\nabla_i u_j) - h_{ij}^w d(\nabla_i w_j), \tag{9.10}$$

where s is the entropy per unit volume and μ_ρ is the chemical potential per unit mass. The fields h_{ij}^u and h_{ij}^w conjugate to $\nabla_i u_j$ and $\nabla_i w_j$ satisfy

$$h_{ij}^u = \left.\frac{\partial \varepsilon}{\partial \nabla_i u_j}\right|_{s,\rho,\mathbf{g},\nabla_i w_i} = \left.\frac{\partial f}{\partial \nabla_i u_j}\right|_{T,\rho,\ldots}, \tag{9.11}$$

$$h_{ij}^w = \left.\frac{\partial \varepsilon}{\partial \nabla_i w_j}\right|_{s,\rho,\mathbf{g},\nabla_i u_i} = \left.\frac{\partial f}{\partial \nabla_i w_j}\right|_{T,\rho,\ldots}, \tag{9.12}$$

where $f = \varepsilon - Ts$ is free energy density. The thermodynamic relation. [Eq. (9.10)] and the current equations [Eqs. (9.5)–(9.9)] can be used to calculate the time and spatial derivatives of the entropy:

$$\begin{aligned} T\partial_t s = &-\nabla_j(j_j^\varepsilon - \mu_\rho g_j - v_i\sigma_{ij} - h_{ij}^u X_i^u - h_{ij}^w X_i^w) \\ &-\mathbf{g}\cdot\boldsymbol{\nabla}\mu_\rho - (\sigma_{ij} + h_{ji}^u)\nabla_i v_j - X_j^u \nabla_i h_{ij}^u - X_j^w \nabla_i h_{ij}^w, \end{aligned} \tag{9.13}$$

and

$$T\nabla_i s = \nabla_i \varepsilon - \mu_\rho \nabla_i \rho - v_j \nabla_i g_j - h_{jk}^u \nabla_i \nabla_j u_k - h_{jk}^w \nabla_i \nabla_j w_k. \tag{9.14}$$

The gradients of the fields h_{ij}^u and h_{ij}^w simplify because the free energy depends only on the gradients of $\mathbf{u}$ and $\mathbf{w}$. Because F depends only on the gradients of $\mathbf{u}$ and $\mathbf{w}$, standard functional derivative identities yield

$$\frac{\delta F}{\delta u_j} = -\nabla_i \frac{\partial f}{\partial \nabla_i u_j} = \nabla_i h_{ij}^u \equiv -\nabla_i \frac{\delta f}{\delta u_{ij}}, \tag{9.15}$$

$$\frac{\delta F}{\delta w_j} = \nabla_j \frac{\delta f}{\delta \nabla_i w_j} = -\nabla_i h_{ij}^w, \tag{9.16}$$

where F is the free energy of Eq. (7.12) and f the free energy density. Equations (9.13) to (9.16) can be used to obtain an equation for the total rate of change of entropy arising both from local production and from flow in and out of a given volume in space. To lowest order in the velocity and fields h_{ij}^u and h_{ij}^w, this equation is

$$\begin{aligned} &T[\partial_t s + \nabla\cdot(\mathbf{v}s + T^{-1}\mathbf{Q})] \\ &\quad = -T^{-1}\mathbf{Q}\cdot\boldsymbol{\nabla}T - (\mathbf{g} - \rho\mathbf{v})\cdot\boldsymbol{\nabla}\mu_\rho - (-p\delta_{ij} - \sigma_{ij} - h_{ji}^u)\nabla_j v_i \\ &\qquad + X_i^w \frac{\delta F}{\delta u_i} + X_i^w \frac{\partial F}{\partial w_i}, \end{aligned} \tag{9.17}$$

where $\mathbf{Q}$ is the heat current which to lowest order in $\mathbf{v}$, etc., is simply $\mathbf{j}^{\varepsilon}$. When the nonlinear terms $h_{ij}^{u}u_{ij}$ and $h_{ij}^{w}\nabla_i w_j$ are neglected, the pressure p appearing in Eq. (9.17) is

$$p = -(\varepsilon - Ts - \alpha\rho) = -(f - \rho\frac{\partial f}{\partial \rho}). \tag{9.18}$$

The integral of the left-hand side of Eq. (9.17) over the volume of the structure must be nonnegative by the second law of thermodynamics. This implies that the right-hand side of Eq. (9.17) must be nonnegative.

In reversible processes, there is no entropy production, and both sides are zero. This condition determines *reactive* or nondissipative currents:

$$\mathbf{g} = \rho\mathbf{v}, \qquad \sigma_{ij} = -p\delta_{ij} - h_{ji}^{u}. \tag{9.19}$$

In irreversible processes, there is entropy production and both sides of Eq. (9.20) must be positive definite. This is accomplished by choosing the "currents" $\mathbf{Q}$, σ_{ij}, X_i^{u}, and X_i^{w} to be linear functions of the fields $\nabla_i T$, $\nabla_i v_j$, $\delta F/\delta u_i$, and $\delta F/\delta w_i$. There are no dissipative parts to the momentum current so that it is totally specified by Eq. (9.19). The couplings must be chosen so as to preserve all symmetries of the system. In particular, dissipative currents can only have terms proportional to fields with the opposite sign under time reversal. Thus the only dissipative contribution to the stress tensor is

$$\sigma_{ij}' = \eta_{ijkl}\nabla_k v_l, \tag{9.20}$$

where η_{ijkl} is the viscosity tensor which, like K_{ijkl}^{u}, is symmetric under interchange of i and j, k and l, and ij and kl.

All of the other currents have the same sign under time reversal, and can, in principle, have contributions from all of the fields except the velocity. Thus, the dissipative heat, $\mathbf{u}$, and $\mathbf{w}$ currents have the form

$$Q_i = \kappa_{ij}\nabla_j T + \Gamma_{ij}^{Qu}\frac{\delta F}{\delta u_j} + \Gamma_{ij}^{Qw}\frac{\delta F}{\delta w_j}, \tag{9.21a}$$

$$X_i^{u} = \Gamma_{ij}^{Qu}\nabla_j T + \Gamma_{ij}^{u}\frac{\delta F}{\delta u_j} + \Gamma_{ij}^{uw}\frac{\delta F}{\delta w_j}, \tag{9.21b}$$

$$X_i^{w} = \Gamma_{ij}^{Qw}\nabla_j T + \Gamma_{ij}^{uw}\frac{\delta F}{\delta u_j} + \Gamma_{ij}^{w}\frac{\delta F}{\delta w_j}, \tag{9.21c}$$

where κ_{ij} is the thermal conductivity tensor, and Γ_{ij}^{Qi}, Γ_{ij}^{u}, Γ_{ij}^{uw}, and Γ_{ij}^{w} are dissipative kinetic tensors. The precise form of these dissipative coefficients depends on symmetry. In three-dimensional systems with cubic or lower symmetry, all of the tensors in Eq. (9.21) are proportional to the unit tensor δ_{ij}. In this case, there are at most six independent coefficients appearing in Eq. (9.21). In quasicrystals, **w** transforms under a different representation of the point group $\mathcal{G}_P$ than do **u** and ∇T, and the cross coefficients Γ^{Qw} and Γ^{uw} vanish. Thus, in icosahedral quasicrystals, Eq. (9.21) becomes

$$Q_i = \kappa \nabla_i T + \Gamma^{Qu} \frac{\delta F}{\delta u_i}, \tag{9.22a}$$

$$X_i^u = \Gamma \frac{\delta F}{\delta u_i} + \Gamma^{Qu} \nabla_i T, \tag{9.22b}$$

$$X_i^w = \Gamma^w \frac{\delta F}{\delta w_i}. \tag{9.22c}$$

In type I incommensurate crystals, the relation between the sublattice displacements and the global displacement and phason variable is not unique. The parameter s in Eqs. (6.16) and (6.18) can be chosen to eliminate the cross coefficient Γ_{ij}^{uw} from Eq. (9.21). In type II systems, the definition of **u** and **w** is essentially unique. In addition, both **u** and **w** transform under the same representation of the point group so that all of the dissipative couplings in Eq. (9.21) are present. It should be noted that $\nabla \cdot \mathbf{Q}$ is a scalar. Thus only the identity representations of u_{ij} and $\nabla_i w_j$ will appear in $\nabla_j[\delta F/\delta u_i]$ and $\nabla_i[\delta F/\delta w_i]$. In cubic systems when both **u** and **w** transform under vector representations, these terms will be proportional to $\nabla^2 \nabla \cdot \mathbf{u}$ and $\nabla^2 \nabla \cdot \mathbf{w}$.

9.4 Hydrodynamics of Icosahedral Quasicrystals

If heat currents are ignored, the linearized hydrodynamic equations for icosahedral quasicrystals are [29]

$$\partial_t \rho + \nabla \cdot \mathbf{g} = 0, \tag{9.23a}$$

$$\partial_t g_i - \nabla_j(\eta_{ijkl} \nabla_k g_l) = -\frac{\delta F}{\delta u_i} - \rho \nabla_i \frac{\delta F}{\delta \rho}, \tag{9.23b}$$

$$\partial_t u_i + \Gamma_u \frac{\delta F}{\delta u_i} - v_i = 0, \tag{9.23c}$$

$$\partial_t w_i + \Gamma_w \frac{\delta F}{\delta w_i} = 0. \tag{9.23d}$$

The viscosity tensor, like the strain elastic tensor, has two independent components which can be expressed as

$$\eta_{ijkl} = (\eta_L - \frac{4}{3}\eta_T)\delta_{ij}\delta_{kl} + \eta_T(\delta_{ik}\delta_{jl} + \delta_{il}\delta_{jk} - \frac{2}{3}\delta_{ij}\delta_{kl}), \quad (9.24)$$

where η_L and η_T are the longitudinal and transverse viscosities. Upon Fourier transformation in space and time, these equations become

$$- i\omega\delta\rho + i\mathbf{q}\cdot\mathbf{g} = 0, \quad (9.25)$$

$$- i\omega\mathbf{u}_T = \frac{\mathbf{g}_T}{\rho_0} - \Gamma^u [\mu q^2\mathbf{u}_T + \mathsf{P}(\hat{q})\mathsf{C}(\mathbf{q})\mathbf{w}], \quad (9.26)$$

$$- i\omega u_L = \frac{g_L}{\rho_0} - \Gamma_u\left[(\lambda + 2\mu)q^2 u_L + \hat{q}\cdot\mathsf{C}(\mathbf{q})\mathbf{w} - iBq\frac{\delta\rho}{\rho_0}\right], \quad (9.27)$$

$$- i\omega\mathbf{g}_T = - \eta_T q^2\mathbf{g}_T - \mu q^2\mathbf{u}_T - \mathsf{P}(\hat{q})\mathsf{C}(\mathbf{q})\mathbf{w}, \quad (9.28)$$

$$- i\omega\mathbf{g}_L = \eta_L q^2 g_L - (\lambda + 2\mu - B)q^2 u_L - \hat{q}\cdot\mathsf{C}(\mathbf{q})\mathbf{w} - i(A - B)q\frac{\delta\rho}{\rho_0}, \quad (9.29)$$

$$- i\omega\mathbf{w} = - \Gamma_w[\mathsf{M}(\mathbf{q})\mathbf{w} + \mathsf{C}_T(\mathbf{q})\mathbf{u}], \quad (9.30)$$

where here and hereafter a bold sans serif letter represents a matrix, the superscript T denotes the transpose matrix, and the subscripts T and L denote parts parallel and perpendicular to **q**. P is a transverse projection operator with components

$$P_{ij} = \delta_{ij} - Q_{ij} = \delta_{ij} - \frac{q_i q_j}{q^2}, \quad (9.31)$$

and

$$\mathsf{C} \equiv \mathsf{K}^{uw}, \quad \mathsf{M} = K_1\mathsf{I} + \mathsf{K}, \quad (9.32)$$

where I is the unit tensor. Note that M and C are both of order q^2.

The speed of the sound modes are isotropic at small wavenumber and can be calculated by setting to zero all dissipative coefficients and the fields **w** in the equations of motion. This is because on the time scales corresponding to propagating sound (that is frequency $\omega \sim$ wavenumber q), **w**, being diffusive, does not respond at all to lowest order in wavenumber. It does have an effect at the next order in wavenumber: it contributes to the damping of sound waves, rendering it anisotropic despite the isotropy of the viscosity tensor (9.24). To see this, simply solve the equations of motion (9.25)–(9.30) for $\omega \sim q$. In this regime, the term Γ^w $\mathsf{K}(\mathbf{q})\mathbf{w}$ in Eq. (9.30) is negligible in comparison to $- i\omega\mathbf{w}$ at small q, so that

$$\mathbf{w} = \frac{- i\Gamma^w\mathsf{C}^{\mathrm{T}}(\mathbf{q})}{\omega}\mathbf{u}. \quad (9.33)$$

Inserting this expression for $\mathbf{w}$ into Eqs. (9.25)–(9.30) leads, after some manipulations (keeping in mind $\omega \sim q$), to the effective Fourier-transformed equations of motion:

$$(-i\omega + \Gamma^{u}\mu q^2)\mathbf{u}_T = \frac{\mathbf{g}_T}{\rho_0}, \tag{9.34a}$$

$$(-i\omega + \eta_T q^2)\mathbf{g}_T + [\mu q^2 \mathbf{I} + (-i\omega)^{-1}\Gamma^{w}\mathbf{PCC}^T]\mathbf{u}_T = 0 \tag{9.34b}$$

for transverse sound and

$$-i\omega\delta\rho + iqg_L = 0, \tag{9.35a}$$

$$(-i\omega + \eta_L q^2)g_L + i(A + 2\mu + \lambda - 2B)q\frac{\delta\rho}{\rho_0} - \frac{\Gamma^{w}}{\omega q}\,\mathrm{Tr}(\mathbf{QCC}^T)\frac{\delta\rho}{\rho_0} = 0 \tag{9.35b}$$

for longitudinal sound where the equilibrium coefficients A and B were introduced in Eq. (7.13). Dissipative cross-coupling of $\mathbf{g}_T$ to ρ and of g_L to $\mathbf{u}_T$, which the substitution (9.33) will generate, have been neglected since they do not affect the attenuations to leading order in q. Note that the term $\mathbf{PCC}_T/i\omega$ in Eq. (9.34b) is of order q^3 (recall $\mathbf{C} \sim q^2$) and is consequently of one order higher in q than the term $\mu q^2 \mathbf{I}$. Likewise, $[\mathrm{T_R}(\mathbf{QCC}^T)/\omega q] \sim q^2$ is one higher order in q than $i(A + \lambda - 2B)q$ in Eq. (9.35b). As a result, both of these terms can affect the dispersion relations only at *second* nontrivial order in q; i.e., they alter the leading damping terms, not the propagating ones. Note that without these terms ($\mathbf{PCC}^T$ and $\mathbf{QCC}^T$) the equations of motion are those of an isotropic solid. This implies that the sound speeds of the full equations are isotropic. The eigenvalues of Eq. (9.34) and (9.35) give the mode frequencies,

$$\omega = c_L q - \frac{i}{2}\left[\eta_L q^2 + \frac{\Gamma^{w}}{\rho_0 c_L{}^2}\frac{\mathrm{Tr}(\mathbf{QCC}^T)}{q^2}\right] \tag{9.36a}$$

for longitudinal sound, with

$$\rho_0 c_L{}^2 = A + \lambda + 2\mu - 2B \tag{9.36b}$$

and

$$\omega_\alpha = c_T q - \frac{1}{2}iq^2\left[(\eta_T + \Gamma^{u}\mu) + \frac{\Gamma^{w}}{\mu q^4}\lambda_\alpha\right], \qquad \alpha = 1,2 \tag{9.37a}$$

for the two polarizations of transverse sound with

$$\rho_0 c_T{}^2 = \mu, \tag{9.37b}$$

where the λ_α ($\sim q^4$) are the two nonzero eigenvalues of $\mathbf{PCC}^T\mathbf{P}$. [Equation (9.37a) is of course just Eq.(9.3) with $\gamma_\alpha(\hat{q}) = \eta_T + \Gamma^{u}\mu + (2\Gamma^{w}/\mu q^4)\gamma_\alpha$.] The attenuations of both transverse and longitudinal sound are sensitive to the presence of icosahedral order through the terms λ_1, λ_2, and Tr $\mathbf{QCC}^T$. In fact, λ_1 and λ_2 contain invariants of higher than sixth order in $\hat{q}$, while their sum $\lambda_1+\lambda_2=\mathrm{Tr}\mathbf{PCC}^T$ as well as $\mathrm{Tr}\mathbf{QCC}^T$ are of sixth order in $\hat{q}$. Since there exist *anisotropic* icosahedral invariants of sixth order in $\mathbf{q}$, one might expect the attentuations to be anisotropic, and indeed they are. There are precisely two independent icosahedral invariants of sixth order in $\mathbf{q}$. They can be taken to be $|\mathbf{q}|^6$, which is isotropic, and

$$I_6 = (1+\tau^6)^{-1}\sum_n (\mathbf{G}_n\cdot\mathbf{q})^6, \tag{9.38}$$

which is not. In terms of these,

$$q^2\mathrm{Tr}(\mathbf{QCC}^T) = K_3^2(-9q^6+10I_6). \tag{9.39}$$

The anisotropic part of the viscous damping is probably smaller in magnitude by a factor of 10^{-10} than the isotropic part. It is nonetheless worth remarking that quasicrystals should, therefore, have sound speeds which are isotropic but sound wave attenuations which even at small $\mathbf{q}$ reflect the icosahedral symmetry of the quasicrystal.

The diffusive modes come from $\mathbf{w}$, the energy density ϵ, and the longitudinal part of $\mathbf{u}$. Similar power-counting can be used here, with $\omega\sim q^2$, to simplify the equations somewhat. One is still left with a fairly complicated 5×5 matrix to diagonalize. The w-field is expected to relax very slowly, as remarked previously, with a diffusivity comparable to that for vacancies.

9.5 *Beyond Hydrodynamics*

The phason mode has been observed in scattering experiments from a number of experimental systems.[30,72] Often the mode is a propagating soundlike mode[80] rather than a diffusive mode predicted by the hydrodynamics presented here. The diffusive mode can be converted to a propagating mode if the "momemtum" $\mathbf{g}_w = \rho\mathbf{v}_w$ associated with the phason velocity $\mathbf{v}_w$ is conserved. There is no exact conservation law requiring $\mathbf{g}_w$ to be conserved. However, if the frictional coupling between sublattices in incommensurate systems is small, $\mathbf{g}_w$ obeys an approximate conservation law of the form

$$\partial_t\mathbf{g}_w + \gamma\mathbf{g}_w + \frac{\delta F}{\delta\mathbf{w}} = 0, \tag{9.40}$$

where γ is a friction coefficient. The equation for $\mathbf{w}$ is then modified to

$$\partial_t\mathbf{w} - \mathbf{v}_w + \bar{\Gamma}^w\frac{\delta F}{\delta\mathbf{w}} = 0. \tag{9.41}$$

If the friction coefficient is zero, these two equations predict a conserved phason momentum and a propagating phason mode. For nonzero γ, the phason momentum always decays to zero, and for frequencies $\omega << \gamma$,

$$\mathbf{g}_w = \gamma^{-1} \frac{\delta F}{\delta \mathbf{w}}, \tag{9.42}$$

and Eq. (9.41) reduces to Eq.(9.23d) with $\Gamma^w = \bar{\Gamma}^w + \gamma^{-1}$.

This section has presented a fairly detailed derivation of the hydrodynamics of quasperiodic structures. This theory, like all theories of broken symmetry hydrodynamics, only makes sense if the free energy is analytic in the gradients of all elastic variables. This analyticity can break down for phason strains. In this case, it is still expected that nonequilibrium phason distortions will decay to equilibrium in a characteristic time τ_{phason} that tends to infinity as the wavenumber q of the distortion goes to zero. The precise dependence of τ_{phason} on q has not, however, been worked out even in simple cases.

10. Pinned Phason Dynamics

It has been argued throughout this article that uniform displacements of the phason variable do not change the free energy and, consequently, that only spatially nonuniform phason fields produce free energy changes. The elastic theory and associated hydrodynamics presented in Sections 7 and 9 assumed an analytic expansion of the free energy in phason as well as normal strains. This assumption has been verified (except in very special cases[65]) in all other systems, such as ferromagnets[24] or superfluid helium[23] in which there is a broken continuous symmetry. The invariance associated with translation of the phason variable in quasiperiodic systems is more complicated than that associated with more "traditional" broken symmetries, and it is possible, as shown by Aubry[40,41], for the analyticity assumption to break down even in the simplest quasiperiodic systems. To see how this comes about, consider the discrete Frenkel–Kontorowa model[33,34,81] studied by Aubry and others. This model consists of harmonically coupled balls at positions x_n interacting with a sinusoidal external potential of period l_b. The Hamiltonian for this model is

$$H_{\text{FK}} = \frac{1}{2} K \sum_n (x_{n+1} - x_n - l_0)^2 - V \sum_n \cos\left(\frac{2\pi x_n}{l_b}\right). \tag{10.1}$$

The separation between adjacent balls is l_0 when the strength V of the interaction potential is zero. This model is similar to the one introduced in Section 3 to illustrate properties of one-dimensional quasiperiodic structures. In equilibrium, there is an average separation l_a between balls determined by the competition between spring forces favoring separation l_0 and interactions with the substrate favoring separation of some integral multiple of l_b. Thus, quite rigorously, the position of the n-th ball satisfies

$$x_n = nl_a + u_a + f(nl_a + u_a - u_b), \tag{10.2}$$

where the Hull function $f(y)$ is a function of period l_b. u_a describes displacements of the balls (sublattice a) whereas u_b describes displacements of the periodic substrate (sublattice b). In most treatments of the Frenkel–Kontorowa model, the substrate is rigidly fixed and the phase u_b is set to zero. It will be useful for the present discussion, however, to allow for the possibility of variations in u_b. For this example, as for general one-dimensional quasiperiodic structures, the phason field w is proportional to $u_b - u_a$. When u_b is fixed, $-u_a$ is essentially w.

For small values of V/K, the function $f(y)$ is continuous, and the frequency of the phason mode determined by solving the equations,

$$m\ddot{x}_n = -\frac{\partial H_{\text{FK}}}{\partial x_n}, \tag{10.3}$$

goes to zero with wavevector q as for the hydrodynamic modes discussed in Section 9. Since the FK Hamiltonian is analytic in the displacements x_n, it is analytic in w when f is analytic. This leads to an analytic expansion of the energy in the the phason strain at long wavelengths. Thus the elastic-hydrodynamic theory of Sections 7 and 9 is applicable to the FK model for small V/K. Above a critical value of V/K, $f(y)$ becomes nonanalytic, and the frequency of the phason mode defined by Eq. (10.3) tends to a finite value as $q \to 0$. As a result, an analytic expansion of the energy in the phason strain becomes impossible and the elastic-hydrodynamic theory is inapplicable. The origin of the analyticity breaking is clear. For large V/K, the balls prefer to sit near the bottom of the well of the cosine potential. The strong potential makes values of x_n near the tops of wells inaccessible and leads to a nonanalytic Hull function f. This analyticity breaking is clearly associated with the discrete nature of the balls and would not be predicted by any perturbation theory in phason strains. If all of the balls are translated by the same arbitrary distance u_a, some of them will end up in high energy configurations near the tops of the wells. They will then relax down to lower energy configurations near the bottom of the wells. Thus, though the energy is still invariant with respect to uniform translations of the balls, there are energy barriers separating translated states of the same energy. These energy barriers lead to *pinning* of the phasons.

When f is analytic, the low energy excited states of the FK model are specified by their phason strain. In the pinned regime, the energy will still go to zero with phason strain, but the average linear phason strain does not uniquely determine a state. There are many locally stable states with the same average linear phason strain. The lowest energy states are those with sequences of balls that are closest to ground state sequences. These excited states have been studied in some detail by Vallet and collaborators[82].

To see the effects of analyticity breaking in higher dimensions, it is instructive to study a toy model in which there are rows of atoms parallel to the x-axis. The position of the n-th atom in a row then becomes a function of the vertical coordinate y. Now allow x_n (y) to be determined by Eq. (10.2) with the simple nonanalytic Hull function,

$$f = -\omega \bar{l} \left\{ \frac{nl_a + u_a(y) - u_b(y)}{l_b} \right\}, \tag{10.4}$$

where $\{x\} = x - [x]$, and $[x]$ is the greatest integer less than or equal to x. $\bar{l}$ is a length and ω a number. With the simple Hull function of Eq. (10.4), Eq. (10.2) can be rewritten as

$$\frac{x_n(y)}{\bar{l}} = n + \eta(y) + \omega[n\sigma + \beta(y)], \tag{10.5}$$

where $\sigma = l_a/l_b$, and where

$$\bar{l}\eta(y) = u(y) = \left(1 - \frac{\omega \bar{l}}{l_b}\right) u_a + \frac{\omega \bar{l}}{l_b} u_b \tag{10.6a}$$

and

$$\bar{l}\beta(y) = -w(y) = \frac{\bar{l}}{l_b}(u_a - u_b), \tag{10.6b}$$

are, respectively, the displacement and phason variables. Equation (10.6) corresponds to Eq. (3.7) with $s = \omega\bar{l}/l_b$ and $\alpha = \bar{l}/l_b$. Equation (10.5) implies that the distance between adjacent balls can take on only two values: 1 and $1 + \omega$. When $\omega = \sigma = \tau^{-1}$, where τ is the golden mean, the resulting sequence is the Fibonacci sequence of long and short intervals shown in Fig. 10.1. Each value of β gives a Fibonacci sequence. A change in β will give rise to a new sequence obtained from the old one by interchanging long and short intervals. Thus if $\beta(y)$ increases from zero at $y = 0$ to 0.5 at $y = L$, there will be exactly one change in length for each interval as shown in Fig. 10.1. The changes in length will cause discontinuities or jags as discrete points. Figure 10.1 also shows the effects of $u(y)$ and of

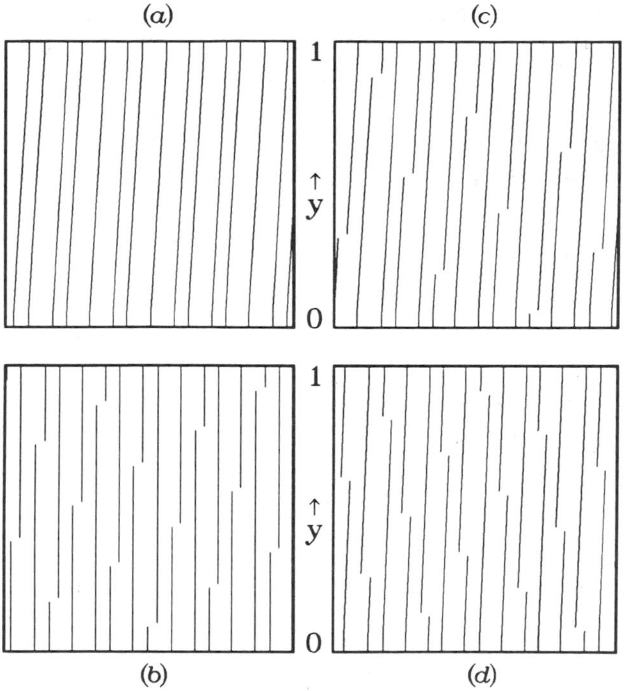

Figure 10.1. Effects of variation of the phase variables in Eqs. (10.4)–(10.6). The lines represent the positions of atoms as a function of y determined by Eq (10.5). At the bottom of each figure is a perfect Fibonacci sequence of long and short intervals determined by Eq. (10.5) with fixed η and β. (a) The effect of a linear variation of u (or η) with y. (b) The effect of a linear variation of w (or β). In the first case, the entire pattern shifts uniformly with y leading to slanted lines. In the second case, jags appear each time $\eta\sigma + \beta(y)$ passes through an integer. There is no slant in the grid. (c) The effect of variation in u_a. (d) The effect of variation in u_b. Note that (c) and (d) have both slant and jags indicating a mixture of strain and phason strain.

$u_a(y)$ and $u_b(y)$ linearly increasing with y. Note that jags are present when u_a and u_b vary, indicating that they have a nonzero projection onto variations of w.

It is natural to assign an energy ε to each jag in Fig. 10.1. Since the number of jags per unit length in the z-direction is proportional to the change $\Delta\beta$ along y, the total energy is

$$E = N_T\,\varepsilon \sim \int dx\,|\,\Delta\beta\,| \sim \int dx\,dy\,|\,\nabla w\,|\,, \qquad (10.7)$$

where N_T is the total number of jags. In higher dimensions, the result generalizes to $E \sim \int d^dx \mid \nabla w \mid$. Thus, the energy goes to zero with the phason strain ∇w but not analytically as predicted by the elastic model.

Penrose tilings [see Figs. 1.1 and 1.2] for pentagonal (icosahedral) quasicrystals can be constructed by taking the dual to Amman quasilattices consisting of grids of parallel lines (planes) normal to the vectors of the vertices of a pentagon (icosahedron) with positions given by the Fibonocci sequence of Eq. (10.5) with $\omega = \sigma = \tau^{-1}$. This method of producing a Penrose tiling is called the dual method. It can be generalized to produce tilings for structures of arbitrary symmetry and rank. The tiles resulting from this construction obey certain matching rules which can be encoded by decorating the tiles as shown in Fig. 1.2. Strain leads to waviness of the Amman grids [Fig. 10.2], whereas phason strains cause jags in the grid similar to those in the one-dimensional grids of Fig. 10.1. Figure 10.2 also shows a dislocation in the Amman quasilattice. The grids in Fig. 10.2 should be compared to the dark lines in the density wave images of Fig. 7.1. Penrose tiles associated with the Amman quasilattices of Fig. 10.2 are shown in Fig. 10.3. Note that strain distorts the tiles from the ideal shape [Fig. 10.3(b)], whereas the jags resulting from phason strain lead to matching-rule violations [Figs. 10.3(c) and 10.3(d)] as explained in Fig. 10.4.

As in the single grid, it is natural to associate an energy with each matching-rule violation. This leads again to an energy proportional to $\int d^dx \mid \nabla \mathbf{w} \mid$. This form of the energy is merely suggestive. No detailed model for the energetics and dynamics of pinned phason strains in quasicrystals has yet been developed. There are some indications that the problem is quite complex, with possibly different forms for the energy associated with the different group representations of the phason strain.[83]

The dual method of producing tilings strongly suggests that the energies associated with nonuniform phason variables might be nonanalytic in the phason strain. Similar results have been obtained using the projection technique.[43,44] The geometry of projection from a five-dimensional space to a two-dimensional space to produce a Penrose tiling or from a six-dimensional space to a three-dimensional space to produce an icosahedral tiling leads to what is equivalent to a nonanalytic Hull function.

Quasiperiodic structures are unique among systems with continuous broken symmetries. They can exhibit two generically different types of energetics and dynamics associated with strains of their elastic variable—continuum elasticity-hydrodynamics and pinned dynamics. In the one-dimensional FK model and presumably in type I incommensurate crystals, there is a transition between these two types of dynamics as a function of some coupling constant (V in the FK model). This transition can also

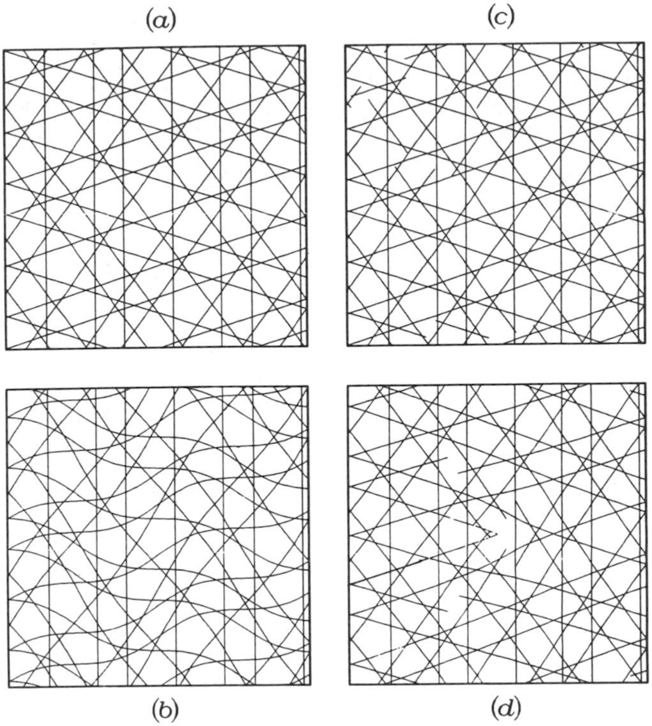

Figure 10.2. Amman pentagrids with spatially varying phases. The Penrose tilings of Figure 10.3 can be obtained from these grids via a generalized dual transformation.[11,12] (a) A portion of a perfect Amman pentagrid leading to the Penrose tiling of Fig. 10.3(a). (b) A distortion of the Amman pentagrid of (a) corresponding to variations in the displacement **u**. (c) An Amman pentagrid containing variations is the phason variable **w** about the value used in (a). The jags in the grids correspond precisely to matching rule violations in the tiling of Fig. 10.3(c). (d) A dislocation in the Amman pentagrid. At large distances from the core, the curvature of the lines and the density of jags fall off inversely with distance.

occur as a function of temperature, at least in mean-field theory.[84] Thus, for these systems, one might imagine a phase diagram in which there is a phase boundary in a temperature-coupling constant plane separating pinned from unpinned phases. Below some critical value of the coupling, the unpinned phase extends down to zero temperature. There are indications discussed above that the phason dynamics of quasicrystals can be pinned at zero temperature as a result of geometry. In this case, one could imagine a phase diagram in which the pinned phase exists at zero temperature for all values of some (as yet undefined) coupling constant.

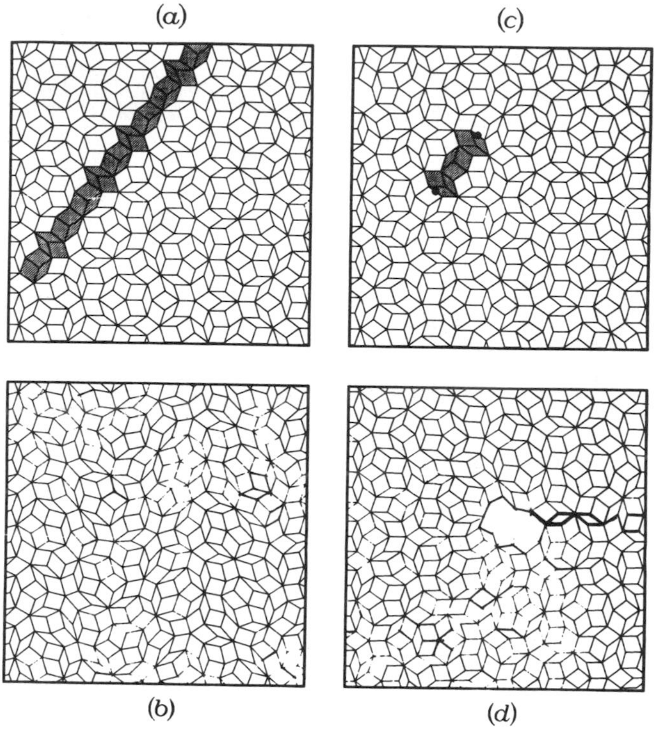

Figure 10.3. Penrose tilings with spatially varying phases. (a) A portion of a perfect Penrose tiling. The shaded unit cells compose a segment of a "worm." (b) A distortion of the tiling of (a) corresponding to variations in the displacement **u**. The unit-cell shapes are distorted but their arrangement is the same as in (a). (c) A tiling containing variations in **w** about the value used in (a). The shaded rhombuses form a flipped segment of a worm [compare to the same region in (a). For details about the relation between worm flips and matching-rule violations, see Fig. 10.4]. The large dots at the ends of the shaded segment indicate edges along which the Penrose matching rules are violated—deviations from the Penrose local isomorphism class. This picture contains several other such mismatches and flipped worm segments. To find them, use Fig. 10.2 as a guide. (d) A dislocation in the Penrose tiling. At large distances from the core, the distortion of the unit cells and the the density of mismatches both become small. although neither can be completely eliminated.

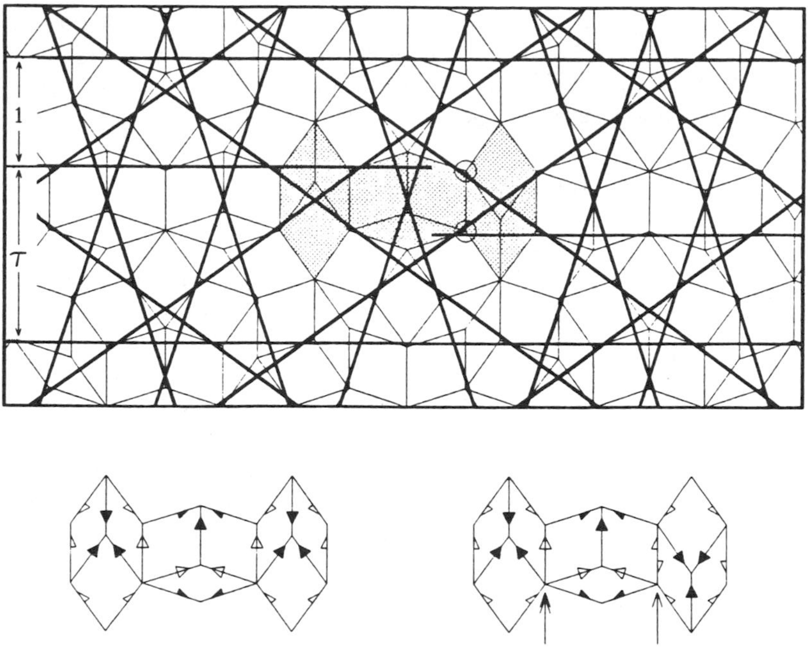

Figure 10.4. Illustration of a "worm" in a Penrose tiling and its signature in the Amman pentagrid. An open circle indicates a type of vertex that is never found in undefected tilings of the Penrose local isomorphism class. The right (left) half of the broken Amman line can be shifted up (down) to restore the ideal pentagrid in the region depicted. The consequent rearrangements of the intersections of the Amman pentagrid cause the right (left) half of the worm to flip when the generalized dual technique is applied. The worm segments depicted below are decorated to illustrate the Penrose matching rules. The rules are that two tiles can join only along edges which have the same color arrow point in the same direction. The segment on the right corresponds to the shaded segment above, where the rightmost hexagon (formed from one "fat" and two "skinny" rhombuses) has been flipped, causing a matching rule violation (indicated by a single-headed arrow). Note, however, that the top and the bottom of the hexagon are decorated in the same way, so that no other mismatches would arise.

A transition to the unpinned phase could occur at some finite temperature. In real quasicrystals, this transition could be preempted by the melting transition.

The physical consequences of pinning are quite complex because pinning implies the existence of a large number of locally stable states into which the system can be frozen. Since transitions to lower energy states require that energy barriers be overcome, the relaxation of phason strains will be slower than if the hydrodynamics of the preceding section applied. The implication of frozen phason strains for scattering experiments will be discussed in the next section.

11. Quenched Phason Strains

Both the hydrodynamics and pinned dynamics of phasons discussed in Sections 9 and 10 predict that nonuniform phason configurations, including those due to dislocations, will relax to complete equilibrium in times τ_{phason} that are much longer than any experimental observation time t_{exp}. This means that any phason strains present in a sample at the moment of preparation remain for indefinitely long times: the phasons are quenched or frozen into initial configurations. Since other variables relax to equilibrium in times much less than t_{exp}, thermodynamic averages should be taken not with respect to the complete equilibrium ensemble but with respect to an ensemble with a fixed, possibly spatially nonuniform phason field $\mathbf{w}_q$. Averages with respect to such a quenched ensemble will be denoted by $< >_w$. For example the scattering amplitudes in the quenched ensemble are

$$< \rho_{\mathbf{G}}(\mathbf{x}) >_w = | < \rho_{\mathbf{G}} > | \, e^{i\mathbf{G}\cdot\mathbf{u}_q(\mathbf{x}) + \mathbf{G}^{\perp}\cdot\mathbf{w}_q(\mathbf{x})}, \tag{11.1}$$

where $\mathbf{u}_q(\mathbf{x})$ is the value of the displacement field in the presence of $\mathbf{w}_q$, and where any effect of $\mathbf{w}_q$ on the intensity $| < \rho_{\mathbf{G}} > |$ has been ignored.

The most striking effect of quenched phason strains is on the positions and widths of X-ray and electron scattering peaks.[60–62] The coherent scattering peak from a quasicrystal with quenched phason strain is

$$I(\mathbf{q}) = \int d^3x \int d^3x' \, e^{-i\mathbf{q}\cdot(\mathbf{x}-\mathbf{x}')} < \rho(\mathbf{x}) >_w < \rho(\mathbf{x}') >_w. \tag{11.2}$$

If the phason strain is spatially uniform, then $\mathbf{w}_q = \mathbf{M}\mathbf{x}$, where $\mathbf{M}$ is a 3×3 matrix. If the displacement field $\mathbf{u}_q$ resulting from $\mathbf{w}_q$ is ignored, the coherent scattering intensity of Eq. (11.2) becomes

$$I = V^2 \sum_{\mathbf{G}} | < \rho_{\mathbf{G}} > |^2 \, \delta_{\mathbf{q},\mathbf{G}+\Delta\mathbf{G}}, \tag{11.3}$$

where $\Delta\mathbf{G} = \mathbf{G}^{\perp}\mathbf{M}$. Thus, a uniform phason strain anisotropically shifts Bragg peaks by an amount proportional to $| \mathbf{G}^{\perp} |$. Since Bragg peaks with the lowest intensity have the largest value[5,13,14] of $| \mathbf{G}^{\perp} |$, the dimmest diffraction peaks will be shifted the most as shown in Fig. 11.1.

Electron diffraction patterns, such as that shown in Fig. 11.2, from small sample volumes[60] where phason strains are expected to be linear

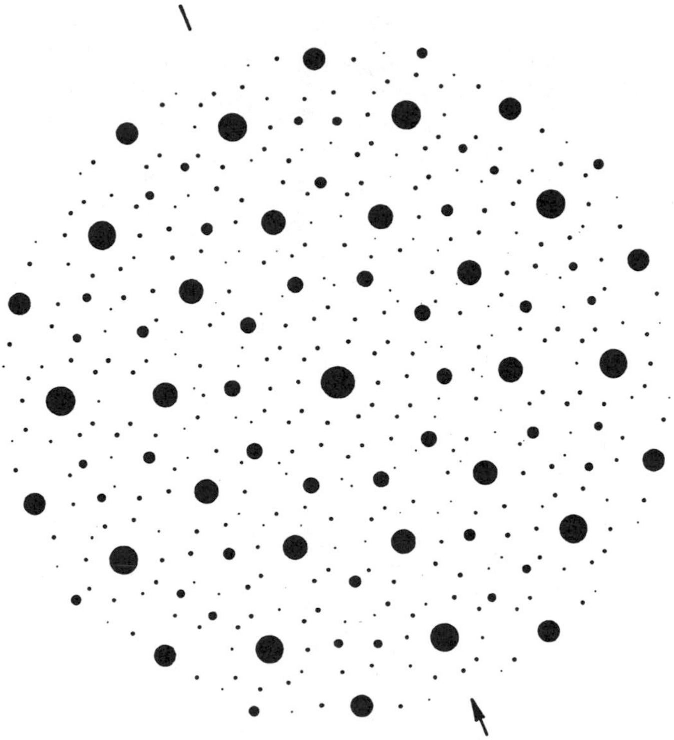

Figure 11.1. A distorted diffraction pattern due to anisotropic linear strain in w. The peaks remain true Bragg peaks (area of spots represents relative intensity). In an ideal quasicrystal, the peaks lie along straight lines. Hold the figure at grazing angle and sight along the arrow to observe the distortion of the pattern.

do show peak shifts proportional to $| \mathbf{G}^{\perp} |$. Patterns taken from larger sample volumes show broadened peaks with unusual shapes[85], such as those shown in Fig. 11.3. These shapes are well explained by averaging the intensity of Eq. (11.3) over a distribution of phason strains that are required to be continuous over the volume of the system.[86] High resolution electron micrographs are real space images obtained by reconstructing the phase information in an electron diffraction pattern. Such micrographs taken from icosahedral quasicrystals[77] look very similar to the density wave images with phason strain shown in Fig. 7.2 and provide further evidence supporting the view that experimental quasicrystals are nearly ideal quasicrystalline systems with frozen-in phason strain.

X-ray powder diffraction probes a much larger volume than does elec-

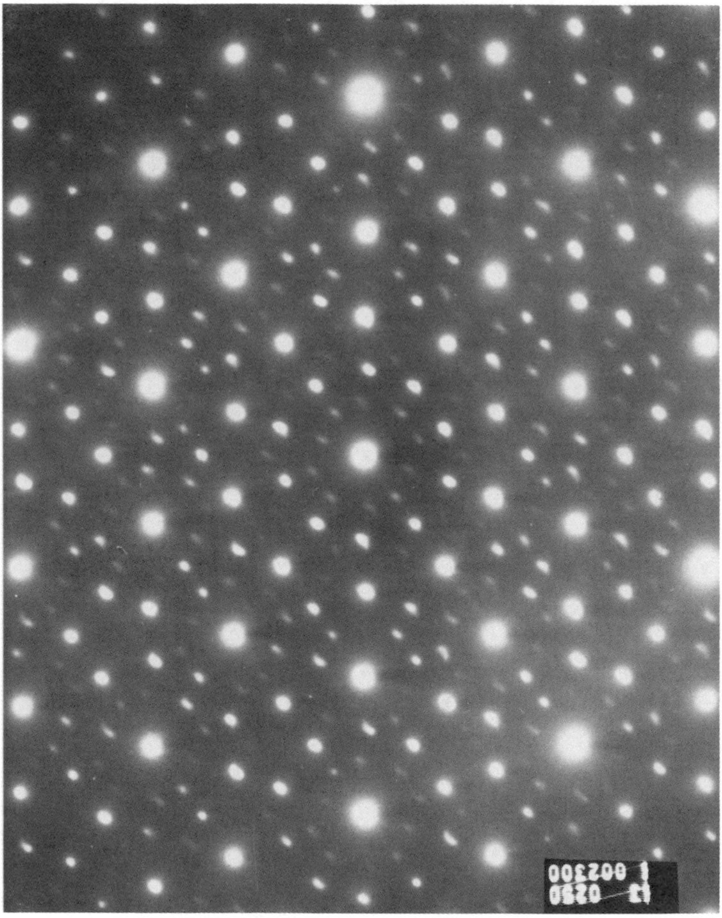

Figure 11.2. An electron diffraction pattern in a fivefold symmetry plane of the icosahedral phase of $Al_{78}\,Cr_{17}\,Ru_5$ which shows particularly large anisotropic shifts similar to those in the theoretically generated Fig. 11.1. This pattern was taken with a well-focussed beam from a relatively small volume.

tron scattering and effectively averages over many possible realizations of the quenched phason variable $\mathbf{w}_q(\mathbf{x})$. In other words, the X-ray scattering intensity is an ensemble average of Eq. (10.3) over some probability distribution for $\mathbf{w}_q$. If the components of $\mathbf{w}_q(\mathbf{x})$ that increase at large $\mathbf{x}$ receive sufficient weight in the probability distribution, the peaks in $I(\mathbf{q})$

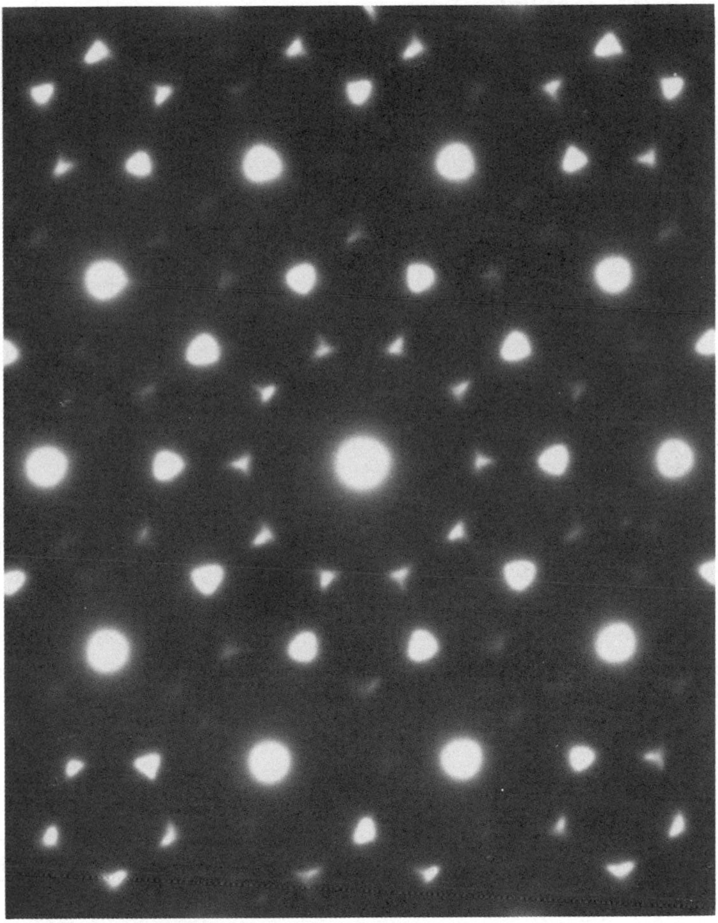

Figure 11.3. An electron diffraction pattern obtained from the same material as in Fig. 11.2 but from a larger scattering volume. Note the broadening and strange shape of the peaks.

will be broadened. If averages with respect to $\mathbf{w}_q$ (**x**) are denoted by square brackets, the scattering intensity becomes

$$I(\mathbf{q}) = \int d^3x \int d^3x' \, [< \rho(\mathbf{x}) >_{w} < \rho(\mathbf{x}') >_{w'}] \, e^{-i\mathbf{q}\cdot(\mathbf{x}-\mathbf{x}')}. \qquad (11.4)$$

The exact form of $I(\mathbf{q})$ depends on the probability distribution for **w**, which is not known in any detail. A reasonable model is that the Fourier transform

$\mathbf{w}_q(\mathbf{q})$ of $\mathbf{w}_q(\mathbf{x})$ is an independent random variable with zero mean and variance proportional to $q^{-(\zeta+3)}$:

$$[w_{qi}(\mathbf{q})w_{qj}(\mathbf{q}')] \sim \delta_{ij}\,\delta_{\mathbf{q},\mathbf{q}'}q^{-(\zeta+3)}. \qquad (11.5)$$

In this case,

$$[w_q^{\;2}(\mathbf{x})] \sim \begin{cases} const. & \zeta < 0 \\ ln\,L & \zeta = 0, \\ L^{\zeta} & \zeta > 0 \end{cases} \qquad (11.6)$$

in a sample of size L. When $\zeta > 0$,

$$[(\mathbf{w}_q(\mathbf{w}) - \mathbf{w}_q(\mathbf{x}'))^2] = A \mid \mathbf{x} \mid^{\zeta}, \qquad (11.7)$$

at large $\mid \mathbf{x} \mid$, where A is a constant. Since $\mathbf{w}_q(\mathbf{x})$ is a Gaussian random variable, $[<\rho_G(\mathbf{x})>_q]$ will be proportional to a nonthermal Debye–Waller factor $\exp(-\tfrac{1}{2}W_q) = \exp(-\tfrac{1}{2}(\mathbf{G}^{\perp})^2[\mathbf{w}_q^{\;2}(\mathbf{x})])$. If $\zeta < 0$, $\exp(-W_q)$ is finite, and the scattering amplitudes will be nonzero. There will, therefore, be Bragg peaks in the scattering intensity but additional diffuse scattering arising from the randomness. This situation is similar to that predicted by a model[21] for quasicrystals based on entropic fluctuations of a projection surface from a higher-dimensional space. If $\zeta > 0$, $\exp(-W_q)$ is zero, and the scattering amplitudes are zero. In this case, the scattering intensity is

$$I(\mathbf{q}) = \int d^3x \sum_{\mathbf{G}} \mid < \rho_{\mathbf{G}} > \mid^2 e^{-i(\mathbf{q}-\mathbf{G})\cdot\mathbf{x}}\, e^{-\frac{1}{2}(\mathbf{G}^{\perp})^2 A \mid \mathbf{x} \mid^{\zeta}}. \qquad (11.8)$$

There are still sharp peaks in $I(\mathbf{q})$ in the vicinity of $\mathbf{G}$, but they are broadened by the randomness. The amount of broadening and the shape of the peaks depend on the value of ζ and more generally on the probability distribution for $\mathbf{w}$. If $\zeta = 1$, the peaks have a Lorentzian-squared line shape with width $\Gamma \sim \mid \mathbf{G}^{\perp} \mid^2$. If $\zeta = 2$, the peaks have a Gaussian shape with $\Gamma \sim \mid \mathbf{G}^{\perp} \mid$. In general, the width scales as $\mid \mathbf{G}^{\perp} \mid^{2/\zeta}$.

The simple model discussed in the last paragraph shows that quenched phason strains can lead to broadening of Bragg peaks with widths that grow monotonically with the magnitude $\mid \mathbf{G}^{\perp} \mid$ of the complementary wavevector as long as contributions from the displacement variable are ignored. Since $\mathbf{u}$ and $\mathbf{w}$ are coupled, a quenched phason field $\mathbf{w}_q(\mathbf{x})$ will induce a quenched displacement field $\mathbf{u}_q(\mathbf{x})$ as discussed above, and the widths should depend on $\mid \mathbf{G} \mid$, but more weakly than on $\mid \mathbf{G}^{\perp} \mid$. Thus, at large $\mid \mathbf{G}^{\perp} \mid$, one would expect widths to grow monotonically with $\mid \mathbf{G}^{\perp} \mid$; at small $\mid \mathbf{G}^{\perp} \mid$, the $\mid \mathbf{G} \mid$ dependence of the width should become important, and the widths should not be monotonic in $|\mathbf{G}^{\perp}|$. The experimental widths of X-ray scattering peaks grow approximately linearly with $|\mathbf{G}^{\perp}|$ at large $|\mathbf{G}^{\perp}|$ but show a slight upward curvature at small $|\mathbf{G}^{\perp}|$. This behavior is observed in experiments.[61,87]

Acknowledgments

This article reviews results obtained in collaboration with P. A. Bancel, P. A. Heiney, D. Levine, S. Ostlund, S. Ramaswamy, J.E.S. Socolar, P. Steinhardt, and J. Toner and could not have been written without their assistance and support. The calculation of the elastic free energy for icosahedral quasicrystals in the appendix was carried out by the author, S. Ostlund, S. Ramaswamy, and J. Toner. The author is grateful to David Wright for a careful and critical reading of an early version of the manuscript and to J.E.S. Socolar for providing many of the figures. The author also acknowledges support from the National Science Foundations under grant Nos. DMR85-20272 and DMR85-19059.

Appendix—Calculation of Elastic Energies

In this appendix, we will outline briefly how the elastic free energy [Eqs. (7.7)–(7.11)] was calculated.

As discussed in Section 7, there are five elastic constants in an icosahedral quasicrystal arising from the $\Gamma_0^{(1)}$ and $\Gamma_0^{(5)}$ representations of u_{ij}, from the $\Gamma_0^{(4)}$ and $\Gamma_0^{(5)}$ representations of $\nabla_i w_j$, and from a mixing of the $\Gamma_0^{(4)}$ representations of u_{ij} and $\nabla_i w_j$. Thus, the elastic free energy has the form,

$$f_{\text{el}}(\mathbf{u},\mathbf{w}) = \frac{1}{2}\lambda(u^{(0)})^2 + \mu \sum_{\alpha=1}^{5} (u_\alpha^{(5)})^2 + \frac{1}{2}K^{(5)} \sum_{\alpha=1}^{5} (w_\alpha^{(5)})^2 + \frac{1}{2}K^{(4)} \sum_{\alpha=1}^{4} (w_\alpha^{(4)})^2 + K^{uw} \sum_{\alpha=1}^{5} u_\alpha^{(5)} w_\alpha^{(5)}, \tag{A1}$$

where $u^{(1)} = \nabla \cdot \mathbf{u}$, $\{u_\alpha^{(5)}\}$ and $\{w_\alpha^{(5)}\}$, respectively, are the set of five independent combinations of u_{ij} and $\nabla_i w_j$ that transform under $\Gamma_0^{(5)}$, and $\{w_\alpha^{(4)}\}$ is the set of four independent combinations of $\nabla_i w_j$ that transform under $\Gamma_0^{(4)}$. Stability requires that

$$\mu > 0, \qquad \lambda > 0, \qquad K^{(4)} > 0, \qquad K^{(5)} > 0, \qquad K^{(5)}\mu > 2(K^{uw})^2. \tag{A2}$$

Here $\{u_\alpha^{(5)}\}$ is simply the five components of the symmetric traceless tensor $\tilde{u}_{ij} = u_{ij} - \frac{1}{3}(\nabla \cdot \mathbf{u})\delta_{ij}$. With a normalization such that $\tilde{u}_{ij}\tilde{u}_{ij} - \Sigma_\alpha (u_\alpha^{(5)})^2$,

$$\begin{aligned} u_1^{(5)} &= (2/3)^{1/2}[u_{zz} - \frac{1}{2}(u_{zz} + u_{yy})], \\ u_2^{(5)} &= 2^{-1/2}(u_{xx} - u_{yy}), \\ u_3^{(5)} &= 2^{1/2}u_{xy}, \\ u_4^{(5)} &= 2^{1/2}u_{xz}, \\ u_5^{(5)} &= 2^{1/2}u_{yz}. \end{aligned} \tag{A3}$$

The calculation of $\{w_\alpha^{(4)}\}$ and $\{w_\alpha^{(5)}\}$ is much less direct. One way to determine these quantities is to use the fact that there is a single term coupling $w_\alpha^{(5)}$ to $u_\alpha^{(5)}$ in the elastic energy so that there is a tensor K_{ijkl}^{uw} satisfying

$$K_{ijkl}^{uw} u_{ij}\, \nabla_k w_l = K^{uw} \sum_\alpha u_\alpha^{(5)} w_\alpha^{(5)}. \tag{A4}$$

Given the transformation properties of $\mathbf{G}_n$ and $\mathbf{G}_n^\perp$, the tensor K_{ijkl}^{uw} must be proportional to

$$J_{ijkl} = (1 + \tau^2)^2 \sum_{n=1}^{6} G_{ni} G_{nj} G_{nk} G_{nl}^\perp , \tag{A5}$$

which can be calculated using Eqs. (4.10) and (6.29). J_{ijkl} is symmetric under interchange of i, j and k, and its independent components are tabulated in Table A1. This table and Eqs. (4.10), (6.29), and (A4) then yield the components of $\nabla_i w_j$ that transform under $\Gamma_0^{(5)}$:

$$\begin{aligned} w_1^{(5)} &= \frac{1}{2\tau}(\tau^2 \nabla_x w_z - \nabla_y w_y + \tau \nabla_z w_x), \\ w_2^{(5)} &= \frac{1}{2\sqrt{3}\tau}[(1-\tau)\nabla_x w_z - (\tau + \tau^2)\nabla_y w_y + (1+\tau^2)\nabla_z w_x], \\ w_3^{(5)} &= \frac{1}{\sqrt{3}\tau}(\tau^2 \nabla_x w_y - \nabla_y w_z), \\ w_4^{(5)} &= \frac{1}{\sqrt{3}\tau}(\nabla_x w_x + \tau^2 \nabla_z w_z), \\ w_5^{(5)} &= \frac{1}{\sqrt{3}\tau}(-\tau^2 \nabla_y w_x - \nabla_z w_y). \end{aligned} \tag{A6}$$

The 9-component vector

$$\tilde{\mathbf{w}} = (\nabla_x w_z, \nabla_x w_y, \nabla_x w_z, \nabla_y w_x, \nabla_y w_y, \nabla_y w_z, \nabla_z w_x, \nabla_z w_y, \nabla_z w_z), \tag{A7}$$

with components $\tilde{w}_l$, $l = 1, \cdots, 9$, can be decomposed into its components that transform according to $\Gamma_0^{(4)}$ and $\Gamma_0^{(5)}$:

$$\tilde{\mathbf{w}}_l = \sum_{\alpha=1}^{4} w_\alpha^{(4)} \tilde{\mathbf{e}}_\alpha^{(4)} + \sum_{\alpha=1}^{5} w_\alpha^{(5)} \tilde{\mathbf{e}}_\alpha^{(5)}, \tag{A8}$$

where the 9-dimensional unit vectors $\tilde{\mathbf{e}}_\alpha^{(p)}$, $p = 4,5$, have components $\tilde{\mathbf{e}}_{\alpha l}^{(p)}$ which satisfy the usual orthogonality and completeness relations

$$\sum_{l=1}^{9} \tilde{\mathbf{e}}_{\alpha l}^{(p)} \tilde{\mathbf{e}}_{\beta l}^{(q)} = \delta^{pq}\, \delta_{\alpha\beta}, \tag{A9}$$

$$\sum_{\alpha,p} \tilde{\mathbf{e}}_{\alpha l}^{(p)}\, \tilde{\mathbf{e}}_{\alpha l}^{(p)} = \delta_{ll'}. \tag{A10}$$

Thus,

$$w_\alpha^{(p)} = \sum_l \tilde{\mathbf{w}}_l \tilde{\mathbf{e}}_{\alpha l}^{(p)}. \tag{A11}$$

With this information,

$$\begin{aligned}
\tilde{\mathbf{e}}_1^{(5)} &= \frac{1}{2\tau}(0,\ 0,\ \tau^2,\ 0,\ -1,\ 0,\ \tau,\ 0,\ 0),\\
\tilde{\mathbf{e}}_2^{(5)} &= \frac{1}{2\sqrt{3}\tau}(0,\ 0,\ 1-\tau,\ 0,\ \tau+\tau^2,\ 0,\ \tau^2+1,\ 0,\ 0),\\
\tilde{\mathbf{e}}_3^{(5)} &= \frac{1}{\sqrt{3}\tau}(0,\ \tau^2,\ 0,\ 0,\ 0,\ -1,\ 0,\ 0,\ 0),\\
\tilde{\mathbf{e}}_4^{(5)} &= \frac{1}{\sqrt{3}\tau}(1,\ 0,\ 0,\ 0,\ 0,\ 0,\ 0,\ 0,\ \tau^2),\\
\tilde{\mathbf{e}}_5^{(5)} &= \frac{1}{\sqrt{3}\tau}(0,\ 0,\ 0,\ -\tau^2,\ 0,\ 0,\ 0,\ -1,\ 0).
\end{aligned} \tag{A12}$$

From these equations, it is straightforward to construct four vectors $\tilde{\mathbf{e}}_\alpha^{(4)}$ orthogonal to the five vectors $\tilde{\mathbf{e}}_\alpha^{(5)}$. We find

$$\begin{aligned}
\tilde{\mathbf{e}}_1^{(4)} &= \frac{1}{\sqrt{3}\tau}(0,\ 0,\ 1,\ 0,\ 1,\ 0,\ -1,\ 0,\ 0),\\
\tilde{\mathbf{e}}_2^{(4)} &= \frac{1}{\sqrt{3}\tau}(0,\ 1,\ 0,\ 0,\ 0,\ \tau^2,\ 0,\ 0,\ 0),\\
\tilde{\mathbf{e}}_3^{(4)} &= \frac{1}{\sqrt{3}\tau}(-\tau^2,\ 0,\ 0,\ 0,\ 0,\ 0,\ 0,\ 0,\ 1),\\
\tilde{\mathbf{e}}_4^{(4)} &= \frac{1}{\sqrt{3}\tau}(0,\ 0,\ 0,\ 1,\ 0,\ 0,\ 0,\ -\tau^2,\ 0),
\end{aligned} \tag{A13}$$

and

$$\begin{aligned}\tilde{\mathbf{w}}_1^{(4)} &= \frac{1}{\sqrt{3}\tau}(\nabla_x w_z + \nabla_y w_y - \nabla_z w_x),\\ \tilde{\mathbf{w}}_2^{(4)} &= \frac{1}{\sqrt{3}\tau}(\nabla_x w_y + \tau^2 \nabla_y w_z),\\ \tilde{\mathbf{w}}_3^{(4)} &= \frac{1}{\sqrt{3}\tau}(-\tau^2 \nabla_x w_x + \nabla_z w_z),\\ \tilde{\mathbf{w}}_4^{(4)} &= \frac{1}{\sqrt{3}\tau}(\nabla_y w_x - \tau^2 \nabla_z w_y).\end{aligned} \tag{A14}$$

The w–w part of the elastic energy density ($F_{ww'} = \int d^d x f_{ww'}$) is thus

$$f_{ww'} = \frac{1}{2}K^{(5)} \sum_l \tilde{\mathbf{w}}_l^2 + \frac{1}{2}(K^{(4)} - K^{(5)}) \sum_{l,l'} \tilde{\mathbf{w}}_l M_{ll'} \tilde{\mathbf{w}}_{l'}, \tag{A15}$$

where $M_{ll'} = \sum_\alpha \tilde{\mathbf{e}}_{\alpha l}^{(4)} \tilde{\mathbf{e}}_{\alpha l'}^{(4)}$. Equations (7.7)–(7.11) are obtained from the above with $K^{(5)} = K_l$ and $K^{(4)} - K_5 = K_2$, and $K^{uw} = K_3$ with $K_1 > 0$, $K_2 + K_l > 0$, and $K_1\mu > K_3^2$.

Table A1. Independent Components of J_{ijkl}

	xxx	yyy	zzz	xxy	xxz	yyx	yyz	zzx	zzy	xyz
x	0	0	$2\tau^2$	0	2τ	0	$-2\tau^3$	0	0	0
y	0	$-2\tau^2$	0	$2\tau^3$	0	0	0	0	-2τ	0
z	$-2\tau^2$	0	0	0	0	-2τ	0	$2\tau^3$	0	0

Exactly the same procedure was used to obtain the elastic energy in the alternate coordinate system [Eq. (4.9) used in Ref. 26. The independent components of $L_{ijkl} = 5(1+\tau^2)^{-2} J_{ijkl}/2$ are tabulated in Table A2.

Table A2. Independent Components of L_{ijkl}

	xxx	yyy	zzz	xxy	xxz	yyx	yyz	zzx	zzy	xyz
x	1	0	0	0	1	−1	−1	0	0	0
y	0	−1	0	1	0	0	0	0	0	−1
z	0	0	−2	0	1	0	1	0	0	0

With the aid of Table A2, the basis vectors $\tilde{\mathbf{e}}_\alpha^{(p)}$ can again be constructed:

$$\tilde{\mathbf{e}}_1^{(5)} = (0, 0, 0, 0, 0, 0, 0, 0, 1),$$

$$\tilde{\mathbf{e}}_2^{(5)} = \frac{1}{\sqrt{3}}(1, 0, 0, 0, 1, 0, 1, 0, 0),$$

$$\tilde{\mathbf{e}}_3^{(5)} = \frac{1}{\sqrt{3}}(0,\ 1,\ 0,\ -1,\ 0,\ 0,\ 0,\ -1,\ 0), \tag{A16}$$

$$\tilde{\mathbf{e}}_4^{(5)} = \frac{1}{\sqrt{3}}(1,\ 0,\ 1,\ 0,\ -1,\ 0,\ 0,\ 0,\ 0),$$

$$\tilde{\mathbf{e}}_5^{(5)} = \frac{1}{\sqrt{3}}(0,\ -1,\ 0,\ -1,\ 0,\ 1,\ 0,\ 0,\ 0),$$

$$\tilde{\mathbf{e}}_1^{(4)} = \frac{1}{\sqrt{3}}(1,\ 0,\ -1,\ 0,\ 0,\ 0,\ -1,\ 0,\ 0),$$

$$\tilde{\mathbf{e}}_2^{(4)} = \frac{1}{\sqrt{3}}(0,\ 0,\ 1,\ 0,\ 1,\ 0,\ -1,\ 0,\ 0), \tag{A17}$$

$$\tilde{\mathbf{e}}_3^{(4)} = \frac{1}{\sqrt{3}}(0,\ 1,\ 0,\ 0,\ 0,\ 1,\ 0,\ 1,\ 0),$$

$$\tilde{\mathbf{e}}_4^{(4)} = \frac{1}{\sqrt{3}}(0,\ 0,\ 0,\ 1,\ 0,\ 1,\ 0,\ -1,\ 0).$$

Equations (A16) and (A17) determine the components $\tilde{\mathbf{w}}_\alpha^{(4)}$ and $\tilde{\mathbf{w}}_\alpha^{(5)}$ that transform under $\Gamma_0^{(4)}$ and $\Gamma_0^{(5)}$. The elastic energy quoted in Ref. 26 was obtained by writing

$$f_{ww} = \frac{1}{2}\left(\frac{2}{3}K^{(5)} + \frac{1}{3}K^{(4)}\right)\sum_l \tilde{\mathbf{w}}_l^{\,2} + \frac{1}{2}\left(\frac{1}{3}K^{(5)} - \frac{1}{3}K^{(4)}\right)\sum_{lm} \tilde{\mathbf{w}}_l(\tilde{\mathbf{e}}^{l(5)}\tilde{\mathbf{e}}_m^{\ (5)} - 2\tilde{\mathbf{e}}_l^{\ (4)}\tilde{\mathbf{e}}_m^{\ (4)})\tilde{\mathbf{w}}_m. \tag{A18}$$

References

1. D. Shechtman, I. Blech, D. Gratias, and J. W. Cahn, *Phys. Rev. Lett.* **53,** 1951 (1984).
2. See for example, Niel W. Ashcroft and N. David Mermin, *Solid State Physics* (Holt, Rinehart and Winston, New York, 1976); C. Kittel, *Introduction to Solid State Physics,* 4th ed. (John Wiley and Sons, New York, 1971).
3. See for example, Michael Tinkham, *Group Theory and Quantum Mechanics* (McGraw-Hill, New York, 1964).
4. P. J. Steinhardt, D. R. Nelson, and M. Ronchetti, *Phys. Rev. Lett.* **47,** 1297 (1981); *Phys. Rev. B* **28,** 784 (1983).
5. D. I. Levine and P. J. Steinhardt, *Phys. Rev. Lett.* **53,** 2477 (1984).
6. D. Levine and P. J. Steinhardt, *Phys. Rev. B* **34,** 596 (1986).
7. (a) R. Penrose, *Bull. Inst. Math. Appl.* **10,** 266 (1974); (b) Martin Gardner, *Sci. Am.* **236,** 110 (1977).

8. R. Amman (unpublished).
9. A. L. Mackay, *Sov. Phys. Cryst.* **26,** 517 (1981); *Physica* **114A,** 609 (1982).
10. R. Mosseri and J. F. Sadoc, in *Structure of Non-Crystalline Materials,* edited by P. H. Gaskell, J. M. Parker, and E. A. Davis (Taylor and Francis, New York, 1983).
11. P. Kramer and R. Neri, *Acta Cryst. A* **40,** 580 (1984).
12. N. de Bruijn, *Ned. Akad. Weten. Proc.* Ser. A **43,** 39 (1981); **43,** 53 (1981).
13. V. Elser, *Phys. Rev. B* **32,** 4892 (1985); *Acta Cryst. A* **42,** 36 (1985).
14. M. Duneau and A. Katz, *Phys. Rev. Lett.* **54,** 2688 (1985).
15. P. A. Kalugin, A. Yu. Kitaev, and L. C. Levitov, Zh. Eksp. Teor. Fiz. Pisma **41,** 119 (1985).
16. J.E.S. Socolar, P. J. Steinhardt, and D. Levine, *Phys. Rev. B* **32,** 5547 (1985).
17. J.E.S. Socolar and P. J. Steinhardt, *Phys. Rev. B* **34,** 617 (1986).
18. F. Gahler and J. Rhyner, *J. Math. Phys. A* **19,** 267 (1986).
19. For recent reviews of quasicrystal research, see *J. Physique* Coll. **3, 47** (1986) and References 20 to 22.
20. *Selected Reprints on Quasicrystals,* edited by P. J. Steinhardt and S. Ostlund (World Scientific, Singapore, 1987).
21. C. L. Henley, to be published in *Comments on Solid State Physics.*
22. Paul Heiney and Peter Bancel, to be published in *Rev. Mod. Phys.*
23. K. Khalatnikov, *An Introduction to the Theory of Superfluid Helium* (W. A. Benjamin, New York, 1965).
24. L. D. Landau and E. M. Lifshitz, *Electrodynamics of Continuous Media* (Pergamon, New York, 1960).
25. P. G. de Gennes, *The Physics of Liquid Crystals* (Clarendon, Oxford, 1974).
26. D. Levine, T. C. Lubensky, S. Ostlund, S. Ramaswamy, P. J. Steinhardt, and J. Toner, *Phys. Rev. Lett.* **54,** 1520 (1985).
27. P. Bak, *Phys. Rev. Lett.* **54,** 1517 (1985).
28. P. Bak, *Phys. Rev. B* **32,** 5764 (1985).
29. T. C. Lubensky, S. Ramaswamy, and J. Toner, *Phys. Rev. B* **32,** 7444 (1985).
30. T. C. Lubensky, S. Ramaswamy, and J. Toner, *Phys. Rev. B* **33,** 7715 (1986).
31. P. A. Kalugin, A. Yu. Katayev, and L. S. Levitov, *J. Phys. Lett.* (Paris) **46,** L-601 (1985).
32. T. C. Lubensky, to be published in the *Proceedings of the NATO Conference on Structural Incommensurability, Boulder, Colorado, July 1986,* edited by Noel Clark.
33. For reviews of incommensurate systems, see P. Bak. *Rep. Prog. Phys.* **45,** 587 (1981).
34. V. L. Pokrovsky and A. L. Talopov, *Theory of Incommensurate Crystals, in Sov. Science Reviews* (Horwood, Zurich, Switzerland, (1985), and
35. *Incommensurate Phases in Dielectrics,* Vols. 14.1 and 14.2 of *Modern Problems in Condensed Matter Sciences,* edited by R. Blinc and A. Levanyuk (North-Holland, Amsterdam, 1986).
36. A. W. Overhauser, *Phys. Rev. B* **3,** 3173 (1971); P. A. Lee, T. M. Rice, and P. W. Anderson, *Solid State Commun.* **14,** 703 (1974).

37. P. C. Martin, O. Parodi, and P. S. Pershan, *Phys. Rev. A* **6,** 2401 (1972).
38. D. Forster, *Hydrodynamic Fluctuations, Broken Symmetry and Correlation Functions* (W. H. Benjamin, Reading, Mass. 1975).
39. W. L. McMillan, *Phys. Rev. B* **12,** 1187, 1197 (1975); V. A. Golovko and A. P. Levanyuk, *Zh. Eksp. Teor. Fiz.* **81,** 2296 (1981) [*Sov. Phys. JETP* **54,** 1217 (1981)]; R. Zeyer and W. Finger, *Phys. Rev. Lett.* **49,** 1883 (1982).
40. S. Aubry, in *Solitons in Condensed Matter Physics,* Vol. 8 of *Springer Series in Solid State Physics,* edited by A. Bishop and T. Schneider (Springer, Berlin, 1978), p. 264.
41. M. Peyrard and S. Aubry, *J. Phys. C* **16,** 1593 (1983); L. de Seze and S. Aubry, *J. Phys. C* **17,** 389 (1984).
42. J.E.S. Socolar, T. C. Lubensky, and P. J. Steinhardt, *Phys. Rev. B* **34,** 3345 (1986).
43. D. M. Frenkel, C. L. Henley, and E. D. Siggia, *Phys. Rev. B* **34,** 3649 (1986). [Kalugin, Kitaev and Levitov (unpublished) have found a loophole in the theorem of this paper. There are apparently special lattices in which atoms move continuously in response to shifts of the phason variable. For these lattices, there should be an elastic rather than a pinned phason free energy at zero temperature. The author is indebted to C. L. Henley for bringing this result to his attention.]
44. P. Bak, Brookhaven preprint.
45. T. Jansen, *J. Physique Coll.* **47,** C3-95 (1986).
46. N. D. Mermin and S. Troian, *Phys. Rev. Lett.* **54,** 1524 (1985); S. M. Troian, *J. Physique* Coll. **47,** C3-271 (1986); S. M. Troian and N. D. Mermin, *Ferroelectrics,* **66,** 127 (1986).
47. M. V. Jarić, *Phys. Rev. Lett.* **55,** 607 (1985).
48. O. Biham, D. Mukamel, and S. Shtrikman, *Phys. Rev. Lett.* **56,** 2191 (1986); preprint.
49. D. Mukamel, this volume.
50. See also, S. Sachdev and D. R. Nelson, *Phys. Rev. B* **32,** 4595 (1985).
51. P. M. deWolff, *Acta Cryst. A* **30,** 777 (1974).
52. P. Bak, in *Scaling Phenomena in Disordered Systems,* edited by R. Pynn and A. Skjeltorp (Plenum, New York, 1986); *Phys. Rev. Lett.* **56,** 861 (1986); *Scripta Met.* **20,** 1199 (1986).
53. P. Bak, this volume.
54. T. Janssen, *Acta Cryst. A* **42,** 262 (1986).
55. C. L. Henley, *J. Non-Cryst. Solids* **75,** 91 (1985).
56. M. V. Jarić, *J. Physique* Coll. **47,** C3-259 (1986).
57. M. V. Jarić and Udayan Mohanty, *Phys. Rev. Lett.* **58,** 230 (1987); M. V. Jarić, *Proc. Xth Intl. Workshop on Condensed Matter Theories,* edited by P. Vashishta (Plenum, New York, 1986).
58. M. Kléman, Y. Gefen, and A. Pavlovitch, *Europhys. Lett.* **1,** 61 (1986).
59. J. Bohsung and H.-R. Trebin, *Phys. Rev. Lett.* **58,** 1204 (1987).
60. T. C. Lubensky, J.E.S. Socolar, P. J. Steinhardt, P. A. Bancel, and P. A. Heiney, *Phys. Rev. Lett.* **57,** 1440 (1986).

61. P. M. Horn, W. Malzfedt, D. P. DiVencenzo, J. Toner, and R. Gambino, *Phys. Rev. Lett.* **57,** 1444 (1986).
62. J. D. Budai, J. Z. Tischler, A. Habenschuss, G. E. Ice, and V. Elser, preprint.
63. See for example the entire issue of *J. de Chimie Physique* **80** (1983).
64. L. D. Landau and E. M. Lifshitz, *Theory of Elasticity* (Pergamon, New York, 1975).
65. G. Grinstein and R. A. Pelcovitz, *Phys. Rev. Lett.* **47,** 856 (1981).
66. D. S. Rokhsar, D. C. Wright, and N. D. Mermin, *Phys. Rev. B* **35,** 5487 (1987) and Cornell preprint.
67. S. Alexander, *J. Physique* Coll. **47,** C3-143 (1986).
68. P. Bancel, P. A. Heiney, P. W. Stephens, A. I. Goldman, and P. M. Horn, *Phys., Rev. Lett.* **54,** 2422 (1985); J. W. Kahn, D. Shechtman, and D. Gratias, *J. Mat. Res.* **1,** 13 (1986).
69. See for example, H. Eugene Stanley, *Introduction to Phase Transitions and Critical Phenomena* (Oxford University Press, New York, 1971) and P. Pfeuty and G. Toulouse, *Introduction to the Renormalization Group and Critical Phenomena,* translated by G. Barton (Wiley, New York, 1977).
70. See for example, P. G. de Gennes, *Superconductivity of Metals and Alloys* (W. A. Benjamin, New York, 1966).
71. The breaking of tenfold symmetry can be seen along the ten "spokes" and at the center of the "tenfold" symmetric "cartwheel" pattern of Ref. 7(b), p. 119.
72. M. V. Jarić in *Proc. XV International Colloq. on Group Theoretical Methods in Physics,* edited by R. Gilmore and D. H. Feng (World Scientific, Singapore, 1987).
73. P. De and R. Pelcovitz, Brown University preprint.
74. J. Toner, unpublished.
75. T. Ramakrishnan and M. Yussouff, *Phys. Rev. B* **19,** 2775 (1979).
76. Frank R. N. Nabarro, *Theory of Crystal Dislocations* (Clarendon, Oxford, 1967).
77. K. Hiraga, M. Hirabayashi, A. Inoue, T. Masumoto, *Sc. Rep. RITU A,* **32,** No. 2, 209 (1985).
78. H. Mori and H. Fujisaka, *Prog. Theor. Phys.* **49,** 764 (1973); H. Mori, H. Fujisaka, and H. Shigematsu, *Prog. Theor. Phys.* **51,** 109 (1974); S. Ma and G. F. Mazenko, *Phys. Rev. B* **11,** 4077 (1975); I. E. Dzyaloshinskii and G. E. Volovik, *Ann. Phys.* (N.Y.) **125,** *67 (1980).*
79. S. Pantelides, private communication.
80. J. P. Pouget, G. Shirane, J. M. Hastings, A. J. Heeger, N. D. Miro and A. G. MacDiarmid, *Phys. Rev. B* **18,** 3465 (1978), I. V. Heilman, J. D. Axe, J. M. Hastings, G. Shirane, A. J. Heeger, and A. G. MacDiarmid, *Phys. Rev. B* **20,** 751 (1979).
81. Y. I. Frenkel and T. Kontorowa, *Zh. Eksp. Teor. Fiz.* **8,** 1340 (1938).
82. Francois Vallet, doctoral Thesis, University of Paris VI (1986); F. Vallet, R. Schilling, and S. Aubry, *Europhys. Lett.* **2,** 815 (1986).
83. J.E.S. Socolar, unpublished doctoral thesis, University of Pennsylvania.
84. G. M. Mazzucchelli and R. Zeyher, *Z. Phys. B* **62,** 367 (1986).
85. P. A. Bancel and P. A. Heiney, *J. Physique,* Coll. **47,** C3-341 (1986).
86. Joshua E. S. Socolar and David C. Wright, *Phys. Rev. Lett.* **59,** 221 (1987).
87. P. A. Bancel, P. A. Heiney, P. M. Horn, and F. W. Gayle, University of Pennsylvania preprint.

Index